Cases on Tour Guide Practices for Alternative Tourism

Gulsun Yildirim
Recep Tayyip Erdogan University, Turkey

Ozlem Ozbek
Bandirma Onyedi Eylul University, Turkey

Ceyhun Caglar Kilinc
Akdeniz University, Turkey

Abdullah Tarinc
Akdeniz University, Turkey

IGI Global
PUBLISHER of TIMELY KNOWLEDGE

A volume in the Advances in Hospitality, Tourism, and the Services Industry (AHTSI) Book Series

Published in the United States of America by
IGI Global
Business Science Reference (an imprint of IGI Global)
701 E. Chocolate Avenue
Hershey PA, USA 17033
Tel: 717-533-8845
Fax: 717-533-8661
E-mail: cust@igi-global.com
Web site: http://www.igi-global.com

Library of Congress Cataloging-in-Publication Data

Names: Yildirim, Gulsun, 1982- editor.
Title: Cases on tour guide practices for alternative tourism / Gulsun Yildirim, Ozlem Ozbek, Ceyhun Caglar Kilinc, and Abdullah Tarinc.
Description: Hershey, PA : Business Science Reference, 2020. | Includes bibliographical references and index. | Summary: "This book examines cases on technological innovations and new trends in tour guide practices"-- Provided by publisher.
Identifiers: LCCN 2019058353 (print) | LCCN 2019058354 (ebook) | ISBN 9781799837251 (hardcover) | ISBN 9781799837268 (paperback) | ISBN 9781799837275 (ebook)
Subjects: LCSH: Tourism--Management--Case studies.
Classification: LCC G155.A1 C317 2020 (print) | LCC G155.A1 (ebook) | DDC 910.72--dc23
LC record available at https://lccn.loc.gov/2019058353
LC ebook record available at https://lccn.loc.gov/2019058354

This book is published in the IGI Global book series Advances in Hospitality, Tourism, and the Services Industry (AHTSI) (ISSN: 2475-6547; eISSN: 2475-6555)

British Cataloguing in Publication Data
A Cataloguing in Publication record for this book is available from the British Library.

For electronic access to this publication, please contact: eresources@igi-global.com.

Advances in Hospitality, Tourism, and the Services Industry (AHTSI) Book Series

Maximiliano Korstanje
University of Palermo, Argentina

ISSN:2475-6547
EISSN:2475-6555

Mission

Globally, the hospitality, travel, tourism, and services industries generate a significant percentage of revenue and represent a large portion of the business world. Even in tough economic times, these industries thrive as individuals continue to spend on leisure and recreation activities as well as services.

The Advances in Hospitality, Tourism, and the Services Industry (AHTSI) book series offers diverse publications relating to the management, promotion, and profitability of the leisure, recreation, and services industries. Highlighting current research pertaining to various topics within the realm of hospitality, travel, tourism, and services management, the titles found within the AHTSI book series are pertinent to the research and professional needs of managers, business practitioners, researchers, and upper-level students studying in the field.

Coverage

- Destination Marketing and Management
- Leisure & Business Travel
- Hotel Management
- Cruise Marketing and Sales
- Tourism and the Environment
- Sustainable Tourism
- Food and Beverage Management
- Customer Service Issues
- Health and Wellness Tourism
- International Tourism

IGI Global is currently accepting manuscripts for publication within this series. To submit a proposal for a volume in this series, please contact our Acquisition Editors at Acquisitions@igi-global.com or visit: http://www.igi-global.com/publish/.

The Advances in Hospitality, Tourism, and the Services Industry (AHTSI) Book Series (ISSN 2475-6547) is published by IGI Global, 701 E. Chocolate Avenue, Hershey, PA 17033-1240, USA, www.igi-global.com. This series is composed of titles available for purchase individually; each title is edited to be contextually exclusive from any other title within the series. For pricing and ordering information please visit http://www.igi-global.com/book-series/advances-hospitality-tourism-services-industry/121014. Postmaster: Send all address changes to above address.

Titles in this Series

For a list of additional titles in this series, please visit: http://www.igi-global.com/book-series/advances-hospitality-tourism-services-industry/121014

Handbook of Research on Resident and Tourist Perspectives on Travel Destinations
Patrícia Pinto (Faculty of Economics and Research Centre for Tourism, Sustainability and Well-being (CinTurs), University of Algarve, Portugal) and Manuela Guerreiro (Faculty of Economics and Research Centre for Tourism, Sustainability and Well-being (CinTurs), University of Algarve, Porugal)
Business Science Reference • © 2020 • 400pp • H/C (ISBN: 9781799831563) • US $285.00

Global Niche Tourism and Opportunities for the Medical Tourism Industry
Manjeet Singh (Central University of Jammu, India) and Subbaraman Kumaran (Alliance University Bangalore, India)
Business Science Reference • © 2020 • 300pp • H/C (ISBN: 9781799834274) • US $215.00

Accelerating Knowledge Sharing, Creativity, and Innovation Through Business Tourism
Muhammad Waseem Bari (Government College University, Faisalabad, Pakistan) Sadia Shaheen (Government College University, Faisalabad, Pakistan) and Meng Fanchen (Beijing Institute of Technology, China)
Business Science Reference • © 2020 • 300pp • H/C (ISBN: 9781799831426) • US $225.00

Contemporary Management Approaches to the Global Hospitality and Tourism Industry
Abraham Pius (Arden University, UK) Husam H. Alharahsheh (University of Wales Trinity Saint David, UK) and Adenike A. Adesanmi (University of East London, UK)
Business Science Reference • © 2020 • 290pp • H/C (ISBN: 9781799822042) • US $225.00

Global Entrepreneurial Trends in the Tourism and Hospitality Industry
Debasish Batabyal (Amity University, Kolkata, India)
Business Science Reference • © 2020 • 300pp • H/C (ISBN: 9781799826033) • US $225.00

Handbook of Research on the Impacts, Challenges, and Policy Responses to Overtourism
Cláudia Ribeiro de Almeida (University of Algarve, Portugal) Alfred Quintano (Independent Researcher, Malta) Moisés Simancas (University of La Laguna, Spain) Raquel Huete (University of Alicante, Spain) and Zélia Breda (University of Aveiro, Portugal)
Business Science Reference • © 2020 • 447pp • H/C (ISBN: 9781799822240) • US $285.00

Economic, Educational, and Touristic Development in Asia
Bryan Christiansen (Global Training Group, Ltd, UK) and Hakan Sezerel (Anadolu University, Turkey)
Business Science Reference • © 2020 • 241pp • H/C (ISBN: 9781799822394) • US $215.00

IGI Global
PUBLISHER of TIMELY KNOWLEDGE

701 East Chocolate Avenue, Hershey, PA 17033, USA
Tel: 717-533-8845 x100 • Fax: 717-533-8661
E-Mail: cust@igi-global.com • www.igi-global.com

Editorial Advisory Board

Table of Contents

Detailed Table of Contents

Mobbing in workplace damages businesses in many ways and negatively affects organizational culture, business relationships, the quality of work tasks, and the quality of life of employees at the organization. In the tourism industry particularly and in the service industry in general, employees' nice behaviors to the customers play a significant role at the value-added part of the product. One of the important factors that will affect the quality of the service that visitors experience in a destination is the tourist guides. While the tourist guides are service providers, the satisfaction of the tourist guides with the working conditions is very important for the destinations and businesses in terms of image and profitability. This study examines the mobbing behaviors that tourist guides encounter in their work environments. Through the interviews with the tourist guides, it is aimed to determine the mobbing behaviors and resources that the tourist guides encounter in their complex and unusual working environments.

Despite advances in the study of tour guiding, there has been limited exploration of the work and life of tour guides, and even fewer studies seeking the views and perspectives of tour guides themselves. This study employed interpretive qualitative phenomenology. Twenty-nine Jordanian tour guides were interviewed to seek their perspectives on their work, motivations for becoming a guide, and the impact of their work on their private lives. The data analysis and interpretation revealed themes focusing on the mechanics of a working day, insurance expenses, unpredictable income, job anxiety, perceptions of guides themselves and by others, and the impact of tour guiding on their private lives. The outcomes of this study provide new understandings of the work-life experiences of tour guides from their perspective, and implications for the tourism industry. The study may be used as a steppingstone for further empirical investigation that could be replicated in different countries and cultural contexts.

The main purpose of this research is to find out tourist guides' contributions to sustainable tourism. It is planned to apply qualitative research method by using interpretive content analysis in phenomenological design. "Using the positive effects of green and sea on people" and "using spices that will interest the group, for example anise" are some of the answers describing tourist guides' leadership (instrumental) roles. "Talking to the villagers and transfer a lot of information to group" and "explain the nature of the region and explaining how the people living there adapt to this nature" are some of the answers of tour guides describing tourist guides' leadership (social) roles. "Giving information about what is famous in the Dardanelles region and go to the tasting" and "if there has been a family who makes tomato paste in the region certainly to let them to get experience" are some of the answers of tour guides describing tourist guides' mediator roles.

This chapter examines the role of student guides as mediators between the institutional mission and heritage of their university and visitors to the historic campus. Drawing on a longitudinal study undertaken at two historic universities in the west, the authors establish that a small cadre of elite, competitively-chosen guides at these institutions perform a role of openness and democracy on behalf of the increasingly complex and hybrid modern university. The chapter considers how student guides are able to navigate their own pride at such privileged engagement and how this privilege impacts on tours offered to visitors, where campus tours become a negotiation based on internal and external influences and are constructed and reconstructed according to the imagined or actual demands of different tour groups. By managing risk at the point of employment, and by encouraging free reign in tours, there is limited risk involved to host universities as student guides offer an informed, personalised heritage experience to both domestic and international tourists.

Sport tourism can be viewed as the amalgam of specific sporting activities, being undertaken by people in various forms of competitive interplay or other interaction, that occur in the places where might be instrumental to the sport and/or tourist experience. It is one of the fastest growing market segments in tourism and travel industry and is receiving increased attention for its social, environmental, and economic effects on destinations. It is estimated that there will be a larger market potential than ever for sport-related tours and tourism. On the other hand, it is well known that tour guides are tourism professionals that lead their guests through the many parts of any destination and they are responsible for planning, organizing, and execution of the tours as a very critical component of the tourism value chain. In this chapter, the basic functions and services of tour guides for sport activities and sport tourism will be discussed and the basic objective is to provide detailed information about this enhancing field of tourism for young generation tour guides.

Tourism investments are concentrated in the southern region of Jordan (Petra and Aqaba). This chapter argues that if food tourism delivers its sustainable benefits, it could be a vehicle for regional tourism development by strengthening the regional food on offer so that deprived areas are regenerated. Villages in Ajloun are selected for the investigation as they are well-known for food production. These villagers have inherited the talent for preparing traditional Jordanian food. This study is the first scientific study conducted to compare two villages in Ajloun, namely Orjan village and Ibbin village for logic justifications: 1) Provide the best example illustrating the best practices of tour guides in Jordan who unconsciously apply the principles of alternative tourism that lead to a sustainable tourism development. 2) This chapter aims to identify many problematic issues such as whether tour guides apply the best practices that are the concrete reflection of sustainable tourism principles.

The importance of tea tourism in India is immense as it has a dual impact by creating a regional tea market and securing jobs for tea labor. Therefore, proper guidance for the tea tourists is essential regarding the tea tourism destinations by addressing the issues, challenges, and opportunities in promoting local culture. Through the public-private partnership, tea tourism should be included in the mainstream tour packages, and local tea customs can be conveyed through a variety of activities promoting sightseeing and other forms of entertainment and tourism experiences. A proper planning is required for making tea tourism more attractive and developed in India. Development of tea tourism sustaining the environment and preserving the heritage and culture will benefit the Indian regions by creating employment opportunities and boosting the rural economy and thereby alleviate the insurgency and other socio-economic problems. It is expected to contribute to the literature on tourist guiding and the promotion of tea tourism and the tourist guides within tea garden boundaries in India.

Gastronomic activities like tasting of local cuisines, visiting of apple orchards, tea gardens, fruit marts, and attending cookery classes have always been an integral part of tour itinerary in Himachal Pradesh. Tourists rejoice it as a means of entertainment as well as cultural learning. The growing trend of gastronomy tourism and development of specific tourism policy under the theme of 'Agro/Organic Tourism' is certainly going to increase the demand for 'gastronomic tour guides' in coming years.

Hence the present research tries to analyse the tour guides' perception about Gastronomy Tourism in Himachal Pradesh, their background knowledge, and opinion about the branding of Himachali Cuisine. Responses from 120 tour guides were collected through a survey instrument via convenience sampling. The analysis depicts the requirement of specialized food guide training programs, attracting young and women tour guides into the profession and emphasis on branding of Himachali Cuisine through listing of ethnic restaurants, preparation of gastronomic map, and organisation of culinary tours and events.

Tourism is a major economic pillar for many countries; it enables them to increase gross domestic product, improves the population lifestyle, and preserves natural, cultural, and religious tourism assets. Tourism appears as a magic tool that could change the future of a developing country. However, tourism requires qualified human resources who can deliver high-quality service. Tour guides assume a great responsibility as they communicate with tourists during their sojourn. Their mission is even more critical if they have to accompany tourists to sacred tourist destinations where tourists are engaged in religious activities. The best practices of guides embedded with tourist behaviors are expected to enhance a peaceful world and provide a good touristic experience.

As religious tourism developed, some stakeholders have focused on the religious guide as a spiritual leader of the pilgrimage to supply religious knowledge and deepen the spiritual experience of the participants. In fact, the customers of religious tour operators seem to recognise religious guides as the important element in selecting a religious tour, and religious tour operators actively promote these religious figures as a significant element in their tour services. This chapter, therefore, considers the social context of religious guides in the Islamic religious tourism industry by mapping them in the structure of the marketplace from the perspective of the flow of 'religious capital'. Particular focus is placed on clarifying the flow of the religious guide's religious capital in the religious tourism industry, and the social networking emerging from the religious tourism industry through considering the novel use of financial and physical resources in the marketplace for religious tourism.

The growing interest in Halal tourism activity represents a prolific lens for scholars and practitioners in the tourism arena to study its concept, its role, and its potential to affect the sustainability of Halal tourism industry. The success of tourism industry is highly dependent on traveler experience with respect to customer service, experience, facilities, comfort, and delight. The notion of Halal tourism is connected and geared to the adoption of Islamic principles and practice in all aspect of tourism activities such as in

Halal hotel, Halal restaurant, Halal spa, Halal package tour, Halal transportation, Halal warehouse, and Halal retail. Halal tourism market reached USD 181 billion in year 2018 and is expected to grow every year. The number of Muslim travelers to Southeast Asian countries has rapidly increased as Southeast Asia (SEA) is known as a dynamic environment with a total Muslim population around 240 million.

Tourist guidance training/education is different in many countries of the world. Tourism types or tourism choices affect the specialization areas of tourist guides. Tourist guides that should be one of the components of an ecologically based integrity are not yet included in all routes. This area, which is not included in the tourist guidance curriculum, has been proposed as a training module for specialization in this study. In this context, the aim of the study is to contribute to the field and literature by creating the infrastructure of the cultural routes education program as a module. The study is a heuristic research. The interview form used in the research was formed by utilizing expert opinions on qualitative data collection methods based on literature review. According to the results of this research, this module is divided into seven main topics.

Art is the most important element that directs people to a destination and gives them an original experience in a developing and changing understanding of tourism. Art tourism motivates people to travel, facilitate communication with different cultures, enhance social pride, and contribute significantly to the region's economy. Regions with artistic and touristic attractions are active with art tourism, and they have an art destination brand. In this study, Ayvalik's potential for art tourism is investigated. At the same time, art events, art venues, and artist atelier/studio in Ayvalik are presented with numerical and visual examples. Ayvalik is developed as a new art route within the framework of art tourism and offers various suggestions for its tourist guides.

Augmented reality (AR) offers an interactive experience of the real-world environment when an object of the real-world is augmented by computer-generated perceptual information and relevant artefacts. This is a conceptual chapter based on the review of available literature. Also, resources on the internet have also been accessed and reviewed. On the context of the Diffusion of Innovation theory, this research aims to outline AR guiding for in an airport used for tourist aviation. Biman Bangladesh Airlines, the national flag carrier of the country, is the example where this study also explains the possible challenges and benefits that AR guiding facilities can possibly have. This research outlines two specific areas of management and marketing issues are analysis on the way to implement such guiding. Findings show

Preface

INTRODUCTION

The shift from mass tourism movements to alternative tourism types is essential for the development and sustainability of tourism in a global sense. Alternative tourism types show a wide variety today according to the curiosity and interests of the tourists with tourism marketing centered on the tourists. Art tourism, halal tourism, gastronomy tourism, sports tourism, and cultural tourism are just a few of them. Tourists participating in this and more alternative tourism activities need to get detailed information about the alternative tourism type. The primary purpose of tour guidance, which is one of the essential parts of tourism, is to provide tourists with in-depth information about the destination, to ensure that it is memorable and to give answers that satisfy the curiosity/interest of the tourists.

The purpose of creating this book is to provide an overview of tour guide applications for alternative tourism types, to express the problems experienced in tour guide applications, to introduce tour guide applications in different kinds of tourism and to contribute to improving the quality of tour guide service. The book focuses on the role of tour guidance in alternative tourism types and presents a collection of 16 chapters about the problems and trends in developing tour guidance. The book will reveal the role of tour guidance by providing a wide range of information about alternative touristic tour guidance in many countries. It aims to be the leading source of information on tour guidance in alternative tourism types to the stakeholders of the tourism industry, such as academics, tour guides, and travel agencies. It also aims to develop tour guidance practices by sharing new research information, findings, and results with these stakeholders. Today, it seeks to be a reference pressure for both new alternative tourism types and those who want to have information about tour guidance applications (policymakers, academics, tourism enterprises, and tour guides).

ORGANIZATION OF THE BOOK

Cases on Tour Guide Practices for Alternative Tourism offers a collection of 16 sections prepared by leading experts in tour guidance in alternative tourism activities as a whole.

The first part is about mobbing under the guidance of a tourist, "Mobbing in Tourism Industry: The Case of Tourist Guide" by Eren and Nebioğlu. As a result of the study, the authors explained the leading causes of mobbing in two titles: personal and operational reasons. Besides, the authors stressed that it is vital to reduce or completely get rid of the effects of mobbing so that the productivity of the workforce does not decrease, the communication with tourists is high, the guide is able to withstand the intense and tiring working conditions psychologically, and the image of the country/region is not adversely affected.

In the second chapter, "Tour Guides' Perspectives on Their Work and Life: A Case Study of Jordanian Tour Guides" prepared by Aloudat, Black, and Everett, the views and perspectives of the interpretive qualitative phenomenology and tour guides were investigated. As a result of the study, the authors found that while tour guides were satisfied in terms of earnings, they were irregular and wavy due to seasonality, international, and regional political conditions. The third chapter, Dönbak's study titled "Tourist Guides as Representatives of Sustainable Tourism: Evidence From Turkey" is about the contribution of tourist guides to sustainable tourism. In this study, instrumental and social leadership role and mediator role of tourist guides within the framework of sustainable tourism, cultural values, natural environment, flora and fauna were evaluated.

The fourth chapter, "Student Guides as Mediators of Institutional Heritage and Personal Experience", examines the role of student guides as mediators between the institutional mission and heritage of their university and visitors to the historic campus. The authors highlighted both the intermediary and the carrier role when introducing the student guides to the campuses covered by inheritance tourism. They also highlighted four areas, such as student guides facilitating physical access, mediating tourists and campus members, strengthening understanding about the place visited and encouraging empathy for the place visited. The next chapter is "Tourist Guidance in Sport Tourism" prepared by İçöz and İçöz, the essential functions and services of tour guides for sport activities and sport tourism will be discussed, and the primary objective is to provide detailed information about this enhancing field of tourism for young generation tour guides.

The sixth chapter is the study of "Tour Guide's Best Practices in Northern Region in Jordan: Land of Olive Oil and Indigenous Food" by Khreis and Benghadbane, which deals with the subject of food tourism. In the study, the authors focused on addressing food tourism in alternative tourism in the southern region of Jordan (Petra and Aqaba) within the framework of sustainable tourism. As a result of the study, the authors emphasized that the guides are not sufficiently trained, that the tour timing is well adjusted, they lack in terms of customer satisfaction, and that the locals are open to tourism development, as well as the restrictions of tourists on dressing and drinking alcohol.

The seventh chapter is the study titled "Tea Tourism and the Importance of Tea Tourists' Guidance in India" prepared by Solak and Amin on tea tourism. The study investigated the correct tour guidance for the sustainability and development of tea tourism, which has an essential place in India. As a result of the study, the authors stressed that tea tourism would increase employment opportunities, contribute to the rural economy, and alleviate socio-economic problems and, in this context, the importance of tea tourism guidance. The next chapter, titled "Analysing the Tour Guides' Perception About Gastronomy Tourism in Himachal Pradesh", written by Sahoo, aimed to measure perceptions about Himachali Pradesh gastronomy tourism and analyze the branding information with the views of tour guides. As a result of the research, the author suggested that a particular food guide training program for tour guides is required.

The ninth chapter, titled "Tour Guide's Best Practices in Christian and Islamic Religious Sites: Study Case Jordan", prepared by Khreis, Chapuis, and Shunnaq, discussed the best guidance practices for groups of tourists believing in different religions. The authors concluded in the research that tour guides are competent to manage the experience, the tour guide is professional enough, the tour is moderately successful, making a moderate contribution to the spiritual value of the religious tour. The next chapter, titled "The Social Significance of Religious Guides: The Development of the Flow of Religious Capital in the Islamic Religious Tourism Industry" by Yasuda, considers the social context of religious guides in the Islamic religious tourism industry by mapping them in the structure of the marketplace from the perspective of the flow of "religious capital". As a result of the study, the author suggested that religious guides are an important center of religious capital flow to create an Islamic cultural space in society.

The eleventh chapter, titled "Tourist Guiding in Halal Tourism: Case of Southeast Asia" by Rahman, focuses on the guiding services provided to tourists in halal tourism in Southeast Asia. As a result of the study, the author emphasized that halal tourist guidance should be improved to support the needs of Muslim tourists. The next chapter is titled "As an Alternative: Cultural Routes Education for Tourist Guides – A Module Suggestion From Turkey" prepared by Tanrisever. The aim of the study is to contribute to the field and literature by creating the infrastructure of the cultural routes education program as a module. As a result of the research, the author developed a module consisting of seven main topics: (i) Technical education, (ii) Flora and fauna, (iii) Trekking leadership, (iv) Regional education, (v) Sustainability, (vi) Communication, and (vii) Nutrition.

The thirteenth chapter, the aim of the study titled "Ayvalik as an Art Route for Tourist Guides Within the Frame of Art Tourism" prepared by Uguz is to develop a new tour route in Ayvalik within the framework of art tourism and offer a suggestion to tour guides. As a result of the research, the author suggested that Ayvalık has many values in terms of art tourism and that all stakeholders related to tourism and art should support art tourism.

The fourteenth chapter is the aim of the study titled "Technology Supported Tourist Guiding in Aviation Tourism" prepared by Hassan, to outline AR guiding for in an airport used for tourist aviation. The findings of this research establish that AR guiding in the most recent times is adopted by the "Early Majority " who are the followers and read the reviews given by the previous adopters on new services or products.

The fifteenth chapter is the purpose of the study titled "Technological Innovation Research With Guided Tour: Recent Trends and Future Directions" by Tarinc, Karaman, Sunar and Sayin, to reveal the technological applications used today in tour guidance and technological developments that will guide the tour guide in the future. In the study, information was given about the technologies used today by the guides and tourists and the technological innovations and trends planned and planned to be used in the future touristic tours.

The last chapter of the book, The aim of the study titled "The Future of Tourism Guidance in the Scope of Industry 4.0 and Next-Generation Technologies", prepared by Topsakal, Bahar, and Yuzbasi, is to examine tourism guidance within the scope of the super-smart tourists of the future., to determine the usage areas of next-generation technologies in the field of tourism guidance, and to give recommendations for tourism guidance in this regard. As a result of the study, the authors suggested that both the education of tourism guides and the standards of tourism guidance should be harmonized with new generation technologies.

Chapter 1
Mobbing in Tourism Industry: The Case of Tourist Guides

Ramazan Eren
https://orcid.org/0000-0002-6619-5360
Akdeniz University, Turkey

Oguz Nebioglu
Alanya Alaaddin Keykubat University, Turkey

EXECUTIVE SUMMARY

Mobbing in workplace damages businesses in many ways and negatively affects organizational culture, business relationships, the quality of work tasks, and the quality of life of employees at the organization. In the tourism industry particularly and in the service industry in general, employees' nice behaviors to the customers play a significant role at the value-added part of the product. One of the important factors that will affect the quality of the service that visitors experience in a destination is the tourist guides. While the tourist guides are service providers, the satisfaction of the tourist guides with the working conditions is very important for the destinations and businesses in terms of image and profitability. This study examines the mobbing behaviors that tourist guides encounter in their work environments. Through the interviews with the tourist guides, it is aimed to determine the mobbing behaviors and resources that the tourist guides encounter in their complex and unusual working environments.

INTRODUCTION

Mobbing in workplace damages businesses in many ways and negatively affects organizational culture, business relationships, the quality of work tasks and the quality of life of employees at the organization. Mobbing is the concept that sometimes referred to as "bullying" is a common behavior in professional life all around the world and due to its importance businesses and academics pay attention to the issue. Bullying, physiological terror, physiological harassment and mobbing these terms can be used interchangeably and describe the behaviors performed by an individual or by a group that targets an individual in a workplace and which can be humiliating, deterrent and unfair.

DOI: 10.4018/978-1-7998-3725-1.ch001

Like other countries in the world, the development of the tourism industry and increasing the touristic revenue are primary concerns for Turkey (Eren, 2019). In the tourism industry particularly and the service industry in general, employees' nice behaviors to the customers play a significant role at the value-added part of the product (Chu & Murrmann, 2006; Schneider & Bowen, 1985). One of the important factors that will affect the quality of the service that visitors experience in a destination is the tourist guides. Tourist guides are in contact with the visitor all day, service providers and the visible face of the destination and the businesses. While the tourist guides are service providers, the satisfaction of the tourist guides with the working conditions is very important for the destinations and businesses in terms of image and profitability.

This study examines the mobbing behaviors that tourist guides encounter in their work environments. Through the interviews with the tourist guides, it is aimed to determine the mobbing behaviors and resources that the tourist guides encounter in their complex and unusual working environments. In order to reach to the purposes, this paper is systematical presents the definition of mobbing concept, identify the negative behaviors can be called as mobbing in the workplace, and the consequences of mobbing with the focus in the tourism industry and tourist guides, by this way it is aimed to identify the context of the interview to reach the real circumstances of tourist guides' work life.

1. BACKGROUND

1.1 The Conceptional Framework of Mobbing, Antecedents and Consequences

The term Mobbing firstly used by an ethologist Konrad Lorenz to describe an animal group behavior. Smaller birds with a group to protect their nests attack to a single bigger animal. Then this terminology adapted by Peter-Paul Heinemann in 1972 for schoolchildren's behavior that used to describe a destructive behavior of small groups often against a single child (Leymann, 1996). The word mobbing borrowed by Leyman in the 1980s when he found a long-term hostile behavior in workplaces. Leyman (1996) made the systematic definition of mobbing in his article and assumed that the term "bullying" is a different kind of behavior because the word bullying carries connotations like physical aggression and threat. Mobbing behavior at the workplace is characterized by sophisticated behaviors and contains humiliating, and malicious behaviors but physical violence is found rarely (Leymann, 1996).

Leymann (1996) identify mobbing activities in five different categories and gives examples. It is important to mention this negative behavior for conducting the research in an appropriate way.

- Effects on victims' communications. The victim gets more silenced by verbal attacks and threats; management blocks the victim from doing business and talking to managers. These behaviors create an isolated position for the victim, and he/she has fewer dialogues with management and other staff.
- Effects on victims' social life. The workplace is a social environment for coworkers. The victim is a forbidden staff by management and coworkers do not talk, eat or have time with the person.
- Effects on victims' personal reputation. People in the workplace gossip about the victim and ridicule about his/her personal characteristics.
- Effects on victims' job status. People do not want to work with the person; people do not give work tasks or given trivial works tasks.

- Effects on victims' health. The victim does dangerous works, people threaten or attack physically or he/she sexually harassed.

The potential causes of mobbing mentioned in the literature can be organizational reasons, the bullies (Leymann, 1996), social system, processes of social exclusion (Schuster, 1996), managers (Kara, Kim, & Uysal, 2018), negative working conditions (lack of communication Zapf, 1999), leadership (Appelbaum, Semerjian, & Mohan, 2012), and coworkers' inappropriate attitudes (Kitterlin, Tanke, & Stevens, 2016). There can be different reasons that can cause mobbing, and at the same time, more than one factor can be observed.

The research conducted by Yurcu & Akinci (2019) revealed that mobbing has a significant relationship with personal values in a negative way and a highly significant and positive correlation with depression. Robinson & Bennett (1995) stated that mobbing could harm employees and organizational norms and described as a threat to the well-being of all components of an organization.

2. MOBBING AND TOURIST GUIDES

2.1 Tourism Industry in the World

Tourism is a fast and continuous development sector all over the world. According to the World Tourism Organization (2019), it constitutes the third-largest export category in the world after the chemical and fuel industry. Total international tourism exports (international tourism revenues and passenger transportation) have been calculated as 1.7 trillion dollars. This development is reflected in the total number of tourists. The total number of people participating in tourism reaches 1.4 billion. The participation of people in tourism is for different purposes, mostly vacation, health and religion are the main ones.

2.2 Researches About Mobbing in the Hospitality and Tourism Industry

In recent years, it can be said that there is an important interest in the concept of mobbing in the tourism literature. Accordingly, the phenomenon of mobbing has begun to be addressed in many areas of the hospitality and tourism industry. Tuzunkan (2018) states that mobbing can take place in four subdimensions that constitute the tourism sector in his study, which examined the tourism sector with four different case studies. Hotels, catering businesses, the travel industry and tourist guidance.

It is understood that mobbing studies in the context of hotels deal with this phenomenon in different regions McMahon, (2000) examine the effects of mobbing on job satisfaction and intention to quit (Çalışkan & Tepeci, 2008) and the exposure of hotel employees to mobbing. Study findings by McMahon (2000) examined hotels operating in Northern Ireland and the Republic of Ireland in the context of mobbing. While there is sectarian harassment in Northern Ireland; Sexual harassment is more common in the Republic of Ireland. Çalışkan & Tepeci, (2008) carried out by Turkey's research findings on employees of hotels operating in three different regions shows that most encountered mobbing type of a person's quality of life and professional status of the attacks. As mobbing increases, job satisfaction decreases; the intention to quit is increasing. Altıntaş, Yılmaz, & Çetinel (2011) findings, another study was done by operating in a destination in Turkey has reached the conclusion that exposure to bullying employees of the hotel.

It can be said that the mobbing studies carried out on food and beverage businesses mainly address the restaurants and the entire food and beverage industry in general. Research findings by Crawford (1997) show that mobbing is a phenomenon expressed by famous chefs in restaurants. Kitterlin et al., (2016) shows that food and beverage industry employees are exposed to mobbing with verbal abuse, inappropriate jokes, sexual intimidation and ridicule. Similarly, the study of Eren & Aydin, (2019) revealed that university students, who graduated from the department of gastronomy and culinary arts, were exposed to negative behaviors in the food and beverage services sector.

In a study in the field of the travel industry, Bentley et al. (2012) revealed that employees are commonly exposed to mobbing. In addition, the issue of mobbing has been explored on tourist guides in recent years. Erkol Bayram (2016) examined the exposure of tourist guides to mobbing in the context of demographic variables. The study on 422 tourist guides shows that the perception of mobbing is higher in single, national, undergraduate graduates and young guides. Yıldırım & Özbek (2019) also examined sexual harassment behavior towards tourist guides. Research findings on 72 tourist guides indicate that there is a higher tendency to sexual harassment towards male guides, mostly because of tour staff rather than tourists.

As stated above, mobbing is seen as an inevitable phenomenon in all dimensions that make up the tourism sector such as hotels, food and beverage businesses, travel businesses and tourist guides. The findings of the research conducted by Tuzunkan (2018) indicate that the mobbing phenomenon can take place in all areas of tourism with different dimensions of the study in all dimensions. One of these areas is seen as tourist guidance.

2.3 What is a Tourist Guide?

The tourist guide is defined "a person who guides visitors in the language of their choice and interprets the cultural and natural heritage of an area which person normally possesses an area-specific qualification usually issued and/or recognized by the appropriate authority" (WFTGA, 2003). Although the definition may seem simple, tourist guidance is a profession with very complex tasks. In this regard, Cohen (1985) examined the tourist guidance profession in four different dimensions. Accordingly, he states that tourist guides can have four basic roles: instrumental, social, interactional and communicative. The instrumental role refers to guiding tourists and introducing the region. The social role defines the group's ability to act together harmoniously. The interactive role includes roles for promoting the region to be introduced in a way that does not disturb the group. The communicative role, on the other hand, explains determining the focus of interest of the group and providing the correct information correctly.

In addition to this role confusion, it can be said that the work environments of tourist guides are also quite complex. The working environment of tourist guides' is beyond ordinary in many ways. Tourist guides can be employees of a travel agency but mostly work as a freelance in Turkey. Tourist guides usually work with different employers and coworkers. Travel agency representatives, vehicle drivers, museum staff, employees of the food and beverage businesses, and employees of the souvenir shops are coworkers for tourist guides, and tourist guides collaborate with different people in every tour program. This comprehensive research considers the complexity of the working environment to identify the causes and consequences of mobbing that tourist guides are exposed.

2.4 Causes of Mobbing

Studies in the literature consider the causes of mobbing in general as organizational, social and personal reasons (Trujillo Flores, Rivas Tovar, & Vilchis, 2014). Organizational causes that trigger mobbing are a delegation of authority, excessive workload, management weaknesses or uncertainties (Beswick, Gore, & Palferman, 2006; Duffy & Sperry, 2007), gaps in communication networks, work monotony, leadership style, organization culture and organizational change(Appelbaum et al., 2012; Bas, 2011; Zapf, 1999).

It has also been stated that mobbing can also be caused by social reasons such as hostility, jealousy, extreme competition, ambition, group pressure, humiliating someone or making scapegoats, social changes and characteristics of the culture (Beswick et al., 2006; Rayner & Hoel, 1997; Zapf, 1999).

Finally, it is stated that mobbing is a process that is affected by the personal characteristics of both the mobber and the victim (Beswick et al., 2006; Zapf, 1999). Personality, psychological condition, personality disorders, individual characteristics, social skills, demographic characteristics and perceptual differences are presented as personal causes that trigger mobbing (Duffy & Sperry, 2007; Einarsen, 1999).

2.5 Consequences of Mobbing

Mobbing behaviors, which have become an important problem for businesses, have serious consequences for victims, organizations and societies. Mobbing is a corrosive process and effects on victims can be observed in terms of mental health and economics. It is stated that mobbing causes mental health problems (Davenport, N., Schwartz, R.D. and Elliott, 2005; Jacobson, Hood, & Van Buren, 2014; Leymann, 1996), high levels of anxiety and depression (Reknes et al., 2014), burn out (Trépanier, Fernet, & Austin, 2015), post-traumatic stress (Hogh & Dofradottir, 2001) and sleep disturbance(Hansen, Hogh, Garde, & Persson, 2014) on victims. According to Einarsen & Raknes (1997), 23% of employees' who exposed to mobbing behavior, experience psychological problems.

Davenport, N., Schwartz, R.D. and Elliott, (2005) stated that it is possible to see health problems such as depression and heart attack in cases where the victim of mobbing is significantly affected, and this situation can also lead the victim to suicide. Economic effects of mobbing for victims who lost his mental health may be associated with hospital, doctor and therapy costs, as well as a lawyer and court costs.

In addition, it is known that the victim suffers from loss of income in case of being fired or forced to quit (Glambek, Matthiesen, Hetland, & Einarsen, 2014; Jennifer, Cowie, & Ananiadou, 2003) and working in jobs that are less than his capacity (Davenport, N., Schwartz, R.D. and Elliott, 2005). In the organizational and societal level, mobbing reduces the efficiency and performance of the organization and causes permanent institutional problems (Jacobshagen, 2004; Hoel et al., 2011). Research has shown that damaged organizational culture no longer motivates staff, leads to the process of alienation, eliminates job dissatisfaction and organizational trust (Zapf, 1999), reducing commitment to work and organization (Schat & Kelloway, 2000). These negative effects cause an increase in the number of unhappy and unhealthy individuals in the society, resulting in an increase of government spending on health (Davenport, N., Schwartz, R.D. and Elliott, 2005).

3. METHODOLOGY

This research addressed the general problem named as mobbing in the tourism industry from the view of tourist guides. This qualitative research aimed to discover the mobbing behaviors in tourist guides work environment, causes of these negative behaviors and effects on tourist guides. The importance of tourist guides in the tourism industry, the unusual business life of the guides and the lack of previous research on this subject make this study necessary. This qualitative study aimed to answer the following research questions;

Q1: Which mobbing behaviors are tourist guides exposed in their work environments and what/who are the causes of mobbing behaviors?
Q2: What are the effects of mobbing behaviors on the professional and private lives of tourist guides?

In this qualitative research semi-structured, in-depth interviews used to gather primary data. Interview instrument developed through the creation of significant questions in line with the literature. Tourist guides invited for a face-to-face interview and informed about the purpose of the research, and the questions in the interview. Interested, and volunteer twelve tourist guides participated in the research. Participants are tourist guides employed by a travel agency or freelancer in Turkey. The researcher recorded the opinions of the interviewees then the answers transcribed and utilized for further interpretations and analysis.

3.1 Research Design

There has been considerable research carried about mobbing, but less attention has been paid to mobbing behavior in the tourism industry. Because of a lack of detailed researches, and the unique circumstances of the work environment of the tourist guides, this qualitative research followed case study design (Yin, 2003). The most important characteristic of the case study is that it concentrates on a particular situation or a small number of situations. Concentration on the case allows the researcher to identify and view the details and to develop possible explanations about a situation. This case study is a single case study that used an embedded unit of analysis (Yin, 2003). The unit analyses in this research were tourist guides and sub-units identified as (a) tourist guides working in a travel agency and (b) tourist guides working as freelance.

3.2 Population and Sample

The population of this study is tourist guides working in Turkey. These tourist guides were expected to be certificated by the national tourism ministry and have a professional tourist guidance license. The number of participants was twelve. Participants have experience of at least three years. Purposive sample method used and tourist guides who are working in Antalya destination participated to the interviews. Antalya is the top coastal destination in Turkey, and most of the tourists travel with package tours. Antalya decided to be a suitable destination to request tourist guides who are working in a travel agency. The freelancer participants were working in multiple destinations.

Table 1. Demographics of participants

Participant	Age	Education	Year of work experience	Employment
K1	37	PhD.	12	Freelance
K2	37	Master of science	10	Travel Agency
K3	44	Bachelor's degree	14	Travel Agency
K4	30	Master of science	6	Travel Agency
K5	32	Master of science	7	Travel Agency
K6	28	Bachelor's degree	3	Travel Agency
K7	39	Bachelor's degree	12	Travel Agency
K8	42	Bachelor's degree	17	Freelance
K9	48	Bachelor's degree	18	Travel Agency
K10	40	Master of science	12	Travel Agency
K11	37	Bachelor's degree	15	Travel Agency
K12	39	Bachelor's degree	14	Freelance

Table 1 provides information about the demographic characteristics of the participants. The ages of the participants vary between 28 and 48. The majority of the participants are bachelor's degree and there are four master of science graduate and one doctorate graduate participant. The work experience of the participants varies from 3 to 18 years. While 3 tourist guides work as freelancers, others work with a travel agency.

4. FINDINGS

4.1 Thematic Analysis of Research Questions

The interviews revealed six main categories, that consist of fourteen sub-categories. Findings give more detailed information about the nature of tourist guidance, perceptions about mobbing, reasons for mobbing, mobbers, mobbing examples, and consequences of mobbing. Table 2 presents all these categories with brief descriptions.

Table 2. Thematic categories and sub-categories.

Main categories	Sub-categories	Descriptions
The nature of work	Capital	It refers to the intellectual knowledge of the tourist guide.
	Time	Flexible working hours. You don't have a clear work schedule.
	Physical strength	Requires too much physical exertion.
	Risks	Aggregate income risks. You don't know how much money you will earn.
	Hierarchy	You fell less hierarchy in the organizational structure.
	Happiness	The impact of this profession on one's ability to self-actualize help tourist guides to be happy.
Tourist Guides' perceptions about Mobbing Phenomenon.	Etymological approach	It literally means harassed.
	Business relationships approach	Negative social relationships are seen in a hierarchical structure.
Causes of Mobbing	Personal Reasons	The psychological situation of the mobber and the effort of the supervisor to hide his insufficient job knowledge.
	Operational Causes	To be under selling pressure.
Mobbers	Operation managers	Mobbing by operation managers who organize the tour programs.
	Chef Tourist Guides	Mobbing by chef tourist guides.
Examples of mobbing	Mobbing by behaviors	Operations managers and chef tourist guides are using a humiliating communication language.
	Mobbing by job scheduling	Send the victim to the tour/To create tour programs where the guide has to work harder but earn less money.
Consequences of mobbing	Job quitting	In order to avoid the mobbing behavior, tourist guide quits his job.
	Obedience	Tourist guides tend to obey the order of his/her superior (obedience pressure).

4.2 The Nature of Tourist Guidance

The first findings of the study reveal the nature of tourist guidance. In accordance with the evidence, tourist guidance at a very basic level is defined as a profession that displays the accumulation of intellectual knowledge as a person's capital. The quotations below present how respondents with codes K1, K4 and K8 define the nature of tourist guidance, respectively.

'Tourist guidance is a job where the accumulation of intellectual knowledge is a person's wealth, and the effective use of this capital is in direct proportion to the money that can be earned.'

'It is a toilsome job that preserves and provides detailed information on history, culture, natural and other features; it is a job that is done by people who are able to help solve tourists' problems and whose professional criteria are defined by-laws.'

'Tourist guidance is a job where, in addition to criteria defined by laws, there is a requirement to possess such qualities as the ability to improve oneself on a regular basis, a healthy nervous system, the

ability to multi-think and to think fast, to constantly plan and orient oneself in accordance with instant changes during a tour.'

In the following quotations from participants numbered K2 and K11 listed, it is stated that due to the fact that tourist guidance involves intensive physical effort, it is a job that requires durability.

'One of the most difficult aspects of this job is that it requires physical durability. If you do not feel tired if you are durable, you will be able to do this job as long as you want.'

'....actually is a very tiring and backbreaking job because there is no such thing as a day or night; it is difficult to start a family.'

In addition, in the following quotations, respondents numbered K3, K5 and K9 emphasize that the concept of time is another important element of tourist guidance, respectively.

'The most negative aspect of the guidance is its inopportune formation. Daily tours are not a problem but if it's a tour with accommodation, staying away from his/her family for a long period is a problem for the guide.'

'At this job, time is one of the most influential factors. You are never able to rest as long as you want or devote time to yourself or the ones you love.'

A tourist guide might experience difficulty interpreting knowledge and skills relating to archaeological sites or city tours destinations in an effective way due to lack of time. It especially concerns young guides.

Tourist guidance is also considered to be a risky profession. In the following quotations from participants numbered K6, K7, K8 and K9, listed respectively, it is stated that tourist guidance is a risky job.

'There is an income risk. There is a fixed salary; however, tourist guides are always on pins and needles. They are stressed about how to make more money because of tourist guidance, nowadays, involves intensive sales pressure and has evolved into a bonus system.

'A few years ago, I got into an accident and was seriously injured during a tourist guided by me. Because I was not covered by social security for self-employed, I was not able to get any financial aid from the Social Security Institution. There is no awareness of what is going to happen in case of a work accident.'

'There are flaws that occur when it comes to health insurance. This situation frustrates a lot of guides.'

Some of the serious risks for those who fondly do this job are cases when this job is considered seasonal, presence of illegal tourist guides, high concentration of guides speaking certain languages and considering this profession a second job.'

In addition to difficult aspects like these, tourist guidance is a job which does not have a complicated hierarchical structure. In this context, respondents numbered K1, K2, K3, K5, K6, K7 and K9 comment on the situation, respectively.

'Tourist guidance is a job which does not possess a hierarchical structure and where superior-interior relationships are not very common.'

All the participants stated that despite all the difficulties experienced, it is a job that brings happiness.

'I have witnessed how a lot of guides were going to work happily. No guide starts his/her job with a sigh.'

As it is clear from the evidence provided above, tourist guidance is a job that requires more intellectual capital than tangible assets. A tourist guide has to be durable enough for intensive physical activities during the day. The difficult aspects of this job are that the work hours of a guide are by far more flexible than the fixed work hours of a white-collar worker and it is considered a risky occupation. The most positive aspects are that it provides you with an opportunity for self-realization and, it has a simple job hierarchy.

4.3 Views of Tourist guides on the Mobbing Phenomenon

The participants provide various descriptions of mobbing. For example, participant numbered K10 has the following view on mobbing:

'Should we consider the subject from a linguistic point of view, this word seems to originate from the English word 'mob' and, in my opinion, means 'to be overwhelmed by something'.

In the following quotations from participants numbered K7 and K9 listed it is stated that the phenomenon of mobbing is the psychological pressure put by a superior as a result of social interaction that occurs in a hierarchical structure.

'In an establishment with superior-subordinate relationship, intimidating each other is one of the methods of applying psychological pressure. This situation may sometimes lead to the resignation of one of the sides or terrorization.'

'Attempts to make a person wary of his/her job, intentions to sicken him/her displayed by other members of staff or an executive are all kinds of administrative and psychological approaches.'

Below is how respondent numbered K4 states that these attempts of sickening and intimidating are mostly a matter that concerns guides who work for an agency; whereas, with self-employed guides, the phenomenon of mobbing is out the of question:

'There is no establishment whatsoever that would apply mobbing to self-employed guides.'

Concerning this, mobbing stands for sickening and intimidating behavior patterns towards tourist guides displayed at businesses they work for, and these behavior patterns are mostly experienced by tourist guides who work for an agency.

4.4 Causes of Mobbing

The research findings show that mobbing behavior originates from two major sources; personal and operational causes. Personal problems, the desire of a superior to expose his/her domination over a guide and their efforts to conceal their weaknesses can be summarized as personal causes.

In the following quotations from participants numbered K5 and K6, listed respectively, it is stated that mobbing may be caused by personal problems:

'Mobbing may be caused by a personal problem that exists between a tourist guide and a superior. A superior may apply personal pressure on the guide he/she has problems within particular by making things more difficult for him/her. (f.e. work hours, way of controlling work etc.). A person who applies mobbing is a psychopath with a psychological disorder and depression.'

'A superior may want a guide to obeying him/her.'

Likewise, respondent numbered K8 states that mobbing may be applied in an attempt to conceal one's own weaknesses:

'Psychological pressure towards subordinates is applied in order to conceal one's own incompetence.'

Respondent numbered K6 states that in addition to these personal reasons, mobbing may also be applied due to operational reasons:

'Mobbing may also be caused by operational reasons. In certain situations, guides may get into a dispute with superiors over the way a tour must be conducted. When this type of dispute continues, superiors may apply growing pressure on the guides. These conflicts may turn into mobbing.'

4.5 Sources of Mobbing

Since tourist guidance is not a job that requires a complex hierarchy, the number of sources that may cause mobbing is also quite limited. In this context, operations managers and senior guides, as the one's guides work mostly with, are defined as primary sources of mobbing. All the participants stated that operations managers or senior/chief guides might be the sources of mobbing. These statements can be summarized as follows:

'Tourist guidance is not a very suitable job for mobbing. Tourist guides are mostly in contact with chief tourist guides and operations managers. These employees may be the biggest sources of mobbing for tourist guides.'

4.6 Examples of Mobbing

Mobbing can be exemplified in two different ways: as personal behavior and as organizational applications. The quotation below presents how respondent numbered K5 defines personal pressure and colleagues' attitudes towards each other.

'Chief guides disrespectfully treat other guides and make it obvious that they dislike them. Their behavior and manner of speaking may cause guides to feel insignificant.'

All the participants of the study stated that organizational mobbing applications might be present as *'assigning guides to tours where they will make less money, giving them work below/over their capacity, restricting their legal rights and defamation'*. The mobbing that is applied to guides by operations managers, in particular, was exemplified by respondent numbered K6 as follows:

'If operations managers want to apply mobbing to guides, they, first of all, may punish them by not giving them a tour group. Likewise, chief guides, who have good relationships with operations officers, may use those relationships as a tool for mobbing to make sure that guides get tours where they will make less money. I have a few co-workers who work at the same agency as me and who complain that they have been rowing with operations managers for 4-5 years and, thus, have been getting tours that bring less money.'

In addition to this, some examples may cause a guide to losing his/her job. Respondent numbered K2 summarizes these situations as follows:

'Last season, a guide, who had a row with the operations manager, did not fulfil his unjustified request. He was assigned to tours we call 'fatigue duty'. Due to the deliberate reports on 'numerous complaints' about him, he had to resign from his job.

4.7 Consequences of Mobbing

In this regard, there are two most known consequences that have been observed and voiced: either obey or quit your job. Respondent numbered K10 describes this situation as follows:

'Sickening and quitting the job or, vice versa, sickening and obeying. The obedient behavior generally occurs because of the low chance of finding a new job in the middle of the guidance season. Quitting the job occurs when a guide is in a position to find an easier job.'

The participants stated that those of their colleagues who have been exposed to mobbing had low work motivation and were unhappy. Those tourist guides, who get fewer tours, get assigned to tours that bring less profit, get assigned to out-of-town tours with accommodation more often than other guides, feel unhappy.

5. CONCLUSIONS AND SUGGESTIONS

According to the results of the study received, the interviews with the participants revealed six major categories and fourteen subcategories. Although tourist guidance, when compared to a lot of other job groups, is a job group where mobbing occurs less, the results of the study shows that mobbing behavior takes place at this job, too. The findings of the research include causes, sources, consequences and examples of mobbing.

Causes of mobbing can be grouped under two titles: personal and operational causes. Personal causes include mobbing that originates from a guide's co-workers', superior's and other people's personal (social, psychological, etc.) problems. This type of mobbing can be exemplified as exclusion, verbal or physical harassment, sexual harassment, bad attitude, problems with communication (social isolation/detachment), defamation, underestimation of one's skills or achievements. As the participants state it, people who apply mobbing due to personal reasons are '*psychopaths with a psychological disorder and depression.*' Operational reasons, though, are when a guide is generally exposed to mobbing of the agency's operational structure. This type of mobbing can be exemplified as: not charging a guide with a duty, giving them work below/over their capacity, restricting their legal rights, deliberately misinforming them, using wage earners to one's own advantage, freezing ties, eliminating and assigning meaningless tasks.

Mobbing is a situation that can be experienced by individuals of any status, occupation, and education group; however, it occurs more often in job and education groups with a hierarchical structure. This hierarchical structure is one of the significant reasons for the increase in the number of sources of mobbing. Since tourist guidance is not a job that has a hierarchical structure, the number of sources that may cause mobbing is also quite limited. In this context, operations managers and senior guides, as the one's guides work mostly with, are defined as primary sources of mobbing. A tourist guide is exposed to personal and operational mobbing if he/she has a row or disagreement with an operations manager or senior/chief guide. The most common consequences of being exposed to sources and reasons of mobbing for guides are obedience, resignation or psychological problems. The person who loves his job will choose obedience as a result of intimidating behavior. However, this affects guides' motivation, creates a distance between them and their job or makes them do a job they do not like. Violent mobbing actions may result in making a guide resign from his job or may him experience psychological problems.

In order to introduce a region's, locality's or country's culture, tourism, history, environment, nature, social and other values, talk about and show surroundings in accordance with the program, previously specified by the travel agency a guide needs to be highly motivated and he must dedicate himself to his job. For the labor productivity not to decrease, for the communication with tourists to be on a high level, for a guide to be able to psychologically withstand the intensive and tiring working conditions, for the image of the country/locality not to be negatively affected it is utterly important to reduce the effects of mobbing or to completely get rid of them. During training, guides could be educated more about the extensive and complicated process of mobbing, its features, about facing obstacles while performing their jobs, about ways of coping with difficulties they might have while dealing with service providers. Thus, they could be prepared for acts of mobbing that occur in the sector. Besides, this could prevent the increase of acts of mobbing for posterity.

REFERENCES

Altıntaş, V., Yılmaz, Y., & Çetinel, F. (2011). Mobbing in Accomodation Businesses: Sample of Antalya. *E-Journal of New World Sciences Academy*, *6*(4), 439–454.

Appelbaum, S. H., Semerjian, G., & Mohan, K. (2012). Workplace bullying: Consequences, causes and controls (part two). *Industrial and Commercial Training*, *44*(6), 337–344. doi:10.1108/00197851211254770

Bas, E. (2011). A capital budgeting problem for preventing workplace mobbing by using analytic hierarchy process and fuzzy 0-1 bidimensional knapsack model. *Expert Systems with Applications*, *38*(10), 12415–12422; Advance online publication. doi:10.1016/j.eswa.2011.04.022

Bentley, T. A., Catley, B., Cooper-Thomas, H., Gardner, D., O'Driscoll, M. P., Dale, A., & Trenberth, L. (2012). Perceptions of workplace bullying in the New Zealand travel industry: Prevalence and management strategies. *Tourism Management*, *33*(2), 351–360. doi:10.1016/j.tourman.2011.04.004

Beswick, J., Gore, J., & Palferman, D. (2006). Bullying at Work: A Review of the Literature. Retrieved from https://www.hse.gov.uk/research/hsl_pdf/2006/hsl0630.pdf

Çalışkan, O., & Tepeci, M. (2008). Mobbing Behaviors at Hotels and Their Impacts on Employee Job Satisfaction and Intent to Remain. Anatolia. *Turizm Araştırmaları Dergisi*, *19*(2), 135–148. doi:10.17123/atad.09003

Chu, K. H. L., & Murrmann, S. K. (2006). Development and validation of the hospitality emotional labor scale. *Tourism Management*, *27*(6), 1181–1191. doi:10.1016/j.tourman.2005.12.011

Cohen, E. (1985). The tourist guide: The origins, structure and dynamics of a role. *Annals of Tourism Research*, *12*(1), 5–29. doi:10.1016/0160-7383(85)90037-4

Crawford, N. (1997). Bullying at work: A psychoanalytic perspective. *Journal of Community & Applied Social Psychology*, *7*(3), 219–225. doi:10.1002/(SICI)1099-1298(199706)7:3<219::AID-CASP420>3.0.CO;2-Q

Davenport, N., Schwartz, R. D., & Elliott, G. P. (2005). *Mobbing: Emotional Abuse in the American Workplace* (3rd ed.). Civil Society Publications.

Duffy, M., & Sperry, L. (2007). Workplace Mobbing: Individual and Family Health Consequences. *The Family Journal (Alexandria, Va.)*, *15*(4), 398–404. doi:10.1177/1066480707305069

Einarsen, S. (1999). The nature and causes of bullying at work. *International Journal of Manpower*, *20*(1/2), 16–27. doi:10.1108/01437729910268588

Einarsen, S., & Raknes, B. I. (1997). Harassment in the Workplace and the Victimization of Men. *Violence and Victims*, *12*(3), 247–263. doi:10.1891/0886-6708.12.3.247 PubMed

Eren, R. (2019). Turkey's Food Image, Travelers' Revisit Intention and Tourist Expenditures. Anais Brasileiros de Estudos Turísticos - ABET, 9(1,2,3), 1–16.

Eren, R., & Aydin, A. (2019). Perceptions and attitudes of culinary students towards food and beverage industry in Turkey. Journal of Culinary Science & Technology, 1–25. doi:10.1080/15428052.2019.1640155

Erkol Bayram, G. (2016). The research on examination of tour guides mobbing exposure within the demographic characteristics. *The Journal of International Social Research, 9*(47), 1185–1194. doi:10.17719/jisr.2016.1465

Glambek, M., Matthiesen, S. B., Hetland, J., & Einarsen, S. (2014). Workplace bullying as an antecedent to job insecurity and intention to leave: A 6-month prospective study. *Human Resource Management Journal, 24*(3), 255–268. doi:10.1111/1748-8583.12035

Hansen, Å. M., Hogh, A., Garde, A. H., & Persson, R. (2014). Workplace bullying and sleep difficulties: A 2-year follow-up study. *International Archives of Occupational and Environmental Health, 87*(3), 285–294. doi:10.1007/s00420-013-0860-2 PubMed

Hogh, A., & Dofradottir, A. (2001). Coping with bullying in the workplace. *European Journal of Work and Organizational Psychology, 10*(4), 485–495. doi:10.1080/13594320143000825

Jacobson, K. J., Hood, J. N., & Van Buren, H. J. III. (2014). Workplace bullying across cultures: A research agenda. *International Journal of Cross Cultural Management, 14*(1), 47–65; Advance online publication. doi:10.1177/1470595813494192

Jennifer, D., Cowie, H., & Ananiadou, K. (2003). Perceptions and Experience of Workplace Bullying in Five Different Working Populations. *Aggressive Behavior, 29*(6), 489–496; Advance online publication. doi:10.1002/ab.10055

Kara, D., Kim, H., & Uysal, M. (2018). The effect of manager mobbing behaviour on female employees' quality of life. *Current Issues in Tourism, 21*(13), 1453–1467. doi:10.1080/13683500.2015.1078298

Kitterlin, M., Tanke, M., & Stevens, D. P. (2016). Workplace bullying in the foodservice industry. *Journal of Foodservice Business Research, 19*(4), 413–423. doi:10.1080/15378020.2016.1185874

Leymann, H. (1996). The content and development of mobbing at work. *European Journal of Work and Organizational Psychology, 5*(2), 165–184. doi:10.1080/13594329608414853

McMahon, L. (2000). Bullying and harassment in the workplace. *International Journal of Contemporary Hospitality Management, 12*(6), 384–387. doi:10.1108/09596110010343666

Rayner, C., & Hoel, H. (1997). A Summary Review of Literature Relating to Workplace Bullying. *Journal of Community & Applied Social Psychology, 7*(3), 181–191. doi:10.1002/(SICI)1099-1298(199706)7:3<181::AID-CASP416>3.0.CO;2-Y

Reknes, I., Pallesen, S., Magerøy, N., Moen, B. E., Bjorvatn, B., & Einarsen, S. (2014). Exposure to bullying behaviors as a predictor of mental health problems among Norwegian nurses: Results from the prospective SUSSH-survey. *International Journal of Nursing Studies, 51*(3), 479–487. doi:10.1016/j.ijnurstu.2013.06.017 PubMed

Robinson, S. L., & Bennett, R. J. (1995). A Typology of Deviant Workplace Behaviors: A Multidimensional Scaling Study. *Academy of Management Journal, 38*(2), 555–572. doi:10.5465/256693

Schat, A. C. H., & Kelloway, E. K. (2000). Effects of perceived control on the outcomes of workplace aggression and violence. *Journal of Occupational Health Psychology, 5*(3), 386–402. doi:10.1037/1076-8998.5.3.386 PubMed

Schneider, B., & Bowen, D. E. (1985). Employee and Customer Perceptions of Service in Banks. Replication and Extension. *The Journal of Applied Psychology*, *70*(3), 423–433; Advance online publication. doi:10.1037/0021-9010.70.3.423

Schuster, B. (1996). Rejection, Exclusion, and Harassment at Work and in Schools. *European Psychologist*, *1*(4), 293–317. doi:10.1027/1016-9040.1.4.293

Trépanier, S.-G., Fernet, C., & Austin, S. (2015). A longitudinal investigation of workplace bullying, basic need satisfaction, and employee functioning. *Journal of Occupational Health Psychology*, *20*(1), 105–116. doi:10.1037/a0037726 PubMed

Trujillo Flores, M. M., Rivas Tovar, L. A., & Vilchis, F. L. (2014). Mobbing: A theoretical model quantifying factors affecting the role of women executives in the institutions of public education in Mexico. *Contaduría y Administración*, *59*(1), 195–228. doi:10.1016/S0186-1042(14)71249-5

Tuzunkan, D. (2018). Mobbing in Tourism Enterprises. *The Case of Turkish Riviera.*, *13*(4), 1813–1819.

WFTGA. (2003). What is a Tourist Guide? Retrieved January 5, 2020, from World Federation of Tourist Guide Associations website: http://www.wftga.org/tourist-guiding/what-tourist-guide

World Tourism Organization. (2019). International Tourism Highlights, 2019 Edition. doi:10.18111/9789284421152

Yıldırım, G., & Özbek, Ö. (2019). A Study on Sexual Harassment to Professional Tour Guides. *Turist Rehberliği Dergisi*, *2*(2), 72–87.

Yin, R. K. (2003). *Case study research: Design and methods* (3rd ed.). Thousand Oaks: Sage Publications Inc.

Yurcu, G., & Akinci, Z. (2019). The mediating role of mobbing in the relationship between personal value and depression: A case study of hospitality employees. *Journal of Human Resources in Hospitality & Tourism*, *18*(4), 471–503. doi:10.1080/15332845.2019.1626852

Zapf, D. (1999). Organisational, work group related and personal causes of mobbing/bullying at work. *International Journal of Manpower*, *20*(1/2), 70–85. doi:10.1108/01437729910268669

ADDITIONAL READING

Duffy, M., & Sperry, L. (2014). *Overcoming Mobbing: A Recovery Guide for Workplace Aggression and Bullying*. Oxford University Press.

Einarsen, S. (1999). The nature and causes of bullying at work. *International Journal of Manpower*, *20*(1/2), 16–27. doi:10.1108/01437729910268588

Einarsen, S., Hoel, H., & Notelaers, G. (2009). Measuring exposure to bullying and harassment at work: Validity, factor structure and psychometric properties of the Negative Acts Questionnaire-Revised. *Work and Stress*, *23*(1), 24–44. doi:10.1080/02678370902815673

Kara, D., Kim, H., & Uysal, M. (2018). The effect of manager mobbing behaviour on female employees' quality of life. *Current Issues in Tourism*, *21*(13), 1453–1467. doi:10.1080/13683500.2015.1078298

Leymann, H. (1996). The content and development of mobbing at work. *European Journal of Work and Organizational Psychology*, *5*(2), 165–184. doi:10.1080/13594329608414853

Yıldırım, G., & Özbek, Ö. (2019). A Study on Sexual Harassment to Professional Tour Guides. *Turist Rehberliği Dergisi*, *2*(2), 72–87.

KEY TERMS AND DEFINITIONS

Job Quitting: Job quitting refers to leaving the job from the workplace for various reasons.

Mobber: Mobber refers to people engaged in mobbing behavior.

Mobbing: The word mobbing means psychological violence, pressure, siege, harassment, or distress.

Mobbing by Behaviors: It refers to the mobbing's realization within the organization with speech, attitude, and behavior.

Mobbing by Job Scheduling: It refers to the creation of job scheduling in a way that will disturb the tourist guide.

Obedience: It refers to submitting to managers' instructions in the workplace.

Tourism Industry: Tourism industry refers to the units consisting of companies and organizations (transportation, accommodation, entertainment, shopping, etc.) that offer services to the tourist and have common functions.

Tourist Guide: The tourist guide is the person who is obliged to give information by accompanying the tour participants in touristic travels.

Chapter 2
Tour Guides' Perspectives on Their Work and Life: A Case Study of Jordanian Tour Guides

Areej Shabib Aloudat
Yarmouk University, Jordan

Rosemary Black
Charles Sturt University, Australia

Sally Everett
King's College London, UK

EXECUTIVE SUMMARY

Despite advances in the study of tour guiding, there has been limited exploration of the work and life of tour guides, and even fewer studies seeking the views and perspectives of tour guides themselves. This study employed interpretive qualitative phenomenology. Twenty-nine Jordanian tour guides were interviewed to seek their perspectives on their work, motivations for becoming a guide, and the impact of their work on their private lives. The data analysis and interpretation revealed themes focusing on the mechanics of a working day, insurance expenses, unpredictable income, job anxiety, perceptions of guides themselves and by others, and the impact of tour guiding on their private lives. The outcomes of this study provide new understandings of the work-life experiences of tour guides from their perspective, and implications for the tourism industry. The study may be used as a steppingstone for further empirical investigation that could be replicated in different countries and cultural contexts.

INTRODUCTION

Tourism accounts for 10.4% of global GDP and 319 million jobs or 10% of total employment in 2018 (World Travel and Tourism Council, 2019). Yet, relatively little remains known about tourism employment (Baum 2007; Baum, et al., 2016; Tribe & Lewis, 2003). Even fewer studies explore the social

DOI: 10.4018/978-1-7998-3725-1.ch002

aspects of tourism-related employment and the private lives of employees (Brymer, Perrewe, & Johns, 1991; Magnini, 2009; Ross, 1997).

Many studies of tourism employment suggest the industry is characterised by long working-hours, stress, seasonality, low status work, and variable remuneration (Baum et al., 2016; Kusluvan, 2003; Kusluvan & Kusluvan, 2003; Ringer, 2013; Tribe & Lewis, 2003). Consequently, exploration of the work and private lives of those employed in the tourism industry is a valuable avenue of research. Perhaps more than any other occupational group in the tourism industry, we suggest that tour guides experience these aspects more acutely than most as a result of tour cancellations, extended tours, irregular work, delays, long working hours, and irregularity of employment (Alrawadieh, Cetin, Dincer & Dincer, 2019; Yetgin & Benligiray, 2019).

In this study a tour guide is defined as a person, usually a professional, who guides groups (and sometimes individuals) around venues or places of interest such as natural areas, historic buildings and sites, and landscapes of a city or a region; and interprets the cultural and natural heritage in an inspiring and entertaining manner, in the language of the visitor's choice (adapted from the European Federation of Tourist Guide Association (Weiler & Black, 2015). Across the world tour guides are employed in both the private and public sectors and under a wide range of legislation and regulations (Weiler & Black, 2015; World Federation of Tourist Guide Association, 2019).

Tour guides perform many functions (Weiler & Black, 2015) and the literature on tour guiding has generally focused on issues such as the diverse roles of guides particularly the roles of communicator and interpreter, visitor satisfaction and expectations, quality assurance mechanisms and training and education (Christie & Mason, 2003; Scherle & Nonnenmann, 2008; Weiler & Black, 2015; Weiler & Ham, 2002).

Few studies have considered the professionalisation of tour guiding (Ap & Wong, 2001; Sun-Ah Ponting, Wearing & Black, 2003) and the impacts of this employment on the guide's work and private lives, and more critically fewer have sought to explore these issues from the tour guide's perspective (Aloudat, 2017; Salazar, 2005; Weiler & Black, 2015). Thus this study fills an important research gap in the tour guiding literature. The study sought to explore the work-life space of Jordanian tour guides from their perspective and is part of a larger study of Jordanian tour guides that resulted on a model that theorised the worldview of tour guides (Aloudat, 2017). This study also contributes to the literature on the long tradition of Arab (Sobh, Belk & Wilson, 2013) and particularly Jordanian hospitality (Shryock, 2003). The chapter presents the relevant literature, followed by the methodology, findings, discussion and conclusion that highlights avenues for future research.

LITERATURE REVIEW

Tour Guides and Tour Guiding

Over two decades ago, Holloway (1981) and Jafari (1985) noted a paucity of literature on tour guiding. Yet, despite this observation, scholars still express concern at the limited understanding we have of tour guides and the tour guiding industry (Ap & Wong, 2001; Black & Ham, 2005; Huang, Hsu, & Chan, 2010; Scherle & Nonnenmann, 2008; Yu & Weiler, 2006; Weiler & Black, 2015). Literature has predominately focused on the diverse roles that tour guides play which range from being information givers (Cohen, 1985; Pearce, 1984; Pond, 1993), cultural and spiritual brokers (Dahles, 2002; Jennings & Weiler, 2006; Macdonald, 2006; Parsons, Houghe Mackenzie & Filep, 2019; Scherle & Nonnenmann,

2008), ambassadors, public relations representatives, hosts, conduits (Pond, 1993), and educators (Poudel & Nyaupane, 2013). Other studies have examined the attributes and knowledge that tour guides should possess (Ap & Wong, 2001; Black, Ham, & Weiler, 2001; Christie & Mason, 2003) and tour guide performance (Alazaizeh, Jamaliah, Mgonja & Ababneh, 2019; Bowie & Chang, 2005; Chan, Hsu, & Baum; 2016; Huang et al., 2010) that may lead to high levels of tourist satisfaction, sustainable behaviour and repeat visits (Alazaizeh, Jamaliah, Mgonja & Ababneh, 2019; Ap & Wong, 2001; Bowie & Chang, 2005; Scherle & Nonnenmann, 2008; Yu & Weiler, 2006).

With the exception of Salazar's (2005) ethnographic study of Indonesian tour guides there have been few empirical studies that have explored the lived experience of tour guides and even fewer that adopt the tour guide's perspective. Salazar's observation of a small sample of local tour guides in Indonesia found they were key performers in the process of 'glocalization' (how guides present and actively deconstruct local culture for a diverse global market) and hinted at the private lives of the guides themselves.

The Private Life of Workers

Studies of employee working conditions and work/life balance have tended to focus on the physical work environment, the time and the facilities for doing work (Herzberg, et al. 2010), emotional labour (Alazaizeh, Jamaliah, Mgonja & Ababneh, 2019; Van Dijik, Smith & Cooper, 2011), as well as the private lives of individual employees (Cullen, Kordey, Schmidt, & Gaboardi, 2003; Gamor, Amissah, Amissah, & Nartey, 2018). Friedman and Greenhaus (2000) for example examined managerial careers to assess the balance between family and work, debating whether these aspects are 'allies' or 'enemies'. They argued that people deeply involved in their work and career restrict their commitments to family life and vice versa, suggesting it is difficult to find the right balance. Friedman and Greenhaus (2000) found the interaction of work and family resulted in stress and conflict between these two aspects, and noted sacrifice was required at the expense of both work and family.

The private lives of tourism workers remain under-researched (Gamor, et al., 2018; Magnini, 2009; Mulvaney, O'Neill, Cleveland, & Crouter, 2006). Exceptions such as Mulvaney et al's (2006) investigation of hospitality workers that highlight how different groups of hotel workers experience stress. A more recent study in Ghana by Gamor et al., (2018) found hotel employees faced work-family conflicts and family-related issues like lack of spousal support and family demands. Likewise, Di Petro and Bufquin's (2017) study of high levels of staff turnover in the US restaurant industry found links to work status congruence, job satisfaction and organisational commitment.

Despite Baum (2007) highlighting the issues surrounding human resources in the tourism industry, little attention has been given to the working conditions of tour guides who form around three percent of the global tourism workforce (Baum, et al., 2016) and very few studies have sought the voices of the tour guides themselves. In one of the few studies that has sought the perspectives of Jordanian tour guides on their work, Aloudat (2017a) found guides are important observers of tourism performance in a destination as local ambassadors, are cultural brokers, field managers, and potential promotional agents given their day-to-day involvement with tourism processes. Recent studies on Jordanian tour guides however present some conflicting results with Ababneh (2017) suggesting guides face work challenges and the relationship between the guides and their employers lacked structure and professionalism, while Alrawadieh et al. (2019) found guides did not experience significant emotional dissonance and there was no negative impact of emotional dissonance on either quality of work life or life satisfaction.

METHODOLOGY

The use of more qualitative approaches in tourism studies is advocated by researchers to enhance existing knowledge and to gain new insights and understandings of tourism concepts (Riley & Love, 2000). Exclusively scientific positivist methods may have limited application because of their lack of attention to subjective meanings and values which are important aspects of the tourism world (Tribe, 2001). Therefore, post-positivist, qualitative research was deemed appropriate to capture data 'on the perceptions of local actors [tour guides] 'from the inside', [emic] through a process of deep attentiveness, of empathetic understanding (*Verstehen*), and of suspending or 'bracketing' preconceptions about the topics under discussion' (Miles & Huberman, 1994:6). Qualitative research allows exploration of attitudes, motives, interest, responses, conflict, and personality (Tribe, 2001) and brings the researcher into close contact to the actors' perspectives (Denzin & Lincoln, 1994).

As there is a paucity of knowledge on the work-life space of tour guides and particularly from the tour guide's perspective an inductive qualitative approach was adopted for this study that sought to describe the essence of the lived experience of the guides (Charmaz, 2006; Goulding, 2002; Patton, 2002; Locke, 2001; Creswell, 2007)

The primary research question was: How do tour guides perceive their work and its impact on their private lives? The secondary research questions were:

1. What motivates tour guides to enter this career?
2. What are the working conditions and features of a tour guiding career?
3. What are their and others perceptions on their career status?
4. What are the impacts of tour guiding on their private lives?

Semi-structured interviews were used as they offered the flexible exploration of the research topic without adhering to a prescribed schedule of questions, therefore, allowing the researcher to gather first-hand information from the research respondents. Interviews with 29 tour guides were conducted across the two phases of the study, the pilot phase and the main fieldwork phase. The interviews were conducted by one of the authors at a time and place selected by the informants. All interviews were audio-recorded in Jordanian and then transcribed and translated into English by one of the authors who is bi-lingual in English and Jordanian. The time of interviews varied from one to two hours. Details of the participants are summarised in Table 1.

This research was carried out in Jordan as one of the researcher's is Jordanian and has close connections to the Jordanian tour guiding community that assisted in accessing the guides for this study. The participants were all Jordanian tour guides and members of the Jordanian Tour Guides Association (JTGA) which is the professional association representing and registering all tour guides in Jordan. Registration is mandatory. JTGA is a non- profit organisation representing 1200 tour guides. It was established under the Tour Guides Regulations No. 65, 1998 in accordance with article 14 of Tourism Law No, 920, 1998. Services provided by the JTGA include: membership registration and licensing in cooperation with Ministry of Tourism and Antiquities (MOTA), training and an on-line membership database. Other services provided with additional fees include life and health insurance. In Jordan, only registered guides are allowed to work as tour guides. Guides must hold a Bachelor degree and complete examinations organised by the Ministry of Tourism and Antiquities (MOTA), and the Jordan Applied University.

Table 1 Demographic Variables of the Tour Guides in the Study

Acronym	Gender	Age	Years of work experience	Foreign language spoken
Guide 1	M	36	16	English/ French
Guide 2	M	35	13	French
Guide 3	F	42	15	English
Guide 4	M	43	10	English
Guide 5	M	53	15	Germany
Guide 6	M	33	6	English
Guide 7	M	40	6	English
Guide 8	M	32	1	Spanish
Guide 9	M	36	15	French/English
Guide 10	M	38	15	English/ Dutch
Guide 11	M	36	15	English
Guide 12	M	37	13	German/ English
Guide 13	M	45	5	English/ Dutch
Guide 14	M	60	21	German/ English
Guide 15	M	-	5	English
Guide 16	M	40	15	French
Guide 17	M	36	10	French
Guide 18	M	40	17	Italian
Guide 19	M	36	15	French
Guide 20	M	36	15	English/Dutch
Guide 21	M	35	16	French
Guide 22	M	36	15	English/Dutch
Guide 23	M	45	15	English/Dutch
Guide 24	M	37	15	French
Guide 25	M	40	14	French
Guide 26	M	36	15	French/English
Guide 27	M	37	12	German
Guide 28	M	36	13	French
Guide 29	M	36	15	English

Most of the informants were male guides and only one female guide was interviewed reflecting the 65 female guides and 1,113 male guides listed by the Jordan Tour Guide Association (MOTA, 2019). This gender difference may be attributed to cultural constraints affecting the status of women in a conservative country. In Jordan and in almost all Arab countries the woman's primary role is the traditional one of family carer. In addition, while women can gain an education and work, there are restrictions on the kind of work that is considered suitable (e.g. there are barriers to women being drivers, technicians, construction workers, or to do any work that needs physical effort). Tour guiding is an uncommon career for Jordanian women (Aloudat, 2017b). The gendered nature of guiding in Jordan is acknowledged and needs further research however, it was the not the aim of this study to explore this aspect of guiding in Jordan.

The pilot phase involved a focus group with seven experienced tour guides. Selection for participation in the pilot's focus group was based on recommendations from an experienced personal-linked tour guide (the key informant). Following the focus group, two additional tour guides were interviewed to build a broader picture of perspectives and emerging themes. The research themes that emerged after the pilot phase were; motivations for becoming a guide, tour guides' work-life space and the features of a tour guiding career.

The main fieldwork phase was conducted over 12 weeks in Jordan and involved semi-structured interviews with sixteen tour guides in a variety of locations across the country. In the first ten semi-structured interviews purposive sampling was employed to ensure a diversity of guides were interviewed, using the criteria of gender, age, location, work experience, foreign language spoken, and the nationality of tourist groups.

The interviews were conducted in different locations at the convenience of the interviewees and commenced with general questions based on the emergent themes from the pilot phase such as motivations to become a tour guide, how they felt about working as guides, their perceptions of their status in relation to other occupational roles in Jordan and the impact of their work on family and work-life balance.

Thereafter theoretical sampling of a further four tour guides (Strauss & Corbin, 1998:73) was employed whereby the sampling process depended on the emergent concepts from the previous ten interviews. This form of sampling sought additional data to explore the different dimensions of the emergent concepts and to fill any gaps in the analysis. For example, through this form of sampling the researcher sought to sample incidents or events, and not individuals or persons per se, in the light of the emergent concepts.

The data analysis was line-by-line and sometimes word-by-word (Strauss & Corbin, 1998) and aimed at identifying categories and subcategories. Connections between the categories and subcategories were then identified through a process of axial coding which helped the theoretical structure to emerge. This skeleton was then refined through higher level selective coding (Strauss & Corbin, 1998). Constant comparison was employed to generate and analyse the data (Strauss & Corbin, 1998). The process of data collection and analysis continued until no further new evidence emerged and theoretical saturation was achieved (Strauss & Corbin, 1990). The themes that emerged are presented in the Findings section. Reliability was ensured in this study through 'member checking' (Charmaz, 2006). It enables a check on whether the findings reflect the participants' reality and whether researcher interpretations are understandable by participants. In this study, member checking was conducted with three interviewees who were presented with initial findings from the interviews and requested to provide feedback as recommended by Guba and Lincoln (1989).

FINDINGS

The data provided rich insights into how the tour guides in this study perceive their work and work-life space. Overall, two key thematic areas with subthemes emerged that are discussed in this section of the chapter with quotations from the tour guides illustrating the key issues. The analysis revealed two key themes: first, tour guides' perspectives on their motivation for career choice and working conditions; and second, tour guides' and others perspectives on their work and the impact of tour guiding on their private lives.

Tour Guides' Perspectives on their Motivation for Career Choice, and Working Conditions

Motivation for Career Choice

Individuals possess many intrinsic and extrinsic motivations for working (Thomas, 2002). The former are internal factors where attributes relate to an individual's psychological makeup and their personal interests and preferences, and the latter represent the external influences of that desire which include pay and other economic benefits, job security, and autonomy. This study found that guides initially choose guiding as a career for many reasons: the promise of a good income, freedom, enjoyment from meeting people from diverse cultures and using relevant qualifications.

Above all, income was one of the main pull factors. It is difficult to precisely quantify the income of Jordanian tour guides because of their freelance status, but it is estimated they work 25 days a month and can earn 1000 to 2000 Jordanian Dinar (JD) (US $1400- 2800 equivalent) per month. In comparison, a teacher in Jordan earns between 300-400 JD per month (US $ 420-560 equivalent) (Civil Service Bureau, 2018). These calculations are based on a minimum of 50 JD per day for a tour guide. One guide said:

I went to a friend and I asked him to find me a job in the Ministry of Tourism. He arranged a meeting for me with the Minister who told me 'we don't have job opportunities, but I advise you to work as a tour guide and your income will be better than mine'... (Guide 13, male, 5 years' experience, English/ Dutch speaking)

The guides also claimed tipping was an additional source of income. The possibility of tips was a strong economic, as well as an emotional motivator for the guides:

When a tourist gives me my tip I consider it part of my wages and a sign of appreciation for the service that I gave to him ... (Guide 20, male, 15 years' experience, French / English speaking)

Tips differed from one tour to another and according to the guides depended on the nationality of the tour group, on the nature of the tour, the age of the travellers and sometimes the group member's educational background.

Although guides were generally satisfied with their earnings, their income was erratic and fluctuated because of seasonality, international and regional political conditions, and seasonal cancellations (addressed later in the chapter). These types of circumstances can be highly disruptive, as guides often book their work in advance and plan their schedules and budgets. Such cancellations cause financial uncertainty. This is particularly difficult when male guides contribute the main household income. As one guide explained:

the income is excellent, but tourism is not guaranteed. ... it is related to the stability and safety ... For example, in 2003, during the Iraqi invasions, there were a lot of cancellations ... (Guide 6, male, 6 years' experience, English speaking).

Tour guides depend solely on their income from working as guides and have no additional income. This is because under The Jordanian Tour Guides' Law, a tour guide must not practice any other profession (MOTA, 1998). There is also an agreement between the JTGA and the Jordan Inbound Tour Operators Agency (JITOA) to fix the minimum wage to 50 JD per day (US $ 43 equivalent). However, there is no official requirement to enforce this pay rate and some guides ask more in the peak seasons, or if there is high demand for a particular foreign language. Conversely, tour guides may accept lower wages, especially if they are inexperienced or if it is low season.

Income instability due to seasonality and political unrest poses challenges, offering little security given the constraints on the number of working days and their income:

... normally I work 20 days a month, but if there are problems the number decreases in a drastic way in 2003 I didn't work except for one or two weeks. Tourism is an industry which is sensitive to regional and international events (Guide 16, male, 15 years' experience, French speaking)

However, some guides enjoyed the freedom and autonomy of the job. Since they are freelancers, they are not restricted to working in crowded offices or under the scrutiny of managers.

... I am happy in my work. There is stress in it, but you are responsible for yourself; you are the manager ... (Guide 13, male, 5 years' experience, English/ Dutch speaking)

Others commented on the diverse nature of the occupation:

I heard about it [tour guiding] from my colleagues and that there is no routine as in office work ... with computer and papers ...; every week you are with a different nationality and in a different site. It is field work ... (Guide, 18, male, 17 years' experience, Italian speaking)

Working Conditions of Tour Guiding

Three subthemes on working conditions of tour guiding emerged from the data: the mechanics of a working day, insurance expenses, and unpredictable income.

The Mechanics of a Working Day

Guides stated that every working day was different. They might work in one place or travel between different attractions or destinations. Their clients may be an individual, or a group consisting of thirty or forty persons and may come from different occupational groups, as well as originating from different countries or cultures. Guides may work with their clients for one hour, one day, or for a couple of weeks or more.

The work of the guide starts from the moment the agency tells the guide about the group. Guides need to obtain the group's details and these preparations make it is easier for the guide to "tune into" the group and make the tour more interesting and "rewarding" for all the group's members. The following extracts are from Guide (1) who described his preparation:

As soon as the agency tells you about the group or the job, you need to go to them [agency] to get some explanation about the tour and the group itself. ... you just take the documents such as the visa list, etc

...I always look at the visa list to check the age, the gender of the group with or without children (Guide 1, male, 16 years' experience, English/French speaking)

Tour guides act as administrators and negotiate the initial encounter with the group before they become a cultural broker, cicerone, leader, ambassador, promoter, information giver, and customer-service provider (Aloudat, 2017; Weiler & Black, 2015; Cohen, 1985). Having key information on the group in advance gives the tour guides the opportunity to prepare for the trip, particularly when there is a group with specialised needs regarding their culture, the purpose of their trip, their physical conditions, their gender or age.

Guide (3) expressed how important it was to be well prepared when she meets her group. This entails not only keeping her knowledge up-to-date, but even by travelling abroad to nearby destinations to broaden her knowledge. She described the extent of her efforts and her enthusiasm to know more:

I went to Sina [in Egypt] because the tourists who come [to Jordan] know Sinai and I must also know it ... I went to Cappadocia [in Turkey]. I asked myself why the tourists when they came [to Jordan] compared Cappadocia with Petra and its rocks and geography so I went there to know... I know Palestine, Besan, Jerusalem, Haifa and the roads of Jerusalem, so when I talk about Jerusalem on the map that is in Madaba [Jordan] I tell the tourists that tomorrow you will go from this door to this door to see this... They realise that I also know about these things ... (Guide 3, female, 15 years' experience, English speaking)

This degree of resourcefulness, commitment and meticulous planning characterises some employees in the tourism industry such as tour guides, despite their perceived low status occupations. Enjoyment of the tourism sector itself remains an underlying magnet that keeps people in this industry (Carson, Carson, & Lundmark, 2017; Hjalager, 2003; Riley, et al., 2002). Above all, Burns (2010) argues that being in the tourism industry they will learn more, meet people, and experience different cultures. However, as in any other occupation, commitment to work differs from one person to another, depending on the level of their job satisfaction (Hall, 1994). Thorough preparation for the tour is often accompanied by good-time management and guides must plan to arrive earlier than tourists and allow extra time for any delays.

Guide (1) described his first encounter with groups:

The first day I make a half day and I prefer to start it in the afternoon to leave the group member's time to rest and relax. In the morning, I prefer to be at least half an hour early ... As soon as you meet your group you must show them that you are self-confident, that you are knowledgeable, are not afraid and are very friendly. (Guide 1, male, 16 years' experience, English/French speaking)

This guide's views have clear links to Goffman's (1959) concept of self-presentation and the performance dimension, where the person's performance regarding self-presentation is described as an individual's effort to create specific impressions in the minds of other people, as well as emotional labour a concept explored in a few tour guiding studies (Houge Mackenzie & Kerr, 2012; Houge Mackenzie & Kerr, 2014; Alrawadieh et al, 2019). Through their first interactions, tour guides generate an impression they are successful and professional guides. These self-presentational techniques have been called attributive tactics that attempt to convey that a person has these characteristics (Leary, 1995, p. 17). They are demonstrated by people in their daily social interactions, including: self-description (describing oneself in a way that conveys a desired impression to others); attitude expressions (expressing attitudes to connote that one possesses certain attributes); and non-verbal behaviour (facial expressions, gestures, body positions and movements). These techniques are an important career strategy for tour guides and can help in their self-promotion. (Leary, 1995).

Insurance Expenses

As independent contractors, tour guides bear the cost of medical treatment for themselves and their families. Moreover, as self-employed people, they are not entitled to government social security services. These issues caused concern and anxiety amongst the guides in this study as these costs had an economic impact on them and their families.

Guide (2) outlined the economic disadvantages of his career:

... there are a lot of problems in this career: there is no social security, no health insurance and no stable income. It is related very closely to political security. ... During my professional life, I went 2 years without working ... (Guide 2, male, 13 years' experience, French speaking guide)

Guide (11) echoes this insecurity:

They [tourism bodies] must provide health insurance and social security for us. This will affect the performance of tour guides positively ... When you sleep you think: Where will I get money from money? I might be sick? Will the tourists come? and this makes psychological tension ... (Guide 11, male, 15 years' experience, English speaking)

This guide indicated that tour guides would like to have health insurance and social security services, particularly as they may encounter accidents, injuries or sickness as part of their work.

Unpredictable Income

Due to the nature of the tourism industry in general and specifically tour guiding, income is unpredictable and fluctuates. Seasonality poses challenges for tour guides, offering little security as it places constraints on the number of working days. Some studies suggest that to overcome the effects of seasonality, tour guides need to have some other work to fill in the time between seasons such as working in exhibitions and conferences, field promotions, or translating (Pond, 1993), while others open their own tour agencies (Pond, 1993). In this study, as discussed earlier in the chapter, as a result of Jordanian legislation their inability to take on other employment together with other factors such as cancellations created significant anxiety for the guides.

The Jordanian legislation prevents guides engaging in any commercial activities and seeking alternative sources of income, and their employment is affected by the low season or unforeseen events such as wars or natural disasters. Guide (1) stated:

... tour guides have a special case. Why? Because the hotels, if there is no international tourism, they can substitute this from other activities like conferences, domestic tourism, or the Arabs instead of the Europeans. And the transportation companies, they can offer internal trips for Jordanians and external

Muslim pilgrimage trips ... and we can't practice any other job rather than tour guiding it is forbidden by law ... (Guide 1, male, 16 years' experience, English / French speaking)

Cancellation is a common feature of the tourism industry, especially in an area with political unrest such as Jordan. Such cancellations also cause instability in the guide's financial situation as they are often the main household contributor. Guide (14) notes:

... tourism is not guaranteed. There are always cancellations due to political unrest... (Guide 14, male, 21 years' experience, German/ English speaking).

Almost all the guides asserted that their economic situation was unstable because of the political turbulence of the region and one guide commented on wars that have broken out in the Middle East. When he was asked about the number of his working days he said:

Except for the years of wars, the Iraqi war and the Israeli-Lebanese war,) ... we can't count these years, I only worked for 20 days in those years ... but on average I work 250 days a year (Guide 11, male, 15 years' experience, English speaking)

Such uncertainty directly affects the quality of life of the guides and their families that is discussed in the following section.

Tour Guides' Perspectives on Their Work and the Impact of Tour Guiding on Their Private Lives

The second main theme emerging from the data analysis was tour guides' perspectives on their work and the impact of tour guiding on the private lives of the guides. Three sub-themes arose from the data: job anxiety, perceptions of guides themselves and by others, and work-life balance.

Job Anxiety

Tour guiding involves many overlapping responsibilities that must be undertaken in a balanced way (Weiler & Black, 2014). For the guides in this study the overlapping roles of tour guides such as company representative, tour manager and cultural broker often resulted in pressure and worry. These tour guides indicated they needed to find a "workable balance" in satisfying different industry stakeholders, especially the travel agency and the clients. Careful preparation and briefing were among the strategies used by tour guides to enhance their performance, but guides often struggled. Guide (1), described his job anxiety as:

After I know about the group I am prepared, but I still worry. You always worry, even the night before you start your trip there is a lot of pressure. ... you keep worrying till you start and hold the microphone and introduce yourself ... My hands become cold and sometimes I am sweating, because I want to see what the people are like ... At the end of the day when I see the people satisfied, this makes me very happy and I forget all the discomfort ... (Guide 1, male, 16 years' experience, English/French speaking)

All the guides in this study asserted that the first moments of meeting the group are significant in making sure the tourists are satisfied. The first service encounter highlights the personality of the guides, their self-confidence, and their level of knowledge. Guide (19) said:

...You should show a strong character in front of the tourists and not hesitate or appear stressed ... the personality comes first, how you show yourself in front of the group. It is not easy to stand in front of 40 persons at the beginning. During the first 10 minutes they are judging the guide; they watch him, how he talks, how he gives information, sometimes with no consideration of his language skills. (Guide 19, male, 15 years' experience, French speaking)

Tourists' first impressions are crucial as the guide's personality is taken as an indicator of the tour's likely success (Pond, 1993). The interviewees highlighted Pond's (1993) finding that an impression of a guide as being nervous, dishevelled, lacking confidence, unenthusiastic or tired can affect the mood of the tour group. The tour guides in this study claimed they have to present themselves using two kinds of self-presentational tactics: verbal techniques (e.g. their informative and leadership skills, listing their experiences and skills) and non-verbal behaviour (e.g. appearance, facial expressions, body position, and gestures (Noe, Uysal, & Magnini, 2010)

To gain regular employment and earn a reasonable living, guides must also satisfy the expectations of the travel agency that offers them work. This careful balancing act between the agency and the clients seeks to ensure they gain sufficient working days and thus earnings as a freelance guide. Therefore, it is important for tour guides to successfully present themselves to the employers to advance their financial rewards. The self-perception of guides and views of others are key in this endeavour.

Perceptions of Guides Themselves and by Others

It has been argued that job satisfaction is related to the importance of the relative prestige of the occupation (Riley, et al., 2002). Participating guides expressed some dissatisfaction with their occupational status compared to the status of other occupations in Jordan. Although all tour guides in Jordan are required to hold an undergraduate degree which is associated with a good educational level and social status, guides perceived their occupational status was low. Comparing his career with the public sector, one guide (19) observed:

... we were happy with our jobs because our status was good and we made money, but when we compared ourselves with public employees .... we found that those who got public sector jobs have now a special status in the community. We have zero status ... The attitude of the community toward the tour guide is not good ... All jobs have a progress career structure, except tour guiding ... it has a declining career structure, you start from the top when you are young and active, then start to decline... (Guide 19, male, 15 years' experience, French speaking)

This reflects a collective view regarding their position and status compared to people working in the public sector. Guide (19) described guiding as a "top-down" career in contrast with other jobs in which the employees progress systematically up the career ladder. Interviewees indicated that a guiding career offers limited or few opportunities to advance from one level to another as the tasks of tour guides generally remain the same. Unlike other jobs, there is generally no differentiation in terms of incentives or rewards between the guides regardless of their years of experience or their educational background.

Guides felt there is little understanding about their work by some members of the Jordanian community. This adverse image has effects which extend into the tour guides' private and social life. One guide (Guide 23) faced a problem when he wanted to get married, and found it difficult to obtain the consent of his partner's parents:

The view of society of tourism is still negative. For example, when I wanted to get married, I asked for the hands of more than 15 girls, but was refused by their parents because I was working as a tour guide … (Guide 23, male, 15 years' experience, English/ Dutch speaking)

Work-Life Balance

Almost all the guides were married with families. In Jordan, the community consists of tribes or big families, and relationships between the families have, until the last twenty years been relatively strong (Kazaeleh, 2009). Many of the guides as a result of the nature of their work and extended periods away from home felt isolated from their family and society, making them feel like strangers in their own communities. This social dislocation and estrangement is particularly significant in a conservative and traditional culture like Jordan (Shoup, 2007). Like many of the guides Guide 16, was eager to talk about his family and social life:

Social relations are zero ... guiding affects me very much. Firstly, I am often away from home. Each week I am away from home for five days, and my kids are growing up while I am far from them. As for my parents because they are not in the same city it takes months before I visit them. ..., there are many social occasions which I have no idea about … Even my parents forget about me, because they know there is no hope that I will be able to come.

Family life and social relations were disrupted by the irregular nature of tour guiding and the heavy demands on time, yet guides did not want to refuse a day's work. At the micro-level, responses indicate that wives/husbands and children were frustrated and dissatisfied when a parent works as a tour guide. This guide described his family relations as follows:

Personally, during the last season my wife said to me, 'is it possible to add us to your schedule?' … (Guide 11, male, 15 years' experience, English speaking)

DISCUSSION

In this study, the perspectives of tour guides are articulated and emerge as rich narrative vignettes of their work, private lives, and experiences. It is clear these tour guides initially chose their career for many reasons, including the potential to receive a good income, freedom, enjoyment in meeting people and because of their relevant qualifications. These reasons are congruent with other studies that demonstrate the attractiveness of tourism jobs (Carson, et al., 2017; Hjalager, 2003; Riley et al., 2002). The freedom and autonomy makes guiding an attractive career choice and was a reason for job satisfaction, a finding that aligns with previous studies that link job satisfaction to the degree of independence and freedom a person has from hierarchical control (Coon & Mitterer, 2012; Watson, 1995; Hall, 1994). This finding is supported by Alrawadieh et al's (2019) more recent study among Jordanian tour guides that unexpectedly found their quality of work life and life satisfaction were not negatively influenced by the emotional dissonance.

The guides considered tour guiding as low status compared to other occupations in Jordan, confirming other tour guiding studies (Ap & Wong, 2001) that have demonstrated the tourism industry's traditional image of low pay, long hours, and low status. The sample group were university graduates who were willing to work as tour guides even though they perceived it was a low status occupation. This finding contrasts with Pizam (1982) who argued that university graduates would rather stay without work than work in an occupation with a low social status. This finding may be explained by the high unemployment rates in Jordan as it reached 18.2% during the first quarter of 2018 (Department of Jordan Statistics, 2018) therefore, guides would prefer to remain in their jobs regardless of its low status given that job opportunities in the Jordanian labour market are limited.

The guides also felt their occupation was deemed socially undesirable by the broader community, reflecting the community's lack of awareness and understanding about the work of a tour guide. The status of tour guiding expressed by tour guides in this study is akin to those in Ap and Wong's work in Hong Kong (2001) where guiding is regarded as 'unglamorous', especially by parents who do not understand or appreciate the role of tour guides, and think that the job of the tour guide is only to take the tourists shopping.

This study found that the image of tour guiding may be affected by the cultural norms of a destination which builds on the work of researchers such as Watson (1995) who found that career status is affected by the existence of traditional criteria of an occupational status within a society's culture. Therefore, it is difficult to generalise the low status of tourism jobs as this usually relates to the local context of the tourism workers (Szivas & Riley, 1999).

The findings revealed the guides were expected to perform a wide range of roles that supports the existing literature (Weiler & Black, 2015) however, this range of roles and expectations from a number of different stakeholders created pressure and anxieties for the guides. This supports the work of Ababenh on Jordanian guides (2017) who found there were conflicts between the guides and their employers and Yetgin and Benlingiary (2019) who found tour guides in Turkey experienced exhaustion and a sense of inefficacy that had a negative effect on all aspects of the tour guides' occupational commitment levels as a result of economic anxiety and burnout.

The data also revealed challenges for the guide's private lives and achieving a work-life balance especially as they experience seasonal employment and tour cancellations (particularly in the Middle East), a situation common among tourism workers (Commons & Page, 2001; Mansfield & Pizam, 2006). These findings support other similar studies on the difficulties facing hospitality and tourism employees, and particularly guides in resolving the balance between work and private life (Friedman & Greenhaus, 2000; Ababneh, 2017; Gamor *et al.*, 2018, Cleveland, *et al.*, 2007).

CONCLUSION AND FUTURE RESEARCH

The findings of this study make an important contribution to the field of tour guiding by revealing from the tour guides' perspectives the challenges and issues facing guides in their work and private lives. The study illustrates how the profession of tour guiding can affect the private life and status of the guides. In adding new insights into the day-to-day realities of tour guiding, this study addresses the paucity of knowledge regarding the work-life space of tour guides (Riley, et al., 2002; Ross, 1993; Tribe & Lewis, 2003). Critically, the study outcomes provide new understandings of the work-life space of tour guides, from their perspective.

Few studies have explored the occupation of tour guiding through the guides' personal perspectives and therefore this study offers a unique window into the work and private lives of tour guides who are an important, yet undervalued occupational group in the tourism industry and the broader community. This study found that while tour guides initially choose their career primarily to gain a good income and were generally satisfied with their earnings, employment was erratic and fluctuated due to seasonality, and international and regional political conditions. Both guides and the broader Jordanian community perceived that guiding was a low social status profession. Of significance, this study demonstrated that the occupation of tour guiding negatively impacts on the lives of tour guides and their families as they struggle to achieve a satisfactory work-life balance.

The study findings have important implications for the tourism industry and tour guide employers. The industry needs to consider providing more rewards, better working conditions, career path progression opportunities, professional and training development opportunities, and scheduling guides' work to achieve a satisfactory work-life balance (Alazaizeh et al., 2019; Ap & Wong, 2001; Weiler & Black, 2015). Implementation of these strategies is likely to enhance the professionalism and self-image of tour guides, but also result in guides continuing to work in the tourism industry, ensuring employers have professional, experienced and committed guides that deliver quality tours. Although tour guiding standards, qualifications, working conditions and wages vary across the world, the findings of this study will be of interest to national tour guide associations and international bodies such as the World Federation of Tourist Guide Associations.

While the findings are plausible they are limited to the substantive area of Jordan and in addition, all except one tour guide interviewed in this study was male, however, this reflects the high proportion of males working as tour guides in Jordan.

Future avenues of research could seek to explore the differences in perceptions of male and female tour guides regarding their career and work-life balance. In examining the status of the tour guiding in relation to diverse cultural contexts, future studies may also ascertain how experiences are affected by the cultural norms of a specific destination. For instance, exploring the image of tour guiding in Western countries compared to those of Asian countries. Further areas of research could also focus on investigating the public perceptions of the status of tourism careers and the people working in the tourism industry, and comparing the working conditions of freelance guides compared to travel agent employees.

REFERENCES

Ababneh, A. (2017). Tourist guides and job-related problems, analysis and evidence from Jordan. *Journal of Quality Assurance in Hospitality & Tourism, 18*(2), 200–217. doi:10.1080/1528008X.2016.1189867

Alazaizeh, M. M., Jamaliah, M. M., Mgonja, J. T., & Ababneh, A. (2019, November 2). Tour guide performance and sustainable visitor behaviour at cultural heritage sites. *Journal of Sustainable Tourism, 27*(11), 1708–1724. doi:10.1080/09669582.2019.1658766

Aloudat, A. Sh. (2017a). *The worldview of tour guides: A grounded theory study*. Noor Publishing.

Aloudat, A. Sh. (2017b). Being in the minority: The challenges of being a female tour guide in Jordan. *Networked Innovation in Guided: The 5th International Research Forum on Guided Tours.*

Alrawadieh, Z., Cetin, G., Dincer, M., & Dincer, F. (2019). The impact of emotional dissonance on quality of work life and life satisfaction of tour guides. *The Service Industries Journal.*

Ap, J., & Wong, K. (2001). Case study on tour guiding: Professionalism issues and problems. *Tourism Management, 22*(5), 551–563. doi:10.1016/S0261-5177(01)00013-9

Baum, T. (2007). Human resources in tourism: Still waiting for change. *Tourism Management, 50*(6), 1383–1399. doi:10.1016/j.tourman.2007.04.005

Baum, T., Cheung, C., Kong, H., Kralj, A., Mooney, S., Thanh, H., Ramachandran, S., Ruzic, M., & Siow, M. (2016). Sustainability and the tourism and hospitality workforce: A thematic analysis. *Sustainability, 8*(8), 809–830. doi:10.3390u8080809

Black, R., & Ham, S. (2005). Improving the quality of tour guiding: Towards a model for tour guide certification. *Journal of Ecotourism, 4*(3), 178–195. doi:10.1080/14724040608668442

Black, R., Ham, S., & Weiler, B. (2001). Ecotour guide training in less developed countries: Some preliminary research findings. *Journal of Sustainable Tourism, 9*(2), 147–156. doi:10.1080/09669580108667395

Bowie, D., & Chang, C. (2005). Tourist satisfaction: A view from a mixed international guided package tour. *Journal of Vacation Marketing, 11*(4), 303–322. doi:10.1177/1356766705056628

Brymer, R. A., Perrewe, P. L., & Johns, T. R. (1991). Managerial job stress in the hotel industry. *International Journal of Hospitality Management, 10*(1), 47–58. doi:10.1016/0278-4319(91)90006-4

Burns, J. (2010). *Career opportunities in travel and hospitality*. Infobase.

Butler, R. (2001). Seasonality in tourism: Issues and implications. In T. Baum & S. Lundtorp (Eds.), *Seasonality in tourism* (pp. 5–21). Elsevier Science Ltd. doi:10.1016/B978-0-08-043674-6.50005-2

Carson, D., Carson, De., & Lundmark, L. (2017). *Tourism, motilities, and development in sparsely populated areas*. Routledge. doi:10.4324/9781315665665

Chan, A., Hsu, C., & Baum, T. G. (2015). The Impact of Tour Service Performance on Tourist Satisfaction and Behavioural Intentions: A Study of Chinese Tourists in Hong Kong. *Journal of Travel and Tourism Marketing, 32*(1-2), 18-33. org/ doi:10.1080/10548408.2014.986010

Charmaz, K. (2006). *Constructing grounded theory: A practical guide through qualitative analysis*. Sage.

Christie, M., & Mason, P. (2003). Transformative tour guiding: Training tour guides to be critically reflective practitioners. *Journal of Ecotourism*, *2*(1), 1–16. doi:10.1080/14724040308668130

Civil Service Bureau. (2018). *Civil Bureau Bylaw*. Available at: http://www.csb.gov.jo/csb/Legislations/bylaw

Cleveland, J. N., O'Neill, J. W., Himelright, J. L., Harrison, M. M., Crouter, A. C., & Drago, R. (2007). Work and family issues in the hospitality industry: Perspectives of entrants, mangers, and spouses. *Journal of Hospitality & Tourism Research (Washington, D.C.)*, *31*(3), 275–298. doi:10.1177/1096348007299919

Cohen, E. (1985). The tourist guide: The origins, structure, and dynamics of a role. *Annals of Tourism Research*, *12*(1), 5–29. doi:10.1016/0160-7383(85)90037-4

Collins, V. (2000). Becoming a Tour Guide, the Principals of Guiding and Site Interpretation. Academic Press.

Commons, J., & Page, S. (2001). Managing seasonality in peripheral tourism regions: The case of Northland, New Zealand. In T. Baum & S. Lundtorp (Eds.), *Seasonality in tourism* (pp. 153–172). Pergamon. doi:10.1016/B978-0-08-043674-6.50013-1

Coon, D., & Mitterer, J. (2012). *Introduction to psychology gateways to mind and behaviour* (13th ed.). Wadsworth Cengage Learning.

Cooper, C., & Shepherd, R. (2014). The relationship between tourism education and tourism industry: Implications for tourism education. *Tourism Recreation Research*, *22*(1), 34–47. doi:10.1080/02508281.1997.11014784

Costa, C., Panik, E., & Buhalis, D. (2013). *Trends in European Tourism Planning and Organization*. Channel View Publication. doi:10.21832/9781845414122

Creswell, J. W. (2007). *Qualitative inquiry and research design, Choosing among Five Approaches* (2nd ed.). Sage.

Cullen, E., Kordey, N., Schmidt, L., & Gaboardi, E. (2003). *Work and family in the e- work era*. IOS Press.

Dahles, H. (2002). The politics of tour guiding: Image management in Indonesia. *Annals of Tourism Research*, *29*(3), 783–800. doi:10.1016/S0160-7383(01)00083-4

Denzin, N., & Lincoln, Y. (1994). *Handbook of Qualitative Research*. SAGE Thousand Oaks.

Department of Statistics. Jordan Government. (2018). Retrieved from: http://dosweb.dos.gov.jo/product-category/jordan-in-figuers_yearbook/

DiPietro, R., & Bufquin, D. (2018). Effects of work status congruence and perceived management concern for employees on turnover intentions in a fast-casual restaurant chain. *Journal of Human Resources in Hospitality & Tourism*, *17*(1), 38–59. Advance online publication. doi:10.1080/15332845.2017.1328260

Friedman, S., & Greenhaus, J. (2000). *Work and family, allies or enemies? What happens when business professionals confront life choices*. Oxford University Press. doi:10.1093/acprof:oso/9780195112757.001.0001

Gamor, E., Amissah, E., Amissah, A., & Nartey, E. (2018). Factors of work-family conflict in the hospitality industry in Ghana. *Journal of Human Resources in Hospitality & Tourism, 17*(4), 482–501. doi:10.1080/15332845.2017.1328263

Goffman, E. (1959). *The Presentation of Every Day Life*. Penguin Books.

Goulding, C. (2002). *Grounded Theory: A Practical Guide for Management, Business and Market Researchers*. Sage. doi:10.4135/9781849209236

Guba, E., & Lincoln, Y. (1989). *Fourth generation evaluation*. Sage.

Hall, R. (1994). *Sociology of work: Perspectives, analyses, and issues*. Pine Forge Press.

Herzberg, F., Mausner, B., & Snyderman, B. (2010). *The Motivation to Work* (12th ed.). Transaction.

Hjalager, A. (2003). Global tourism careers? Opportunities and dilemmas facing higher education in tourism. *Journal of Hospitality, Leisure, Sport and Tourism Education, 2*(2), 1–12. doi:10.3794/johlste.22.35

Holloway, J. (1981). The guided tour: A sociological approach. *Annals of Tourism Research, 8*(3), 377–402. doi:10.1016/0160-7383(81)90005-0

Houge-Mackenzie, S., & Kerr, J. (2013). Stress and emotions at work: An adventure tourism guide's experiences. *Tourism Management, 36*, 3–14. doi:10.1016/j.tourman.2012.10.018

Houge-Mackenzie, S., & Kerr, J. (2014). The psychological experience of river guiding: Exploring the protective frame and implications for guide well-being. *Journal of Sport & Tourism, 19*(1), 1, 5–27. doi:10.1080/14775085.2014.967796

Huang, S., Hsu, C., & Chan, A. (2010). The tour guide performance and tourist satisfaction: A study of the package tours in Shanghai. *Journal of Hospitality & Tourism Research (Washington, D.C.), 34*(1), 3–33. doi:10.1177/1096348009349815

Jafari, J. (1985). Editor's page. *Annals of Tourism Research, 12*(1), 1–2. doi:10.1016/0160-7383(85)90035-0

Jennings, G., & Weiler, B. (2006). Mediating meaning: Perspectives on brokering quality tourist experiences. In G. Jennings & N. Nickerson (Eds.), *Quality tourism experiences* (pp. 57–78). Elsevier. doi:10.1016/B978-0-7506-7811-7.50010-9

Jordan Tour Guides Association. (2019). Retrieved from: http://www.tourguides.com.jo/index.php?pg=QUJPVVQgSlRHQQ==&sub=T3VyIEhpc3Rvcnk=

Kazaeleh, A. (2009). *Conflict between social values and organizational values in educational administration*. Daralhamed Publication. (in Arabic)

Kusluvan, S. (2003). *Managing employee attitudes and behaviours in the tourism and hospitality*. Nova Science.

Kusluvan, S., & Kusluvan, Z. (2003). Perceptions and attitudes of undergraduate tourism students towards working in the tourism and hospitality industry in a developing economy. In S. Kusluvan (Ed.), *Managing employee attitudes and behaviours in the tourism and hospitality industry* (pp. 76–98). Nova Science.

Leary, R. (1995). *Social psychology series. Self-presentation: Impression management and interpersonal behaviour.* Brown & Benchmark Publishers.

Locke, K. (2001). *Grounded Theory in Management Research.* Sage publications.

Macdonald, S. (2006). Mediating heritage: Tour guides at the former Nazi Party Rally Grounds, Nuremberg. *Tourist Studies, 6*(2), 119–138. doi:10.1177/1468797606071473

Magnini, V. P. (2009). Understanding and reducing work-family conflict in the hospitality industry. *Journal of Human Resources in Hospitality & Tourism, 8*(2), 119–136. doi:10.1080/15332840802269643

Mansfeld, Y., & Pizam, A. (2006). *Tourism security and safety from theory to practice.* Elsiever Butterworth-Heineman.

Miles, M., & Huberman, M. (1994). *Qualitative Data Analysis: an Expanded Sourcebook.* SAGE.

Ministry of Tourism and Antiquities. Jordan Government. (1998). *Legislations.* Retrieved from http://www.mota.gov.jo/Contents/Law_Regulations.aspx

Mulvaney, R., O'Neill, J., Cleveland, J., & Crouter, A. (2006). A model of work-family dynamics of hotel manager. *Annals of Tourism Research, 34*(1), 66–87. doi:10.1016/j.annals.2006.07.002

Noe, F., Uysal, M., & Magnini, V. (2010). *Tourist customer service satisfaction an encounter approach.* Routledge. doi:10.4324/9780203852361

Nyaupane, P., & Timothy, D. J. (2013). Assessing visitors' preferences of various roles of tour guides in the Himalaya. *Tourism Analysis, 18*(1), 45–49. doi:10.3727/108354213X13613720283647

Parsons, H., Houge Mackenzie, S., & Filep, S. (2019). Facilitating self-development: How tour guides broker spiritual tourist experiences. *Tourism Recreation Research, 44*(2), 141–152. Advance online publication. doi:10.1080/02508281.2019.1582159

Patton, M. (2002). *Qualitative Evaluation and Research Methods* (3rd ed.). SAGE.

Pearce, P. (1984). Tourist- guide interaction. *Annals of Tourism Research, 11*(1), 129–146. doi:10.1016/0160-7383(84)90100-2

Pizam, A. (1982). Tourism manpower: The state of the art. *Journal of Travel Research, 21*(2), 5–9. doi:10.1177/004728758202100204

Pond, K. (1993). *The professional guide.* VNR.

Poudel, S., & Nyaupane, G. P. (2013). The role of interpretive tour guiding in sustainable destination management: A comparison between guided and non-guided tourists. *Journal of Travel Research, 52*(5), 659–672. doi:10.1177/0047287513478496

Riley, M., Ladkin, A., & Szivas, E. (2002). *Tourism employment analysis and planning: Aspects of tourism* (Vol. 6). Channel View Publications. doi:10.21832/9781873150382

Ringer, G. (2013). *Destinations: Cultural landscapes of tourism.* Routledge. doi:10.4324/9780203441381

Ross, G. F. (1993). Tourism and hospitality work interest and motivation among potential employees. *Hospitality Research Journal.*, *16*(2), 17–27. doi:10.1177/109634809301600204

Ross, G. F. (1997). Career stress responses among hospitality employees. *Annals of Tourism Research*, *24*(1), 41–51. doi:10.1016/S0160-7383(96)00032-1

Salazar, N. (2005). Tourism and glocalization "local" tour guiding. *Annals of Tourism Research*, *32*(3), 628–646. doi:10.1016/j.annals.2004.10.012

Scherle, N., & Nonnenmann, A. (2008). Swimming in cultural flows: Conceptualising tour guides as intercultural mediators and cosmopolitans. *Journal of Tourism and Cultural Change*, *6*(2), 120–137. doi:10.1080/14766820802364147

Shoup, J. (2007). *Culture and Customs of Jordan*. Greenwood Publishing.

Shryock, A. (2004) The new Jordanian hospitality: house, host, and guest in the culture of public display *Comparative Studies in Society and History*, *46*, 35-62.

Sobh, R., Belk, R. W., & Wilson, J. A. J. (2013). Islamic Arab hospitality and multiculturalism. *Marketing Theory*, *13*(4), 443–463. doi:10.1177/1470593113499695

Strauss, A., & Corbin, J. (1990). *Basics of qualitative research grounded theory procedures and techniques* (2nd ed.). Sage.

Strauss, A., & Corbin, J. (1998). *Basics of qualitative research: Techniques and procedures for developing grounded theory* (2nd ed.). Sage.

Sun-Ah Ponting, S., Wearing, S., & Black, R. (2010) Exploring Practitioner Conceptualisations of Professionalism among Australian Ecotour Guides. In *CAUTHE, Proceedings of the 20th Annual CAUTHE Conference*. University of Tasmania.

Szivas, E., & Riley, M. (1999). Tourism employment during economic transition. *Annals of Tourism Research*, *26*(4), 747–771. doi:10.1016/S0160-7383(99)00035-3

Thomas, K. W. (2002). Intrinsic Motitvation at Work: Building Energy & Commetmints. San Francisco: Berett-Koehler.

Tribe, J. (2001). Research Paradigms and the Tourism Curriculum. *Journal of Travel Research*, *392*(4), 442–448. doi:10.1177/004728750103900411

Tribe, J., & Lewis, A. (2003). Attitudes of the young to careers in hospitality and tourism: Review and recommendations. In S. Kusluvan (Ed.), *Managing employee attitudes and behaviours in the tourism and hospitality industry* (pp. 67–76). Nova Science.

Van Dijik, P. A., Smith, L. D., & Cooper, B. K. (2007). Are you for real? An evaluation of the relationship between emotional labour and visitor outcomes. *Tourism Management*, *32*(1), 39–45. doi:10.1016/j.tourman.2009.11.001

Watson, T. (1995). *Sociology, work, and industry*. Routledge.

Weiler, B., & Black, R. (2015). *Tour guiding research: Insights, issues, and implications*. Channel View Publications.

Weiler, B., & Ham, S. (2002). Tour guide training: A model for sustainable capacity building in developing countries. *Journal of Sustainable Tourism*, *10*(1), 52–69. doi:10.1080/09669580208667152

World Federation of Tourist Guides Association. (2019). *World-wide survey of tourist guides*. Retrieved from: http://wftga.org/membership/surveys-tourist-guiding-worldwide

World Travel and Tourism Council. (2018). *Travel & Tourism Economic Impact 2018*. World Travel and Tourism Council.

World Travel and Tourism Council. (2019). *Travel & Tourism Economic Impact 2019 World*. London UK: WTTC. Retrieved from: https://www.wttc.org/-/media/files/reports/economic-impact-research/regions-2019/world2019.pdf

Yetgina, D., & Benligiray, S. (2019). The effect of economic anxiety and occupational burnout levels of tour guides on their occupational commitment. *Asia Pacific Journal of Tourism Research*, *24*(4), 333–347. doi:10.1080/10941665.2018.1564681

Yu, X., & Weiler, B. (2006). Service ethics for ecotourism guides. In B. Prideaux, G. Moscardo, & E. L. James (Eds.), *Managing tourism and hospitality services: Theory and international applications* (pp. 181–194). CAB International. doi:10.1079/9781845930127.0181

Chapter 3
Tourist Guides as Representatives of Sustainable Tourism:
Evidence From Turkey

Eda Rukiye Donbak
https://orcid.org/0000-0002-1752-0390
Faculty of Tourism, Adıyaman University, Turkey

EXECUTIVE SUMMARY

The main purpose of this research is to find out tourist guides' contributions to sustainable tourism. It is planned to apply qualitative research method by using interpretive content analysis in phenomenological design. "Using the positive effects of green and sea on people" and "using spices that will interest the group, for example anise" are some of the answers describing tourist guides' leadership (instrumental) roles. "Talking to the villagers and transfer a lot of information to group" and "explain the nature of the region and explaining how the people living there adapt to this nature" are some of the answers of tour guides describing tourist guides' leadership (social) roles. "Giving information about what is famous in the Dardanelles region and go to the tasting" and "if there has been a family who makes tomato paste in the region certainly to let them to get experience" are some of the answers of tour guides describing tourist guides' mediator roles.

INTRODUCTION

The tourism industry, which is largely labor intensive (Grobelna, 2017; Kim and Park, 2013), and today one of the most developed industries in the world (Baherli and Sucuoglu, 2015), also it is considered by many countries as a means of economic development (Kuo et al., 2018). Some studies have suggested that tourism industry can contribute to conserving natural areas by supporting development of ecotourism and related fields, such as alternative, sustainable and responsible tourism (Blamey, 1997; Orams, 1999; Randall & Rollins; 2009). Tourist guides are professionals involved in shaping tourists'

DOI: 10.4018/978-1-7998-3725-1.ch003

first impressions of destinations, and the role of memorable travel experiences they provide for each tourist in the development of sustainable tourism is critical (Kuo, et al., 2018).

The purpose of this study is to try to understand how licensed tourist guides from Turkey, define their roles within the scope of sustainable tourism. In order to evaluate the assumed roles, the participants were asked about semi-structured questions in the framework of social and instrumental leadership, mediating between local culture and tourists, ensuring behavior towards protecting the environment and economically supporting the local people.

BACKGROUND

Tour Guiding and Guiding in Turkey

In the definition of tourist guidance according to The International Forum of Travel and Tourism Advocates (IFFTA) is emphasized interpretation role of tourist guide. According to IFFTA a tourist guide is the person who makes the interpretation of the archeology, history, natural environment and cultural knowledge to the tourists. (IFFTA, 2019). Federation of Tourist Guide Association's definition of tourist guidance is focused in a tourist guide's providing services and experience to tourists (FEG, 2019). According to FEG, "*tourist guiding has developed for decades as a distinct profession within the tourism industry alongside other complimentary professions that also contribute to the visitors' overall experience*" (FEG, 2019). World Federation of Tourist Guide Associations (WFTGA), makes the definition of the tourist guiding profession by using their ability of knowledge of foreign languages and criteria of interpreting tourism sources of the tourist guides (WFTGA, 2019).

While making the definitions of the tourist guidance profession of these important institutions, which we can say that they represent tourist guides in international platforms, they have used the foreign language skills of the guides, their ability to transfer information about nature and cultural heritage to the tourist as the basic criteria in their definitions.

Interpretation of tourist guides plays an important role in satisfying tourists and showing destination loyalty, but they also have some kind of teaching roles that enable tourists to analyze the cultural characteristics of the destination they experience (Cheng, et al., 2018). While tourist guides have a critical importance in linking the destination with their visitors, their importance in the development of the local tourism industry in most destinations cannot be denied, and the quality of service provided by guides to tourists not only contributes to the business development of the travel agencies, but also improves the image of the destination (Chang, 2014; Mak, et al., 2011; Huang, et al, 2010). Tourist guides not only ensure the health and safety of tourists, but also help them to enjoy a rich sightseeing experience, and they have an important role in evaluating the tourists' own travels as well as the performance of the guide (Kuo, et al., 2016). While tour guides have an active role in ensuring all the impressions and satisfaction of a tour on tourists, their main responsibility is to lead the group to solve problems, and a secondary role is to make entrepreneurship by avoiding short-term profits such as tips and commissions (Tsaur, et al., 2018; Heung, 2008). According to Serenari, et al., (2012) and Randall & Rollins, (2009) firstly, Cohen's (1985) classification of the tourist guides' roles as pathfinder and mentor on the basis of historical and modern roles influenced the subsequent researchers, for example Weiler & Davis (1993) reformed Cohen's (1985) classification and suggested their role as motivator and environmental interpreters (Serenari, et al., 2012).

In Turkey tourist guidance includes management of tours that provided for the national or foreign tourists in the form of persons or groups are not in the nature of travel agency activities, the tour chosen by the language before the trip shall be introduced to tourists in accordance with the cultural, tourism, history, environment, nature, social or similar values and assets of the country and organized by the travel agencies. Travel program is defined in the written documents of the travel agency and sold to the consumer is carried out and managed by tour guides on behalf of the agency. The first guide profile in Turkey is the interpreter-guide who speaks language with the policies followed after the Crimean War and the cooperation made in the field of education, culture and trade with Europe in the period after the Tanzimat of 1839 (Ahipaşaoğlu, 2006). In the 1965s, with the regulations of the Ministry of Tourism and Promotion and the opening of the guidance departments in the universities after 1995, a university graduate guide profile with an associate degree and a bachelor's degree for the guidance profession was formed (Tolga et al., 2015; Hacıoğlu, 2008). It is thought that the university education received by the tourist guides is important for the development of a country's tourism industry. It is still a sensitive issue to recruit highly motivated employees who are well educated in the aforementioned field and who have professional competence and who are committed to their work to the hospitality and tourism sector (Grobelna, 2017; Wan et al., 2014; Richardson & Butler, 2014; Lucas & Johnson, 2003).

Table.1 Tourist Guide Statistics in Turkey (2019)

Chambers of Licensed Tourist Guides in Turkey	Number of Guide (National Scale)	Number of Guide (Female) (National Scale)	Number of Guide (Male) (National Scale)
ADRO	149	53	96
ANRO	721	276	445
ARO	1632	388	1244
ATRO	551	204	347
BURO	101	37	64
ÇARO	140	46	94
GARO	114	34	80
İRO	4260	1546	2714
İZRO	992	415	577
MUTRO	385	134	251
NERO	558	145	413
ŞURO	111	33	78
TRO	98	37	61
Total	9812	3348	6764

Source; Turkey Tourist Guide Association (TUREB), 2019.

In the Table.1, according to the regions in Turkey numbers of female and male tour guide outlines. According to this, it is seen that male guides are approximately twice the female guides, while the room with the highest number of members is the Istanbul Regional Tourist Guides Chamber (İRO).

Table.2 summarizes the number of licensed national tourist guides according to the languages they guide. According to other languages, the number of those who guided the English language in Turkey is seen to be more. After English, the number of guides licensed from German and French, respectively, is the highest. It is seen that there are very rare number of guides who hold license from Icelandic, Hindi, Slovak, Georgian languages. Among the reasons why the number of licensed guides is so different is the ranking of the countries that send the most tourists in the world. In recent years, the increase in the number of licensed guides in Russian and Chinese languages has been remarkable. After these languages, the increase in the number of licensed guides in Arabic and Portuguese languages is also remarkable.

Table.2 Distribution of Tourist Guides by Language in Turkey

Languages of Tourist Guides in Turkey	Number	Languages of Tourist Guides in Turkey	Number
German	1653	Spanish	766
Arabic	364	Swedish	28
Bosnian	8	Italian	551
Bulgarian	231	Icelandic	1
Czech	6	Japanese	589
Chinese	284	Korean	33
Danish	8	Polish	47
Indonesian	130	Lithuanian	2
Armenian	7	Hungarian	22
Persian	39	Macedonian	8
Finnish	7	Norwegian	32
French	1118	Portuguese	376
Georgian	1	Rumanian	21
Croatian	24	Russian	697
Hindi	2	Serbian	45
Dutch	236	Slovak	3
Hebrew	23	Urdu	9
English	7037	Greek	142

Source; Turkey Tourist Guide Associations (TUREB), 2019.

Tourist Guide's Roles in Sustainable Tourism

Serenari, et al., (2012) emphasizes that the effects of tourism on natural landscape areas are examined intensively, and states that policy-oriented tourism perspective is utilized in the sustainable development of mountainous areas. Nature based tourism is also excepted as one of the most growing segment of tourism industry (Ewert & Shultis, 1997; Randall & Rollins, 2009) and ecotourism is considered a type of tourism that supports the conservation of natural resources while providing economic benefit to the local people while providing the opportunity for local and foreign tourists to learn about the natural life while also recognizing the local culture (Brandon, 1996; Davenport *et al.*, 2002; Emmons, 1991; Honey, 1999; Ormsby & Mannle 2006). Weiler & Ham (2002), emphasized that ecotourism differentiates itself from other types of tourism, in addition to providing tourists with a pleasant and enjoyable time as well as creating awareness about the cultural and natural environment. It is thought that the level of knowledge and awareness of tourists about the natural environment, and the management of attitudes and behaviors that may have an impact on the natural environment through interpretation and training constitute the basis of ecotourism activities (Walker & Weiler, 2017). Minimizing negative impacts from visitors in visited both natural and cultural heritage areas is critical to the ecotourism industry (Lackey & Pennisi, 2019). Ecotourism guidance can be carried out voluntarily or self-employed, not only guiding a specific area, but also being aware of the importance of ecological values and transferring this importance in a positive way to the ecosystem while guiding (Shephard et al., 2000). As tourist guides have a role model in ecotourism for environmentally and culturally sensitive behavior, they are considered to conserving natural resource, to provide education and interpretation and being cultural mediator between clients and local community (Black et al., 2001; Muñoz, 1995). By interpreting of information of tour guides make tourist understand and develop new insights towards environment (Reisinger & Steiner, 2006; Carmody, 2013), cultural and natural heritage of a site (Black & Ham, 2005; Chang, 2014; M. Alazaizeh, et al., 2019) and encourage pro-conservation attitudes of nature (Gray, 1993). According to Cohen's (1985) and Weiler & Davis' (1993) model which describe tourist guides contribution to sustainable tourism takes place in four dimensions (Weiler et al., 2011);

"1. Enhancing visitors' understanding and valuing of the site and its natural and cultural resources, through interpretive guiding;

2. Influencing visitors' decisions about their voluntary onsite behavior, through communicating and role modelling sustainable tourism practices;
3. Monitoring and managing visitors' on-site behavioral compliance, through enforcing regulations and role modelling practices associated with protecting ecological and cultural values;
4. Fostering visitors' post-visit pro-environmental and proconservation attitudes and behaviors, through persuasive communication".

Pereira & Mykletun (2012) used questions in below for evaluating tourist guide's contribution level to sustainable tourism;

"(1) What are the levels of congruence between the guides' roles within several spheres – leadership, mediatory, resource management and economy and the study's established indicators?

(2) To what extent do guides have the knowledge and understanding of sustainability and ecotourism so that their roles enhance ecotourism efforts in the area?".

According to Barre (2013), authenticity plays an important role in the evaluation of sustainable tourism. We can say that experience and authenticity in tourism define different points. Generally, tourism experience is the subject of going out of routine, shared with family or friends, valuable time, fun and learning (Alazaizeh, et al., 2019), while authenticity in tourism covers ethnic, historical or cultural tourism which includes representation of other or history of others (Prentice et al., 1998). Especially from the perspective of Taylor's (2001) approach to sincerity in tourism, it can be stated that tourist guides are a bridge between authenticity and the aims of sustainable tourism (Barre, 2013). A tourist guide can easily transform a visitor's sightseeing activity into memorable learning experience (Carmody, 2013).

It is possible to say that tourist guides have an important role in obtaining tourism experiences and conveying authenticity. Since the beginning of the 2000s, it has been observed that in parallel with the development of the tourism industry, tourist guides have an important role in the sector (Ap & Wong, 2001; Yu, Weiler, & Ham, 2002; Weiler & Yu, 2007; Huang & Weiler, 2010; Tsaur, Dai & Liu, 2018) and the interest of the researchers towards the performance of these roles has increased (Zhang & Chow, 2004; Huang, Hsu, & Chan, 2010; Chang, 2014). Mediating and culture broking are the interpretive roles of tour guides' (Ap & Wong, 2001). In his work, Jensen (2010), states that the roles and functions of tourist guides have very different features, but their interpretative and leadership roles are fundamental for the positive development of a socially positive relationship between the host community and tourists. He also states that, in addition to the tourist guide roles that Jensen (2010) has already mentioned, focusing on the representation roles of tourist guides, they fulfill the role of mediation between local people and tourists. Ababneh (2017) find out in his research that tour guides interpreted cultural site of Jarah without disconnect its local community and landscape however the author suggests; "*Effective heritage interpretation should capture the tourists' attention, meet their experiential needs and acknowledge the holistic value of the heritage resource*". Ratz (2017) suggested that "*guides can limit and control visitors' access to Budapest's cultural resources by carefully deciding on niche themes, designing their tours' itineraries and inviting local participants, both from the business sector and the resident community*".

Method

In this study, qualitative research method was applied to evaluate the role of tourist guides for sustainable tourism in Turkey. Qualitative research is a type of approach to investigate and understand the meaning that individuals or groups attribute to social issues (Creswell, 2014). Yıldırım & Şimşek (2018, p. 41) similarly describe the qualitative research method; "*Qualitative data collection methods such as observation, interview and document analysis are used, and a qualitative process is carried out to present perceptions and events in a realistic and holistic manner in the natural environment*".

Research Group

The research group is eight licensed tourist guides experienced in nature and culture tours and to be determined by snowball method. Demographic and professional information of the tour guides interviewed are given in Table 3. It is seen that the age range of the tourist guides interviewed is between 26 and 54, and the tourist guide with the most guidance experience has been in this profession for twenty-two years, and the tourist guide with the least experience has been doing this profession for three years.

Only four of the eight tourist guides included in the study stated that they had a specialist certificate.

It is seen that three of the youngest of the guides interviewed have a bachelor's degree in tourist guidance.

This result can be considered to said starting to become popular in recent years among young people of the tourist guide profession in Turkey. Five other older tourist guides obtained their licenses after completing one-year ministerial courses.

Table.3 Demographic and Occupational Attributes of Tourist Guides

Tourist Guides Interviewed	Gender	Age	Education	How did you get your guidance license?	How many years have you been a tourist guide?	Do you have any expertise?	Do you have a certificate about your area of expertise?	What languages are included in your license?	Which regions do you tour extensively?	Which chamber of regional tourist guides are you registered in?
TG1	M	28	Bachelor Degree in Tourism Guidance	With university diploma	For four years	Cultural Tours	No	Russian	Black Sea and South East Regions and cşty of Konya and Kars	İRO
TG2	F	39	Bachelor Degree in Tourism	From Ministry of Culture and Tourism	For twelve years	City of Canakkale and surroundings	No	English	Mediterranean, Aegean, Marmara, Black Sea and South East	ÇARO
TG3	F	51	Bachelor 's Degree outside area of Tourism	From Ministry of Culture and Tourism	For nineteen years	Gelibolu Island,	Yes	German	Gelibolu Island, North Eagean Region, City of Canakkale	ÇARO
TG4	F	54	Bachelor Degree in Tourism	From Ministry of Culture and Tourism	For twelve years	City of Canakkale and It's battlefield area	Yes	English	City of Canakkale and surroundings (Ancient cities of Assos, Pergamon and battlefield area	ÇARO
TG5	M	54	Bachelor 's Degree outside area of Tourism	From Ministry of Culture and Tourism	For twenty two years	City of Istanbul and Aegean Regions	Yes	English and Norwegian	Regions of Central Anatolia,Aegean, East Anatolia	İZRO
TG6	F	30	Bachelor Degree in Tourism Guidance	With university diploma	For five years	No	No	English	Agean, Maramara and Central Anatolia Regions	ADRO
TG7	M	40	Bachelor 's Degree outside area of Tourism	From Ministry of Culture and Tourism	For fifteen years	History of Art and Mountaineering	Only Mountaineering	English	City of Istanbul, Nevşehir and Antalya	MUTRO
TG8	M	26	Bachelor Degree in Tourism Guidance	With university diploma	For three years	Specific Tourism Regions	Yes	English	Mediterranean, Aegean, Central Anatolia, South East Anatolia Regions	GARO

Data Collection

The main qualitative data collection methods are basic and supportive data collection methods. The data were obtained through a semi-structured interview technique, which is included in the basic data collection method.

FINDINGS

Evaluation of Research Question 1: Do You Think You Have Social and Instrumental Leadership Roles for Tour Groups?

Firstly, the first research question that related with instrumental leadership has been made easier for the participants to comment and the question has been asked to the participants as follows: Can you use your knowledge about natural life, flora and fauna in case you encounter unexpected situations during the tour or in case of disputes within the group in order to increase the morale of the group and to balance the dynamics of the group? Geographical regions with distinctive features, mountains with tourism attraction, spices in our culture, and traditional cultural elements are specified in the following statements:

TG1; "*I tell mythological events by decorating. I give an example to the Black Sea Region of Turkey*".
TG2; "*I use the positive effects of green and sea on people. Fruits, for example, I give information about olive. I don't talk about a lot of issues related to the healing information to avoid problems.*
TG3; "*I use the Kaz Mountains as a cornerstone of my interpreting. Especially about olive oil. For example, when there are no pencils, olive seeds are used. That wine has been used for thousands of years. I will also give information about pods*".
TG5; "*I use Turkish baths and massages*".
TG7; "*I will, yes. I use spices that will interest the group. For example anise*".

The question was asked to the participants in order to interpret the social leadership that includes the first research question more easily; would you introduce your group to the local people's lives even if you are on nature tours? Ensuring that they meet with local people during the tour, the story of local people settling in that area, how they harmonized their lives with the climate and geographical conditions of the region were specified in the following statements:

TG1; "*Yes, I do. The walking tour I let tourists to talk them*".
TG2; "*Sure, I talk to the villagers. I learn and transfer a lot of information to them*".
TG3; "*Yes, for example, I talk about local dishes, clothing styles*".
TG4; "*Of course, I tell them about how they got there, about that period*".
TG6; "*Of course, yes, when I explain the nature of the region, I explain how the people living there adapt to this nature*".

Evaluation of Research Question 2: Do You Consider Yourself a Representative of the Local Culture?

The second research question that related with mediator role of tour guides has been made easier for the participants to comment and the question has been asked to the participants as follows: Do you consider yourself a representative of the local culture and make room for social programs related to the local culture in your tour schedule? Characteristics of the region visited, content of the tour programs, use of the time period of the tour, presence of tourist testing facilities were specified in the following state:

TG1; "*I can do more in Cappadocia region when I compare with other regions. Generally I do it region by region*".
TG4; "*We can give information about what is famous in the Dardanelles region and go to the tasting*".
TG6; "*Of course, we're trying to get what's famous in the area*".
TG7; "*Yes I do, but sometimes programs come ready and there is no time, but if there is time, I definitely give space*".
TG8; "*Yes, for example, I use handicrafts a lot. For example, if there has been a family who makes tomato paste in the region certainly I let them to get experience*".

Evaluation of Research Question 3: Do You Raise Awareness Among Tourists in Terms of Behavior Changes That Aim to Protect the Long-Term Nature?

The third research question that related with sustainability of natural resources has been made easier for the participants to comment and the question has been asked to the participants as follows: Do you give information to the group members about the functions and protection of the ecosystem? Do you monitor the behavior of the group members in nature and try to reduce their environmental impact?

Level of interest of the tour participants, use of plastics, smoking smoke, visiting protected areas, uncertainty in continuity were specified in the following state:

TG1; "*It depends on the region we go to and the type of tour we go on and the interest of the people I transfer what needs to be done region by region. I indicate that tea trees should not be damaged in Black Sea tours. I even give examples of plastic cups. We do, but I don't know what the people's point of view is on the opposite side*".

TG2; "*If the tour is rich in that, of course. Especially when it comes to fish. I intervene when the water bottle is thrown. I have not done anything until now except for the works such as planting seedlings*".

TG3; "*I pay particular attention to issues related to the situation in the Kaz Mountains and I will not stop reacting before this situation ends. Especially I do not allow smoking in my groups. If I see garbage, I collect it. I usually emphasize and make a statement saying we need nature*".

TG4; "*I think and convey that the participants of the tour may be interested. I definitely mention the issue of Kaz Mountains. At least I give information on environmental protection. As I pass through the Bosphorus, I also see those who throw something and warn*".

TG5; "*Yes definitely, but we have little domination. If there is a damaging situation, I will also be kindly warned*".

TG6; "*Of course, I warn someone who smokes, for example, in the forest*".

TG7; "*Of course, do not touch, tear off the flower, etc. Everyone wants, but it depends on the program. We are trying to do as much as we can. We do as much as we can, it depends on the tour audience*".

TG8; "*Obviously I am very superficial. Because people are uncomfortable when they give detailed information about this issue. I will definitely reduce and examine. When there are participants trying to enter the areas that are forbidden to enter, I will also block and warn directly. I think it's not a fact I can do personally. Because I cannot ensure its continuity*".

Evaluation of Research Question 4: Do You Think You Made Economic Contributions to the Local People in Your Tours?

The fourth research question that related with sustainability of the economic environment has been made easier for the respondents to comment and the question has been asked to the participants as follows: Do you offer product information to local communities in your frequent tour areas about how to diversify their crafts? Village breakfasts, rug production, organic vegetable growing, village cooperatives, souvenirs, local guides were specified in the following state:

TG1; "*Of course. I definitely do. I mutually share how it should be improved. But I do not want to use a local guide on my tours*".

TG2; *"The villagers usually tell me about the characteristics of their nomadic culture, and then I tell them in my tours and I take my tour groups for breakfast in the village of Mavikazık. I usually dominate the regions. So I don't need local guides".*

TG3; *"I am, but I do not know if it is effective. Because he doesn't care about trades. For example, our rugs are very famous but not even for sale. We do not work with local guides. But we usually use it in the Kaz Mountains trekking".*

TG4; *"Yes, I inform women about the local babies they make and advise the heroes to knit their babies. I transfer information about drug-free products made in Gökçeada. I don't but according to the information I got from other friends, the local guides there helped when going to Kaz Mountains".*

TG5; *"For unavailable reasons, it's my own neglect because I don't know about it. But I still try to give you little information on matters that I have knowledge of".*

TG6; *"I usually do it, I say them "you better do that, etc". I transfer information such as. I try to take the guests to such places".*

TG7; *"No because they don't understand and listen".*

TG8; *"Yes I am. I usually give information on how to sell. For example, in the Ayder plateau, cooperatives sell honey and pollen, and the products are both very high quality and more affordable. I direct the groups directly. But I don't support local guides too much. Because the biggest problem of us is those people and they harm our profession".*

CONCLUSION

When the findings of the research carried out by qualitative methods were examined, the majority of the guides in the research performed their social and intermediary leadership roles while benefiting from cultural values as well as their knowledge about the natural environment, flora and fauna. (Research question 1). First of all, this finding is an example of Serenari (2012)'s proposition of historical and modern role categories of guides. According to this proposition, while the pathfinder roles of the guides are in the historical roles class, the leadership roles based on motivation and environmental interpretation are evaluated in the modern role category. Findings obtained within the context of intermediary leadership roles are distinctive features of geographical regions (TG1; *Cappadocia*, TG4; *Dardanelles* and TG1; *Black Sea),* mountains with tourism attraction (TG3 and TG4; *Kaz Mountains*), spices with a large place in culinary cultures (TG7; *Anise*), and traditional cultural elements (TG5; *Turkish Bath and massage*). The findings obtained within the scope of social leadership roles were to ensure that the participants of the tour came together with the local people during the tour (TG1; *conversations* and TG2; *meeting with the villagers*). Social leadership roles include transferring how local people have settled in the past to those who have taken part in the tour (TG4), how the local people have adapted to the geographical and climatic conditions of the region where they live (TG6), and transferring local clothing and food of the local people (TG3). Research findings on social leadership are in line with Barre (2013)'s suggestion that tourist guides contribute to sustainable tourism principles by intermediating between local people and tourists. In addition, it is seen that TG3 and TG4 come to the forefront with the regions in which they have expertise certificates, while TG1 comes to the forefront with the regions where they often tour. This result shows that there may be a relationship between the regions where the guides are constantly touring with the areas where they have a specialist certificate and using these roles.

When the findings obtained from the answers to the second research question were examined, it was understood that most of the guides included in the research considered themselves representatives of the local culture within the context of their mediation roles. This finding is an example of the proposal that Jensen (2010), as stated earlier in his study, arises from the role of guides to represent and have mediation functions between local people and tourists. When the findings obtained within the scope of mediation roles were examined, it was understood that the characteristics of the region visited were determinant for the guides' mediating roles. For example; TG1 states that while he can mediate in the Cappadocia region, it cannot do tours in the Aegean and Mediterranean regions. As the reason for this finding, it can be argued that the type of tourism carried out in the mentioned regions may be due to being far from sustainable tourism principles. Other findings were the content of the tour programs, the use of time on the tour and the availability of opportunities for tourists to experience.

For example, TG8, stated that he exhibited the role of mediation in providing interaction between local people and tourists by chance (TG8; *for example, I use handicrafts a lot, if there is a family making tomato paste, I will definitely make them try it in the region*). Those who state that tour programs are decisive in this regard the statement summarizes the situation; "TG7; *Sometimes the programs are ready and there is no time, but if the time remains, I definitely give place*". Ratz (2017) states that tourist guides can provide a more efficient interaction between the local community and the tourist when they ensure that the program is redesigned, taking into account the shortcomings in the tour programs.

The third research question is for guides to interpret their roles in ensuring the sustainability of natural resources. Most of the guides included in the research stated that they have missions in ensuring the sustainability of natural resources. Alazaizeh et al. (2019) states that licensed tour guides reduce the negative effects of visitors on protected areas by using effective and powerful techniques, while providing them with sustainable environmental behaviors. The level of interest of the participants in the tour about the environmental issues (TG8; *people are disturbed when they give detailed information about this issue*) has been determinant in the roles of the guides. The use of plastic materials (TG1; *I give an example even from plastic cups*), preventing smoking in critical places (TG3; *I do not allow smoking in my groups*), controlled travel in protected areas (TG8; *when there is a participant trying to enter areas that are forbidden to enter, I also have direct obstacles and a warning*) are directly related to intervention in behavior. The uncertainty of the continuity in ensuring the sustainability of natural resources is another finding that has shown that guides attach importance to feedback.

The fourth research question is about the sustainability of the economic environment. When the findings obtained in this context are evaluated, many of the guides included in the research stated that they have a positive contribution to the sustainable economic environment. Finding for the contribution of the guides to the sustainability of the economic environment is to get village breakfasts, to advise on the production of rugs, to introduce organic vegetable growers to tour groups, to buy the products produced in the village cooperatives, to provide model recommendations to gift manufacturers. According to Ratz (2017), the personal relationships of tourist guides with visitors create alternative appeal to the communities and regions they represent, and the guides turn into strategists for the local community and the tourist area.

REFERENCES

Ahipaşaoğlu, H. S. (2006). *Guidance in Tourism*. Gazi Bookstore.

Alazaizeh, M. M., Jamaliah, M., Mgonja, J. M. T., & Ababneh, A. (2019). Tour guide performance and sustainable visitor behavior at cultural heritage sites. *Journal of Sustainable Tourism*, *27*(11), 1708–1724. doi:10.1080/09669582.2019.1658766

Bahçelerli, N. M., & Sucuoğlu, E. (2015). Undergraduate tourism students' opinions regarding the work conditions in the tourism industry. *Procedia Economics and Finance*, *26*, 1130–1135. doi:10.1016/S2212-5671(15)00941-7

Barre, S. (2013). Wilderness and cultural tour guides, place identity and sustainable tourism in remote areas. *Journal of Sustainable Tourism*, *21*(6), 825–844. doi:10.1080/09669582.2012.737798

Black, R., & Ham, S. (2005). Improving the quality of tour guiding: Towards a model for tour guide certification. *Journal of Ecotourism*, *4*(3), 178–195. doi:10.1080/14724040608668442

Black, R., Ham, S., & Weiler, B. (2001). Ecotour guide training in less developed countries: Some preliminary research findings. *Journal of Sustainable Tourism*, *9*(2), 147–156. doi:10.1080/09669580108667395

Blamey, R. K. (1997). Ecotourism: The search for an operational definition. *Journal of Sustainable Tourism*, *5*(2), 109–130. doi:10.1080/09669589708667280

Brandon, K. (1996). *Ecotourism and Conservation: A Review of Key Issues*. World Bank.

Carmody, J. (2013). Intensive tour guide training in regional Australia: An analysis of the Savannah Guides organization and professional development schools. *Journal of Sustainable Tourism*, *21*(5), 679–694. doi:10.1080/09669582.2012.744412

Chang, K. (2014). Examining the effect of tour guide performance, tourist trust, tourist satisfaction, and flow experience on tourists' shopping behavior. *Asia Pacific Journal of Tourism Research*, *19*(2), 219–247. doi:10.1080/10941665.2012.739189

Cheng, Y. S., Kuo, N. T., Chang, K. C., & Chen, C. H. (2019). How a tour guide interpretation service creates intention to revisit for tourists from Mainland China: The mediating effect of perceived value. *Journal of China Tourism Research*, *15*(1), 84–104. doi:10.1080/19388160.2018.1517067

Cohen, E. (1985). The tourist guide: The origins, structure and dynamics of a role. *Annals of Tourism Research*, *12*(1), 5–29. doi:10.1016/0160-7383(85)90037-4

Creswell, J. W. (2014). *Research Design, Qualitative, Quantitative and Mixed Methods Approaches* (4th ed.). Sage Publishing.

Davenport, L., Brockelman, W. Y., Wright, P. C., Ruf, K., & del Valle, F. B. R. (2002). Ecotourism tools for parks. In J. Terborgh, C. van Schaik, L. Davenport, & M. Rao (Eds.), *Making Parks Work: Strategies for Preserving Tropical Nature* (pp. 279–306). Island.

Emmons, K. (1991). Educational planning and ecotourism: Targeting school students. In J. A. Kusler (Ed.), *Ecotourism and Resource Conservation: A Collection of Papers* (Vol. 2, pp. 826–838). Association of Wetland Managers.

Ewert, A. W., & Shultis, J. (1997). Resource based tourism: An emerging trend in tourism experiences. *Parks & Recreation*, *32*(9), 95–103.

FEG. (2019, Nov 11). *European Federation of Tourist Guide Association*. Retrieved from https://www.feg-touristguides.com/about.php

Grobelna, A. (2017). The perception of job-related motivators when choosing a career in the tourism and hospitality industry – a comparative study between Polish and Spanish students. *International Journal of Management and Economics*, *53*(2), 84–106. doi:10.1515/ijme-2017-0014

Hacıoğlu, N., Kaşlı, M., Şahin, S., & Tetik, N. (2008). *Tourism education in Turkey*. Detay Publishing.

Heung, V. C. S. (2008). Effects of tour leader's service quality on agency's reputation and customers' word-of-mouth. *Journal of Vacation Marketing*, *14*(4), 305–315. doi:10.1177/1356766708094752

Honey, M. (1999). *Ecotourism and Sustainable Development: Who Owns Paradise?* Island.

Huang, S., Hsu, C. H. C., & Chan, A. (2010). Tour guide performance and tourist satisfaction: A study of the package tours in Shanghai. *Journal of Hospitality & Tourism Research (Washington, D.C.)*, *34*(1), 3–33. doi:10.1177/1096348009349815

IFTTA. (2019, Nov 11). *The International Forum of Travel and Tourism Advocates*. Retrieved from https://iftta.org/glossary

Kim, H., & Park, E. J. (2013). The role of social experience in undergraduates' career perceptions through internships. *Journal of Hospitality, Leisure, Sport and Tourism Education*, *12*(1), 70–78. doi:10.1016/j.jhlste.2012.11.003

Kuo, N. T., Chang, K. C., Cheng, Y. S., & Lin, J. C. (2016). Effects of tour guide interpretation and tourist satisfaction on destination loyalty in Taiwan's Kinmen battlefield tourism: Perceived playfulness and perceived flow as moderators. *Journal of Travel & Tourism Marketing*, *33*(1), 103–122. doi:10.1080/10548408.2015.1008670

Kuo, N. T., Cheng, Y. S., Chang, K. C., & Chuang, L. Y. (2018). The asymmetric effect of tour guide service quality on tourist satisfaction. *Journal of Quality Assurance in Hospitality & Tourism*, *19*(4), 521–542. doi:10.1080/1528008X.2018.1483283

Lackey, N. Q., & Pennisi, L. (2019). Ecotour guide training program methods and characteristics: a case study from the African bush. *Journal of Ecotourism*. Retrieved from https://www.tandfonline.com/doi/full/10.1080/14724049.2019.1686005

Lucas, R., & Johnson, K. (2003). Managing Students As A Flexible Labor Resource In Hospitality and Tourism In Centraland Eastern Europe and the U. K. In S. Kusluvan (Ed.), *Managing Employee Attitudes And Behaviors In The Tourism and Hospitality Industry* (pp. 153–170). Nova Publishers.

Mak, A. H. N., Wong, K. K. F., & Chang, R. C. Y. (2011). Critical issues affecting the service quality and professionalism of the tour guides in Hong Kong and Macau. *Tourism Management, 32*(6), 1442–1452. doi:10.1016/j.tourman.2011.01.003

Muñoz, O. (1995). Interpretation guidelines. In D. Hawkins, M. Epler Wood, & S. Bittman (Eds.), *The Ecolodge Sourcebook*. Ecotourism Society.

Orams, M. B. (1999). *Marine tourism: Development, impacts and management.* Routledge.

Ormsby, A., & Mannle, K. (2006). Ecotourism benefits and the role of local guides at Masoala National Park, Madagascar. *Journal of Sustainable Tourism, 14*(3), 271–287. doi:10.1080/09669580608669059

Randall, C., & Rollins, R. B. (2009). Visitor perceptions of the role of tour guides in natural areas. *Journal of Sustainable Tourism, 17*(3), 357–374. doi:10.1080/09669580802159727

Reisinger, Y., & Steiner, C. (2006). Reconceptualising interpretation: The role of tour guides in authentic tourism. *Current Issues in Tourism, 9*(6), 481–498. doi:10.2167/cit280.0

Richardson, S., & Butler, G. (2012). Attitudes of Malaysian tourism and hospitality students' towards a career in the industry. *Asia Pacific Journal of Tourism Research, 17*(3), 262–276. doi:10.1080/10941665.2011.625430

Serenari, C., Leung, Y. F., Attarian, A., & Franck, C. (2012). Understanding environmentally significant behavior among whitewater rafting and trekking guides in the Garhwal Himalaya, India. *Journal of Sustainable Tourism, 20*(5), 757–772. doi:10.1080/09669582.2011.638383

Shephard, K. L., & Royston-Airey, P. C. M. (2000). Exploring the role of part-time ecotourism guides in Central Southern England. *Journal of Sustainable Tourism, 8*(4), 324–332. doi:10.1080/09669580008667367

Taylor, J. P. (2001). Authenticity and sincerity in tourism. *Annals of Tourism Research, 28*(1), 7–26. doi:10.1016/S0160-7383(00)00004-9

Tolga, Ö., Korkmaz, H., & Atay, L. (2015). A research on professional attitudes of undergraduate tourist guidance students. *Journal of Travel and Hotel Management, 12*(2), 26–41.

Tsaur, S., Dai, Y., & Liu, J. (2018). SOCO's impact on service outcomes of tour guides: The moderating effect of customers' shopping orientation. *Current Issues in Tourism, 21*(8), 917–933. doi:10.1080/13683500.2015.1118444

TUREB. (2019, Nov 11). *Association of Tourist Guides in Turkey.* Retrieved from http://www.tureb.org.tr/en

Walker, K., & Weiler, B. (2017). A new model for guide training and transformative outcomes: A case study in sustainable marine-wildlife ecotourism. *Journal of Ecotourism, 16*(3), 269–290. doi:10.1080/14724049.2016.1245736

Wan, Y. K. P., Wong, I. A., & Kong, W. H. (2014). Student career prospect and industry commitment: The roles of industry attitude, perceived social status, and salary expectations. *Tourism Management, 40*, 1–14. doi:10.1016/j.tourman.2013.05.004

Weiler, B., & Davis, D. (1993). An exploratory investigation into the roles of the nature-based tour leader. *Tourism Management, 14*(2), 91–98. doi:10.1016/0261-5177(93)90041-I

WFTGA. (2019, Nov 11). *World Federation of Tourist Guide Associations.* Retrieved from http://www.wftga.org/tourist-guiding/what-tourist-guide

Yıldırım, A., & Şimşek, H. (2018). *Qualitative research methods in the social sciences.* Seçkin Publishing.

ADDITIONAL READING

Cheng, T., Wu, H. C., Wang, J. T., & Wu, M. R. (2019). Community Participation as a mediating factor on residents' attitudes towards sustainable tourism development and their personal environmentally responsible behavior. *Current Issues in Tourism, 22*(14), 1764–1782. doi:10.1080/13683500.2017.1405383

Christie, M. F., & Mason, P. A. (2003). Transformative tour guiding: Training tour guides to be critically reflective practitioners. *Journal of Ecotourism, 2*(1), 1–16. doi:10.1080/14724040308668130

Potter, A. E. (2016). "She goes into character as the lady of the house": Tour guides, performance, and the Southern plantation. *Journal of Heritage Tourism, 11*(3), 250–261. doi:10.1080/1743873X.2015.1100626

Powell, R. B., & Ham, S. H. (2008). Can ecotourism interpretation really lead to pro-conservation knowledge, attitudes and behaviour? Evidence from the Galapagos Islands. *Journal of Sustainable Tourism, 16*(4), 467–489. doi:10.1080/09669580802154223

Qiu, Q., Zheng, T., Xiang, Z., & Zhang, M. (2020). Visiting intangible cultural heritage tourism sites: From value cognition to attitude and intention. *Sustainability, 12*(1), 132. doi:10.3390u12010132

Rokenes, A., Schumann, S., & Rose, J. (2015). The Art of guiding in nature-based adventure tourism – how guides can create client value and positive experiences on mountain bike and backcountry ski tours. *Scandinavian Journal of Hospitality and Tourism, 15*(1), 62–82. doi:10.1080/15022250.2015.1061733

Weiler, B., & Yu, X. (2007). Dimensions of cultural mediation in guiding Chinese tour groups: Implications for interpretation. *Tourism Recreation Research, 32*(3), 13–22. doi:10.1080/02508281.2007.11081535

Yamada, N. (2011). Why tour guiding is important for ecotourism: Enhancing guiding quality with the ecotourism promotion policy in Japan, *Asia Pacific Journal of Tourism Research*, 16(29, 139-152.

KEY TERMS AND DEFINITIONS

Authenticity in Tourism: While it is outside of special interest tourism or nature-based tourism for the realization of hobbies, it refers to the experience that can be obtained within the types such as culture, history, ethnic tourism representing others or the past.

Conservation Attitude of Nature: It expresses the environmentalist educational comments of tourist guides and the protective attitudes expected to occur in tourists.

Ecotourism: It allows tourists to experience nature experiences while providing them with authentic experiences in a planned manner, while making direct economic contributions to the local community in the destination.

Interpretation Role of Tourist Guide: It refers to transferring information to tourists by interpreting information in order to affect the impression and attitude of the tourists.

Mediator Role of Tourist Guide: It refers to the culture of the local people in the destination and the cultural mediation between the tourists.

Memorable Learning Experience in Tourism: It refers the happiness and excitement created by the sense of innovation perceived as a result of the social interaction of the tourists with the local people thanks to the professional performances of the tourist guides.

Sustainable Tourism: It refers to managing the tourism structuring of a region within a whole of principles considering the future developments of its natural and cultural environment.

Chapter 4
Student Guides as Mediators of Institutional Heritage and Personal Experience

Simon C. Woodward
School of Events, Tourism and Hospitality Management, Leeds Beckett University, UK

Elizabeth Carnegie
Newcastle Business School, Northumbria University, UK

EXECUTIVE SUMMARY

This chapter examines the role of student guides as mediators between the institutional mission and heritage of their university and visitors to the historic campus. Drawing on a longitudinal study undertaken at two historic universities in the west, the authors establish that a small cadre of elite, competitively-chosen guides at these institutions perform a role of openness and democracy on behalf of the increasingly complex and hybrid modern university. The chapter considers how student guides are able to navigate their own pride at such privileged engagement and how this privilege impacts on tours offered to visitors, where campus tours become a negotiation based on internal and external influences and are constructed and reconstructed according to the imagined or actual demands of different tour groups. By managing risk at the point of employment, and by encouraging free reign in tours, there is limited risk involved to host universities as student guides offer an informed, personalised heritage experience to both domestic and international tourists.

INTRODUCTION

Whilst there is an established corpus of work on the role of the tourism guide in general (see for example Cohen, 1985; Dahles, 2002; Salazar, 2005; Salazar, 2006; Gelbman & Maoz, 2012; Tucker & Carnegie, 2014; Mesaritou, Coleman, & Eade, 2016; Farkic, 2018), the topic of students as tour guide, presenter and interpreter of university heritage and ambition remains under-researched. Indeed, the subject of campus tourism in its broadest sense is something that has received limited attention as an area of aca-

DOI: 10.4018/978-1-7998-3725-1.ch004

demic investigation, even though in many destinations university campuses can form an important part of the tourist space (see for example Connell, 1996; Ming 2007; Shaw & Fincher, 2010; Schofield & Fallon, 2012; Woodward, 2013).

In the UK, towns like Durham, Cambridge and Oxford promote themselves as cultural destinations mainly because of the heritage associated with their universities and colleges, whilst in the USA the presence of Harvard University, Auburn University and the University of Virginia (UVa) in Cambridge, Mass; Auburn, Alabama and Charlottesville VA are key reasons for tourists to visit those locations. Other destinations where universities form a major part of the tourism experience include the streets around the Sorbonne in the Latin Quarter of Paris, France; Heidelberg in Germany with its iconic student prison (studentenkarzer) and Alcalá de Henares in Spain. In his overview of the importance of the historic university campus in contemporary tourism, Woodward states that "*the historic buildings associated with centuries of learning provide an iconic and attractive physical presence that helps the destination build its sense of place*" (2013:265). In a review of the contribution of Higher Education Institutions (HEIs) to city branding, Popescu argues that "*a 'consumer' of the university, either student or academic staff becomes, in most cases, a 'consumer' of the city in which the university is located*" (2012:494). It is contested in this chapter that the reverse can also be true, namely that a visitor to a university or college town can also become a 'consumer' of the HEI in either its tangible or intangible form, and that to make the most of, the visitor or consumer needs an informed guide.

This paper seeks to fill part of the existing knowledge gap on campus tourism and particularly on the role of tourist guides by exploring the role of student guides as intermediaries between campus visitors on the one hand and the host university on the other. The situation can at times be challenging, since a university is on the one hand an elite organization (i.e. a community of people) but also a physical space on the other, and thus can present multiple identities to the visitor with no previous affiliation to, or knowledge of, the institution and its estate. The findings discussed within this chapter are drawn from a longitudinal research exercise carried out over several years at the University of Virginia in Charlottesville, USA (hereafter UVa) and at the University of Durham in the North East of England. It should be noted that both of these study locations are located within World Heritage Sites, which brings an additional layer of heritage value into the equation.

BACKGROUND

As indicated above, the role of campus tourism in the contemporary tourism experience has received comparatively limited attention in the literature. Connell (1996, 2000) and Woodward (2013) explore in outline the varying functions that universities can play within a destination's product, with Ming (2007) exploring the Chinese experience in particular. More specifically, the role of the university and college campus as public space, and thus part of the wider destination product, has been reviewed by, *inter alia*, Gumprecht (2003, 2007) and Bromley (2006) whilst the contribution of students (and their presence in a destination) to placemaking in their host communities has been explored by Shaw & Fincher (2010) and Pereira Brando Albino (2015).

Universities are powerful place-based organisations, with a global reach in research terms, staff and students. Often major landowners, universities can be viewed as contributing to or exacerbating troubles of communities such as a lack of affordable housing for local people, through their own development strategies intended to co-exist side by side with, rather than as part of, the communities in which they are located (Bromley, 2006; Pereira Brando Albino, 2015). Universities, and particularly elite institutions, however can play an important part in building and marketing their host cities as tourism destinations as well as in the delivery of the tourist experience itself (Connell, 1996; Connell, 2000; Gumprecht, 2003; Ming, 2007; Woodward, 2013) This in itself requires a situation where multiple stakeholders must become involved in the planning and delivery of services to visitors, including students themselves (Woodward, 2013).

Gumprecht's assertion that the university campus "*is a symbol of the college as a place apart*" (2007:96) reflects the tension that is sometimes found in relationships – institutional and physical - between so-called 'town and gown'. This topic is explored by O'Mara (2012) who considers the US experience of universities (and particularly elite institutions) as catalysts for urban regeneration and by Pereira Brando Albino (2015) in her study of university: town relationships in Exeter (UK) and Coimbra (Portugal). Both of these studies have considered the case of what they term 'elite' universities – that is, institutions renowned for a tradition of academic excellence. Because of their longevity, such educational establishments often have single buildings or assemblages on campus that are recognised as being of particular architectural significance and of appeal to tourists. For instance, more than 140 buildings associated with the University of Oxford and its constituent colleges are listed Grade I (the highest level of protection available to the built environment in England) (Oxford City Council, 2017).

Another aspect of 'elitism' of relevance here is that of World Heritage Site (WHS) status, a recognition of the outstanding universal value of selected tangible heritage assets (Orbasli, 2008) and what Smith (2006) would term as representing the Authorised Heritage Discourse – that is, assets that are valued in heritage terms because they represent a western perspective on aspects of the past considered important. There are several university complexes inscribed on the UNESCO World Heritage List including the University of Virginia, Charlottesville, VA (inscribed 1987); the University and Historic Precinct of Alcalá de Henares in Spain (inscribed 1998) and the Campus Central de la Ciudad Universitaria (UNAM) in Mexico City (inscribed 2007). One can add to the list parts of the University of Durham in Northern England, which occupies many of the core buildings within the Durham Castle and Cathedral WHS (inscribed 1986) including the castle itself, which is home to University College and hence a place of residence for undergraduate and postgraduate students. Parts of the University of Greenwich in London occupy buildings located within the former naval college that are now part of the Maritime Greenwich World Heritage Site (inscribed 1997).

Thus, as well as maintaining an elite status as institutions of higher education, these universities share a further signifier of elitism, namely their status as globally significant heritage assets deemed to be of Outstanding Universal Value. For instance, the rationale for inclusion on the World Heritage list of the 16th century Alcalá de Henares is that it was the first planned university city in the world and the original model for the *Civitas Dei* (City of God) (University and City Council of Alcala de Henares & Community of Madrid, 1997). Both of the study locations under consideration here - one in the UK and one in the USA - are not only elite universities within their respective national higher educational contexts, but are also inscribed on the World Heritage list and are thus deemed to be of Outstanding Universal Value within a global heritage context. There are thus multiple layers of heritage 'value' surrounding such institutions that can, without interpretation, be confusing to the casual visitor. How student guides seek to mediate between these layers is the topic of this chapter.

It is conventional for universities around the world to recruit student 'ambassadors' from amongst their under-graduate and post-graduate cohorts to support admissions and academic staff on open days targeting applicants and their families. Such tours are variously seen as marketing activities (Klein, 2004);

as rituals (Magolda, 2000; 2001) and as experiences (Hoover, 2009). However, tours for prospective students are not, however, the primary area of interest for this paper. Rather, the role of student guides in facilitating campus visits by 'casual' visitors and both national and international tourists is investigated as a conscious outward-facing act of universities seeking to engage within external global markets as an extension of intellectual hospitality (Kaufman, 2001 in Lynch et al, 2011).

As is the case at any heritage visitor attraction, it is desirable that anyone visiting a university campus as a leisure activity is offered the opportunity to understand and engage with its heritage significance and values. So how do student guides make meaningful both the tangible historic environment of the university, but also the intangible values associated with academic life there? Does this become even more complex when the destination is also a World Heritage Site with its associated 'baggage' of Outstanding Universal Value? What are student motivations for taking on such a role for their *alma mater*? And finally, who decides what messages are communicated? This latter point is particularly significant given an increasing emphasis on reputation management and corporate branding in the global higher education sector (Hemsley-Brown & Goonawardana, 2007; Stephenson & Yerger, 2014; Mampaey, et al., 2015).

Guides are deemed to have a role in place making and as Dahles (2002:783) notes, guides are expected to have a 'specific body of knowledge and a high standard of tourist guiding' in their role as mediator or mentor building bridges between host and guests. As Salazar notes (2005:631) guides need to be mindful of the 'different expectations of specific groups'. In the context of this study it is also important to note the transient nature of students themselves who are a mobile body, often moving into and then away from the University area once their studies are complete. What this means is that there is the potential for a lack of continuity within the student guide systems, which has to be managed by the relevant university to ensure that the baton is passed over the years without loss of quality or integrity. Each successive cohort of student guides will have interests and issues that reflect the times. Through their commitment to specific narratives of place, students at UVa for instance have been instrumental in shaping the way formal narratives are debated within the university, particularly as regards slavery.

For visitors, university tours offer privileged access to elite organisations which might otherwise be denied them. Salazar (2005: 640) determines that it is '*the human contact, the close encounters with people…the feeling one has of actively participating in the lived life*' that tourists remember about tours. Historic tours taken by student guides, themselves insiders of the organization, offer privileged access both to the physical place but also a perceived authentic engagement with the lived lives of a small band of elite students, who share their internal and private world with visitors. That the tour guides that form part of this study are drawn from the most privileged student groups in the main – many live within the prestigious rooms on the lawn at University of Virginia or inhabit the castle within the University of Durham, further enforces this sense of real and intimate engagement.

Student guides have a strategic and necessary role within the more formal context of elite universities welcoming tourists within an increasingly globalised tourist (Salazar, 2005) and student markets. As mentioned earlier, student guides can effectively challenge 'path dependant' narratives of organisations and equally may actively seek to raise issues that do not form part of the Authorised Heritage Discourse (AHD) (Smith, 2006). Their tours often respond to and reflect tourist interests and concerns and tourists may come from various parts of the world (Salazar, 2005) or are drawn from migrant communities within the broader locality. Thus, guides reclaim, reshape and openly debate aspects of history and present-day university life, albeit from their perspective as ambassadors and intermediaries.

To summarise so far, this chapter explores how student guides at two elite institutions of higher education construct and mediate a portfolio of stories and messages about their university that reflect not only the Authorised Heritage Discourse (AHD) of that institution, but also more personal perspectives and values. The authors contend that these narratives are additionally shaped by guides' responses to tourists and thus are changed to suit the real or imagined needs of each new visiting group. In this way, it is argued that narratives of place, institution, personal identity are constructed around the need to be mindful of the ways in which the university, and indeed the student guides, are viewed by both domestic and international tourists.

STUDY METHODS

Initial Data Gathering

The findings reported here are drawn from a longitudinal survey that commenced in 2011 and which is investigating university student engagement with World Heritage values. To gain an initial understanding of their engagement with heritage as a concept as expressed by WHS status, the programme of primary research commenced with an online questionnaire survey of students connected to colleges and departments at the University of Durham with a presence on the city's peninsula (i.e. the students either lived or studied within the area covered by the UNESCO WHS inscription). That initial research delivered an analytical sample of just over 200 usable responses, and analysis raised a number of key themes that merited further investigation including the sense of privilege felt by some students at being able to live and study within such a historically and culturally significant place as well as their pride at being associated with a historic institution.. Many students also reported the pleasure they received from sharing their experiences of the University's historic buildings and associated structures within the World Heritage site with visitors – family and friends – when they came to visit. Indeed, more than 80% of respondents to our initial survey confirmed that they always took friends and family visitors into the heart of the World Heritage site in Durham so that they could experience it for themselves.

To investigate these topics in more depth, a programme of focus groups was held with students from Durham University including representatives of those resident within University College (the college at Durham that occupies the castle), others who lived elsewhere on campus but who studied in buildings on the Peninsula (i.e. within the WHS) and a third group who both lived and studied on parts of the campus away from the WHS. At this point it became clear to the researchers that Durham student feelings of pride and privilege could quite conceivably be shared by students at other elite, historic universities, as they seemed to be based on an acknowledgement that the traditions of the University were a particularly important a factor in supporting their sense of 'belonging' to that institution. Moreover, a chance remark by one focus group participant about how much he enjoyed his role as a student guide, escorting visitors around the castle in Durham, identified the potential of this topic as a particular route for further investigation.

A review of UNESCO's World Heritage list confirmed that the University of Virginia (UVa) at Charlottesville in the USA, as well as being similar in terms of age, is also of similar status to Durham University in terms of WHS recognition and in terms of its position within the Higher Education framework in its country. Thus, the study was extended to include additional primary research in Virginia where language would not be a barrier and where there were also some geographical similarities in the setting (see Table 1). The value of this in methodological terms has been to broaden the scope of inquiry across two HEIs and thus it allows us to explore some of the conceptual and practical issues around the role of student tour guides within a wider context.

Case Study Institutions

Table 1 presents key facts about the two institutions and their host communities with more detailed presented in the discussion below.

Table 1. Key facts about the two case study locations

	University of Durham, UK	UVa, Charlottesville, USA
Date University established	1832[1]	1819[2]
Current student population	17,500 (2015 data)[3]	22,500 (2017 data)[4]_
University ranking	Top 10 in UK; top 100 in World[5]	No.2 Public University in the USA; top 100 in USA[6]
Date of WHS inscription	1986[7]	1987[8]
Area covered by WHS inscription	9 ha[9] (includes Cathedral and associated ecclesiastical buildings)	11 ha (UVa campus only)[10]
Population of host city	48,000 (2011 census)[11]	46,600[12]
Number of tourists to the destination per annum	4 million (2015 data)[13]	Data unavailable
Socio-economic context of the destination	42% of population lives in areas considered to be in the top 30% of deprived communities nationally[14] Main employers are local government, Higher Education and other public sector service jobs[15]	27% of population below poverty line[16] Main employers are the University, local government and healthcare[17]

The University of Durham (also known as Durham University) was founded by Act of Parliament in 1832 and given a Royal Charter in 1837 (Pocock, 2013). Closely linked to the Cathedral until the early 20th century (indeed, the Dean and Chapter were strongly represented on the University's board of governors until 1909), the University occupies many buildings associated with the Cathedral, including Durham Castle, which for centuries was the home of the Bishop of Durham until ownership was transferred to the University after the accession to the throne of Queen Victoria (Roberts, 2003). The Castle, Cathedral and many associated buildings on the peninsula immediately to the south of the city centre were inscribed on the World Heritage list in 1986. Key reasons for inscription were the high quality of the architecture of both Castle and Cathedral and associated buildings and their association with the political power of the Prince Bishops of Durham who governed what was an almost independent state in the North of England between the late 11th century and 1603, when James VI of Scotland succeeded to the English throne as James I and a buffer zone between the two countries was not required (Roberts, 2003).

The Castle is occupied by University College, the oldest of the 16 colleges that provide residential and pastoral services to Durham students. Around 150 University College students live in the Castle itself, with others occupying residences on the periphery of the World Heritage Site or elsewhere within the city. Other buildings occupied by the University that form part of the WHS include a 16th century

Grammar School (now the University's music school); Cosin's Hall (now home of the Institute of Advanced Study); Abbey House (home of the Theology Department); the Pemberton Rooms (classrooms and offices) and Cosin's Library and the Exchequer Building (both now part of the Palace Green library housing rare books) (Pevsner & Williamson, 1985). Thus it is not only students belonging to University College who will engage with Durham's World Heritage buildings on a regular basis – those from other colleges studying theology, music or consulting rare books in Palace Green library will also use the buildings, whilst all Durham students first matriculate and then graduate at special ceremonies held within the Cathedral.

The University of Virginia (UVa) was established in 1819 by Thomas Jefferson, the 3rd President of the United States and principal author of the American Declaration of Independence. Associations with Jefferson are the source of considerable pride on campus though, as will be explored later, his role as a slave owner and the use of slaves to build many of the original campus buildings tend to be overlooked or minimised in most corporate and official documentation promoting UVa. Located on the outskirts of the town of Charlottesville, some120 miles south-west of Washington DC, the original campus (referred to even today as the Academical Village) was designed by Jefferson around a central Rotunda rather than a church, reflecting his belief that scientific enquiry should be separated from religion (Dabney, 1981). The core of the campus was inscribed on the World Heritage list in 1987 along with Jefferson's home Monticello which lies on the outskirts of Charlottesville, to the south-east of the town centre.

Stretching south-south eastwards of the Rotunda on either side of a vast green space called the Lawn are a series of classical buildings containing the original class rooms and student accommodation designed by Jefferson. These residences are much sought after and allocated only to 54 final year undergraduates who have excelled in their studies and service to the University, further reinforcing the elite nature of the cohort of students whose temporary home is at the core of the UVa campus.

In summary, both universities are historically important, have at least part of their footprint within a WHS, occupy popular tourist zones, are economically dominant within the region and attract regular visits from tourists keen to explore their heritage in all its manifestations.

Phase Two Data Gathering

A second stage of qualitative data gathering was initiated to cover both institutions. This phase of research included observation of guide behaviour on formal, tourist-focused tours of both universities; researcher participation in guided tours of both institutions; additional focus groups with students at both institutions; semi-structured interviews; one to one with interviews with student guides at both UVa and Durham and finally semi-structured interviews with representatives of both HEIs. Secondary source material such as tour leaflets, web materials and student notes and responses to formal guiding materials was also analysed to shape interview questions and focus group sessions. All interview notes were collated, reviewed and key themes established.

The key aims for this part of the wider project were to determine how elite student guides gain an understanding of how guides shape narratives to increase the democratic reach of universities as tourist spaces, and whether they generally support or subvert institutional 'path-driven' values of their university. To achieve this, a programme of semi-structured interviews was carried out with guides from both study locations to explore the process undertaken to be accepted as 'elite' guides, why they seek to be guides and the importance they place on their role as intermediaries between institution and visitor. A key area of questioning examined whether their pride in their institution and their privileged position as students at elite universities, chosen to fulfil an elite public-facing role, shapes their approach to the narratives they personally develop for both domestic and international tourist audiences.

As Alicia from UVa acknowledged:

"I do see the significance on a global scale....when I get tourists that come here from all over the world, that's when I really see the importance of this place and why it is as well-known as it is".

Before going on to present the key findings which discuss how students shape tours and the way they engage with tourist audiences, the authors first explore the process involved in becoming a guide at each university, as this understanding is essential to place their activities within the broader context of the Universities as heritage tourist attractions.

STUDY FINDINGS: RECRUITMENT AND OPERATIONALISATION FOR STUDENT-LED TOURS

Guides and the Process of Selection for at Each University

As previously discussed, being a tour guide at a historic university offers students the opportunity to share their knowledge, to become a public face of their university, and in the words of several guides, 'give something back' to the institution they are proud to attend. At UVa, Grace decided to become a tour guide because: "*I walked about the grounds and I saw how beautiful they were and the quality of the people that were here and I learned a bit of the history. I realised that I really wanted to be that voice to show the people what it was like*". Guides, whether paid or not, (Durham pays, Virginia students are volunteers) are competitively chosen and our study determined that the majority are women, mostly history majors, drawn from the top students in each university.

University of Durham Selection Procedure

Until 2014, student guides working on tours of the castle were recruited through the World Heritage Site Visitor Centre which is located within University premises on Owengate in the heart of the WHS. Any student studying at Durham was eligible to become a guide. However, in mid-2014, towards the end of the initial fieldwork stage of this research project, procedures for recruiting guides for the tours of Durham Castle altered, as did the delivery of such tours. This was a conscious decision by the Master of University College who perceived the previous, more *ad hoc* process managed through the World Heritage Centre to be a reputational risk for the College.

In a process led by the then newly-appointed Curator of The Castle and the Museum of Archaeology a number of important steps were taken. Firstly, the number of Castle tour guides was reduced from around 30 down to 10, with the intention of providing a more coherent product. Secondly, recruitment of guides was also removed from the WHS Centre to the Curator based within Castle, with priority being given to University College students and alumni. In other words, those with an immediate personal relationship to the castle, the College and its heritage. Interestingly, Avi, a PhD student originally from the US said that because she's not British "*the castle is kind of mine, but I'm not connected to it*", suggesting that even within the guide cohort there are 'layers of belonging' to the institution.

Curatorial staff from the University's Oriental Museum and Palace Green Library were also trained to be able to offer back-up guide support during exam time and vacations when fewer student guides are available to staff the tours. A further change was made to the guides' attire with students no longer wearing academic gowns as was previously the case. These were perceived to be a symbol of the elite student life and thus a visual reinforcement of so-called 'town and gown' differences. Guides are now required to wear a branded jacket (see Figure 1). Finally, the practice of tipping was discouraged and guides were paid a higher hourly rate, and were also compensated financially for attending compulsory training sessions.

Figure 1. Student guide at the castle occupied by University College, Durham

In spring 2017, the student guide cohort for University College tours numbered 10 female and 4 male guides, with an average length of guide service of 1 year 5 months. The guides were split equally between undergraduate and postgraduate students from disciplines ranging from Archaeology and Classics to Natural Sciences and Aeronautical Engineering. Most though not all undergraduates and postgraduates were also members of the college (previously, guides had come from across the University). As at UVa (see below) there is perceived to be a status aspect of being a castle guide and interviewees reported that it can be competitive, particularly for the 'best' shifts. There is also reported to be some competition between guides over the quality of their tours.

The route of the tours around the castle was also altered, partly for conservation reasons in order to reduce the amount of footfall on certain fragile staircases. Content was also altered to provide a chronological perspective on the castle's history. The new tour content was restructured to tie the building into key political events affecting the city and the county, thus reinforcing to some extent the AHD perspective on its 'heritage' as espoused in the original and revised WHS nomination documents presented to

UNESCO. Guides are required to follow the core themes dictated to them in their briefing notes, but are allowed to digress into areas of personal interest relevant to the Castle's heritage and their own relationship with it. In part this reflects the final theme offered in the formal tour, where participants end up in the Great Hall and are regaled with stories about its role as a place of entertainment for both communities who have called the castle home in centuries past including the Prince Bishops who occupied the castle for more than 400 years and, since the 1830s, university students. More recently, a children's trail has been developed and the Curator is currently working on foreign language tours. Current visitor numbers on guided tours are around 30,000 per annum with a further 2,000 or so school children visiting on separate organised tours.

In short, the developments in terms of the delivery of castle tours at Durham in the last 5 years reflect an increasingly managerialist approach to university administration, with the increased professionalisation of the product representing a move by University College to exert more control over public access to its estate within the World Heritage Site.

University of Virginia Selection Procedure

The University of Virginia has two formal guide services. One functions on a similar basis to Durham University's 'back-up' service, with Rotunda administrative staff undertaking tours during the summer months when students are not on campus and the student guide services, an elite competitively selected voluntary body, the latter of which forms the basis of discussions here. Both tours are free for tourists to take. Elizabeth, PhD candidate at the time of writing, worked initially in the Rotunda when waiting to take up a place as an undergraduate. She continued to work as a paid guide throughout her studies. She notes, "*We are not paid much but we are not volunteers but it is a menial job so it is not high paid but it is very convenient for me*". Ironically, being able to offer the same tours as a competitively selected volunteer guide is considered a prestigious rather than menial activity.

UVa has around 110 historic tour guides, predominately White American women, with each guide required to give 11 tours a semester or an average of one per week. Every semester 300 students undertake the selection process to become a historic tour guide. Sixty-five are interviewed with some twenty-six students chosen from this number to get to the next stage of offering a 15-minute trial tour. Although there is a 650-page package of information successful guides need to learn and refer to, prospective guides base their fifteen-minute trial tour on a fact sheet about the University of Virginia. Tours are conducted on the Lawn, and need to be succinct, offering an edited down version of events that reflects the students' own interests. Thus even at interview stage, students are able to tailor, to an extent, the authorised narrative about the University and its heritage to reflect their own interests.

Successful probationary guides then spend a semester in training learning everything about the UVa 'family' including "*Thomas Jefferson, Thomas Jefferson's father, old traditions long gone that you read about*". (Grace). Once training is complete, they are able to take on the responsibility of running tours of their own.

FINDINGS AND DISCUSSION

Opening Remarks

The following discussion considers the key motivations for students to become a historic tour guide at their host institution and examines how guides experience the role. There is a particular focus on pride and privilege and how 'elite' is determined within institutions and negotiated amongst the students themselves. Thereafter, the authors reflect on how specific guides address often contentious, and potentially 'unauthorised' versions and visions of the history of these WHS elite universities, whilst remaining within the institutional framework that employs them.

Why be a Guide?

Students and academic staff/ faculty are the regular inhabitants of university space and are an expected or desired backdrop to any campus tour. Both UVa and Durham University and promote student guides as offering an 'authentic' experience, with other tours seen as back-up or suitable alternatives when students are not on campus. In other words, such back-up tours inevitably will lack some of the personal anecdotes brought to the tours by current students. Nonetheless, both universities are actually more accessible as tourist spaces at such times, particularly during summer vacations. Jo (English, PhD student, Durham) believes "*it is better to be a guide over the summer when the undergraduate students are not around so there are fewer of them and more tourists, so money is better*".

All guides interviewed for this project stressed they offer their own experiences and insights, drawing on official training materials but amending presentations in response to both past and contemporaneous encounters with tourist. In other words, there is an overt acknowledgement from the outset of the staged authenticity of the backstage experience (MacCannell, 1973; Urry 1990). For a university to achieve the twin pillars of providing intellectual hospitality and evidencing the nature of democratic university life (Kaufman, 2001 in Lynch et al, 2011), students need to share the key values of their place of study but must also feel free to challenge or even criticise them from within. Within the context of this research, this can be evidenced by elite student guides having pride in being a student at a highly regarded WHS university, twinned with the understanding that this role is privileged, that they are privileged to be ambassadors for their respective institution and that they have been given the authority also to shape narratives and take views on history and contemporary university life that furthers debate and promotes inclusiveness. This pride and willingness to share the heritage of the place with others was evident in interviews at UVa: Alicia commented that "*I realised I really want to be the voice to show the people what it was like*" and Suzanne noted "*I knew I wanted to talk about a school that I felt had given me so much*".

These principles underpin the student guide role at both universities and are tested during the selection process. Interviewees often confirmed that their motivations for becoming a guide was more from a public relations and personal pride agenda than profit. Whilst those who were paid as guides were happy to be earning, they all voiced a genuine desire to raise what they saw as 'historic truths', as well as reflect present day values of their host institution. At UVa in particular, guiding offers social benefits within a high-status network and community. Although as noted earlier many of the guides are history majors, this is not exclusively the case and we were told that guiding offered transferrable skills. Avi, a PhD candidate at Durham and an international student, reported that she became involved partly because of the opportunity to earn some income, partly to share her own relationship with the institution and its buildings with visitors but also to practice her communication skills, stating that "*it is a lot of fun to see people's reactions to me, and to the site*". This demonstrates that for some guides there is an entirely pragmatic reason to taking on this role, seen here as the personal benefits of income generation and skills development.

The extraordinarily competitive selection process at UVa, where a large number of hopefuls seek to be guides, means failure is likely and success in being accepted on the team is perceived as a significant personal achievement. Like most applicants, Suzanne did not succeed in her first attempt to become a guide but appreciates that the competitive selection process ensures that those chosen take particular pride in the role. As she told the interviewer:

"trying out in your first semester, first year, you haven't really been at this university long enough to understand it so to make an impression on them you have to be very exciting and maybe to focus on the history more than your personal experience because you haven't had a lot of personal experience of the university yet. You have only been here about three weeks if you try our first semester first year. My second semester, I made my tour the many hats of Thomas Jefferson and I started off at Pavilion 7 and talked about the founding and Thomas Jefferson as the politician and then I went to a garden and talked about Thomas Jefferson and his love of horticulture in the garden and then I went to Pavilion 3 and talked about him as an architect and all the architecture of the university and finally I stopped at Pavilion 2 and talked about Thomas Jefferson as an educator. They want to see that you have a theme. That's the defining thing. Every tour guide develops their own tour so every tour is different".

All the UVa students interviewed noted that acting as a historic tour guide increased their own knowledge about the WHS and the relationship with Thomas Jefferson and nearby Monticello, and thus their pride in their university and the consequent sense of privilege. Almost all indicated their engagement with guiding has helped them define their own relationships with the university's heritage, although in some instances that presented challenges. For instance, guides leading UVa tours are increasingly proactive in addressing slavery aspects of the past as the University itself begins to grapple with this topic (Bromley, 2018; McCance, 2019).

Different Audiences

Guides from both institutions reported tailoring tours to specific audiences, influenced by their size, profile and assumed interests, and will happily respond and revise their planned tour to take accounts of questions. All take pride in being able to personalise their tours and guides at UVa in particular are keen to craft tours based on aspects of Jeffersonian heritage, a reflection of course of the statesman's association not just with UVa but with the foundation of the United States themselves. Alicia (UVa) stated "*because Thomas Jefferson is such an integral part of these conversations it seems like maybe to the students and all of the visitors there is more of an emphasis on Thomas Jefferson the man, versus the architecture of the place*". Another guide from UVa, Grace, demonstrated that her pride in place is clearly linked with the sense of continuity of place at the University of Virginia. She notes:

"In my introduction to the tour I'll ask tourist if they have been to Monticello and a lot of time they do; they either come from there to here or from here to there and I mention UVa in conjunction with Monticello is a WHS and I always like to mention that UVa is the only WHS that is still in use for its original

purpose so the library, woodlands are still lived in, professors and students still live in the lawn rooms and something I think is a really cool fact to know as tourists then understand how important this place is and to know that it is still used for its original purpose. I honestly believe there is not a school in the entire world like it".

Durham guides are able to develop more diverse tours although there is always a split between the physical environment and tangible heritage and the intangible, and between the personal aspect of being a student living there and the public aspects of the site's history such as the role of both castle and cathedral in the governance of the region for centuries, as well as the castle's more recent academic function. A key theme from the University's perspective is the continuity of occupation of the building for nearly 1,000 years and the fact that it remains a living, working place.

At Durham, castle visitors are particularly interested in student relationships with the building – both 'normal' visitors but also students from elsewhere and potential applicants. Jo from Durham University said she was happy to talk about even the most mundane of topics, such as what it is like to eat your meals in a 14th century hall originally built for the Prince Bishops of Durham, but which to many visitors reminds them of the 'Hogwarts' dining hall seen in the Harry Potter movies, inspired by the J K Rowling novels about the eponymous boy wizard. Indeed, the practicalities and experience of living in a castle is a repeated topic of discussion at Durham. Jo admitted being asked questions about the type of food served in the dining hall and was it 'normal food', whilst both Avi and Jo reported receiving Harry Potter-related questions on most tours they escort, with Avi going so far as to venture that for some demographics (though she wouldn't identify them by name) the Hogwarts-type experience is the only thing that seems to engage them about the building and she felt that tourists perceive the two environments – one real, one imagined – to be very similar.

Personal Responses to the Site's Contested Heritage

This topic was particularly evident in discussions with guides at UVa, where in recent years some members of the guide cohort have consciously sought to develop tours that address more challenging aspects of contested history, and in particularly slavery and Jefferson's position as a slave owner. Suzanne suggested that that this is now being actively promoted as a theme "*so there's a couple of things we're encouraged to hit on, and one of the things is slavery at the university and integration of the university*". This of course reflects the recent moves by the University's President to acknowledge more publicly the relationship that the institution and its founder had with slavery during its formative years (Bromley, 2018; McCance, 2019).

For Grace this is an important public service part of her role as "*every single tourist that I have ever taken on my tours and talked to about slavery has seemed to appreciate that and are nodding their heads and even have sad expressions on their face which makes me happy as it makes me sad, and when I see people being sad about what I say it makes me feel I'm doing a good job*"

By encouraging guides to discuss some of the aspects of Thomas Jefferson's life such as his relationship with Sally Hemmings, and the fact that the students loaned their slaves to build the college and adjacent gardens, the university is moving beyond the formal conservative values to allow the dissonant heritage discourse to be discussed. Grace believes this is necessary to challenge what some term as a 'cult' of Jefferson at UVa.

"Lectures often start or end with a quote. There are quotes in the gym. They put him (Jefferson) as God but you know in general most guides really want to address the fact of slave owning on their tours because I think we are all very aware. It is a very sad part of our history and we want to address it in our tours and every single tourist that I have ever taken on my tours and talked about slavery to has seemed to appreciate that".

Suzanne presents the interesting point that:

"We're still finding out more, it's sad how little we know. We talk about how when you look at the lawn, it of course looks very beautiful and Jefferson had a very specific image for what he wanted the lawn to look like and he spent a lot of time on the architecture, but you know Charlottesville being very hilly if you walk around it's not going to have any natural terracing as you see on the lawn and that was all done by manpower, by slaves moving the land so I usually take my tours out into a garden where if you look back from the garden you can't see the rotunda you can't see the lawn and that's where the slave quarters were. I absolutely feel comfortable talking about, I know it can be sensitive still especially when you talk about Thomas Jefferson and Sally Hemmings. There's still tourists who will deny there was any relationship between Thomas Jefferson and Sally Hemming even although there's pretty much definitive DNA evidence". Student guides are thus at the forefront of tackling previously difficult heritage on behalf of their host institution.

Elizabeth acknowledges that:

"Slavery is coming out more and more. There has been a push, especially from African American enthusiasts and African-American members of the university community that's brought this to light". She believes it was not ignorance but a conscious desire "to hush it up, maybe pretended it never happened. It was the elephant in the room no one wanted to discuss. They did contract slaves that did a lot of work here so now they're starting to acknowledge that and they're now starting to bring to light applications of African-American people who had applied to the university but that were rejected. They never said it is because you were black, they just said 'We do not think your application is suitable for the university at the time".

Grace feels passionately that it is important to discuss slavery as a natural part of her historic tours.

"I take them to a garden at first and then I talk about how this is a very beautiful place and how it wasn't always this way. Originally, this was a workplace where animals were kept and the gardens were and also where slaves lived. And there were slaves at the UVa and it was a very big part of Thomas Jefferson's life therefore it was also a big part of the university's life. I will mention that Thomas Jefferson was slave owner who owned 600 slaves throughout his lifetime and then I will go through like the history of African Americans at the university briefly first talking about what slaves did in the construction of the university doing the actual construction work and building of the Rotunda and the terrace and the lawn. UVa was actually the first university to purchase a slave as in owned by the university which was unheard of at the time. Prior to that all the slaves that worked here were contracted out by individual people so were hired through professors or hotel keepers or the students".

It is important to note, within the context of a discussion about the ability of student guides to present contested narratives about their host institution within the spirit of academic inquiry, that increased pressure over a period of time by the student body at UVa for the university to acknowledge the role of slave labour led eventually to the construction in 2019 of a Memorial to Enslaved Laborers, commemorating 200 years since the first slave labour was used on the site (Bromley, 2019).

Being free to shape tours round elements of the contentious past reinforces changed (or changing) attitudes in the present, and indeed the need to be viewed as inclusive, diverse and welcoming institution within both the formal university and in the student body. This supports town and gown relationships as well as offering tourists from all over the world a sense of 'warts and all' history. Grace (a UVa guide) determines that guiding is structured around university values of "*focusing on the history and **altering** the history by ensuring that we are very much up to day with current trends in academic research*" (our bold). And one of those trends in current research is allowing students to openly discuss issues reinforcing both the university's history and its current purpose in line with the opinion of Gelbman and Maoz to the extent that student guides become '*mediators and commentators of local cultures*' and as the voice of their university, are *drawing on tourism strategically in order to address issues of national significance*' (2012:109).

CONCLUDING REMARKS

Primary research conducted by the authors and presented here has demonstrated that the student guide is both an intermediary and a carrier of the institutional values of the historic university, given their competitively selected role as the 'elite of the elite'. The contribution of this research to the corpus of existing work on heritage tourism and indeed on tourist guides lies in its exploration of a previously invisible aspect of the tourism experience in many historic destinations, where the elite values of World Heritage Site status and top-ranked universities collide with tourist expectations of a 'regular' heritage experience at these places, and where the people mediating their experience and understanding of the location tend to be young, educated and drawn from the elite population in that place.

The historic campus tour is an established part of the engagement strategies of both universities and must be seen to stand alone within the destination. Thus, although nearby Monticello at UVa and the cathedral at Durham are part of their respective World Heritage Sites and also of the wider tourism offering in those places, there are no formal links in tourism terms although links may be made within the tours. As such the student guides, by making these links, informally strengthen the place-based narratives beyond their own university and into the wider tourism experience.

Tours are constructed and shaped by the 'elite' guides' own interests and values, and by those of the domestic and international tourists who visit. Ultimately, through the formal recruitment process and careful selection of guides, the peer to peer system at UVa in particular where the baton is passed on through networks, the testing of historic knowledge at selection and the support systems in place, both institutions are able to ensure the guides are and remain ambassadors rather than protagonists, regardless of whether they feel they are challenging systems from within.

Being a historic tour guide is a privileged role at both universities. The competition required to take on the role and the responsibility placed upon these young people stimulates a sense of privilege that is acknowledged by the guides themselves. All take pride in their role, feel connected to their university and recognise that they are playing a part in reputation management and organisational and self-marketing, whilst at the same time constructing tours that offer visitors the experience they consider they are seeking based on assumptions around demographics and questions from the group. Tours led by 'elite' guides, ensure that sense of privileged access is reinforced to visitors and down through each successive year, as student move through their studies.

Some guides engage empathetically within their visitors via the subject matter they choose to develop, reinforcing the ideas of Tucker (2016). At UVa, guides felt it was especially important to discuss shameful aspects of the university's past, as both a mark of respect and evidence of cultural change in the present-day university values, again reinforcing the ideas of Winter (2013). At the University of Virginia, Jefferson - the man and his life - is generally far more important a topic for exploration than is the built environment of the WHS, whereas at Durham the footprint of the university and the castle seemed the main focus of interest. However, in both places thee current lived experience in these special, historic places is also celebrated and discussed as much as their historical functions. The university is the dominant force in both cases, with WHS values secondary and not always even mentioned in tours.

In both institutions, students are free to create new narratives within their tours that reflect their own interests, even when these narratives appear to challenge the formal narratives and conservative values that are inevitably fixed points for elite universities. Student guides thus perform an open exchange of information with tourists in the spirit of cultural diplomacy (Bryce and Carnegie, 2013). By allowing and even encouraging students to shape their own historical tours for tourists, the universities under discussion here are performing and highlighting the democratic nature of university life, breaking away from its own path dependency. Student guides represent the university values, whilst at the same time challenge them and making themselves appear both as individuals and as a 'body politic', approachable. All tours are a negotiation, continuously created and recreated. As Alicia (UVa) noted: *"You are never going to get the same tour twice and they always question as well"*.

REFERENCES

Bromley, A. E. (2018). *New Commissioners will focus on untold stories of UVa in the age of segregation.* Retrieved from https://news.virginia.edu/content/new-commissioners-will-focus-untold-stories-uva-age-segregation

Bromley, A. E. (2019). *Work on memorial to enslaved laborers gets underway.* Retrieved from https://news.virginia.edu/content/work-memorial-enslaved-laborers-gets-underway

Bromley, R. (2006). On and off campus: Colleges and universities as local stakeholders. *Planning Practice and Research, 21*(1), 1–24. doi:10.1080/02697450600901400

Bryce, D., & Carnegie, E. (2013). Exhibiting the 'Orient': Historicising theory and curatorial practice in UK museums and galleries. *Environment & Planning A, 45*(7), 1734–1752. doi:10.1068/a45359

Cohen, E. (1985). The tourist guide. The origins, structure and dynamics of a role. *Annals of Tourism Research, 12*(1), 5–29. doi:10.1016/0160-7383(85)90037-4

Connell, J. (1996). A study of tourism on university campus sites. *Tourism Management, 17*(7), 541–544. doi:10.1016/0261-5177(96)89217-X

Connell, J. (2000). The Role of Tourism in the Socially Responsible University. *Current Issues in Tourism, 3*(1), 1–19. doi:10.1080/13683500008667863

Dabney, V. (1981). *Mr Jefferson's University: A History*. University of Virginia Press.

Dahles, H. (2002). The politics of tour guiding. Image management in Indonesia. *Annals of Tourism Research, 29*(3), 783–800. doi:10.1016/S0160-7383(01)00083-4

Farkic, J. (2018). Outdoor guiding as hospitality work. *Annals of Tourism Research, 73*(C), 197–199. doi:10.1016/j.annals.2018.04.004

Gelbman, A., & Maoz, D. (2012). Island of peace or island of war: Tourist guiding. *Annals of Tourism Research, 39*(1), 108–133. doi:10.1016/j.annals.2011.05.008

Gumprecht, B. (2003). The American College Town. *Geographical Review, 93*(1), 51–80. doi:10.1111/j.1931-0846.2003.tb00020.x

Gumprecht, B. (2007). The campus as a public space in the American college town. *Journal of Historical Geography, 33*(1), 72–103. doi:10.1016/j.jhg.2005.12.001

Hemsley-Brown, J., & Goonawardana, S. (2007). Brand harmonization in the international higher education market. *Journal of Business Research, 60*(9), 942–948. doi:10.1016/j.jbusres.2007.01.019

Hoover, E. (2009). Golden Walk" Gets a Makeover from an Auditor of Campus Visits. *The Chronicle of Higher Education, 55*(26).

Kaufman, E. (2001). *The Delirium of Praise*. Johns Hopkins University Press.

Klein, A. (2004). Reinventing the campus tour. *University Business, 7*(11), 51–53.

Lynch, P., Molz, J. G., Mcintosh, A., Lugosi, P., & Lashley, C. (2011). Theorizing hospitality. *Hospitality & Society, 1*(1), 3–24. doi:10.1386/hosp.1.1.3_2

MacCannell, D. (1973). 'Staged Authenticity: Arrangement of Social Space in Tourist Settings'. *American Journal of Sociology, 79*(3), 589–603. doi:10.1086/225585

Magdola, P. M. (2000). The Campus Tour: Ritual and Community in Higher Education. *Anthropology & Education Quarterly, 31*(1), 24–46. doi:10.1525/aeq.2000.31.1.24

Magolda, P. M. (2001). What our Rituals Tell us about Community on Campus: A Look at the Campus Tour. *About Campus: Enriching the Student Learning Experience, 5*(6), 2–8. doi:10.1177/108648220100500602

Mampaey, J., Huisman, J., & Seeber, M. (2015). Branding of Flemish higher education institutions: A strategic balance perspective. *Higher Education Research & Development, 34*(6), 1178–1191. doi:10.1080/07294360.2015.1024634

McCance, M. (2019). *UVA Today Series Presents Stories Explored by Commission on Segregation*. Retrieved from: https://news.virginia.edu/content/uva-today-series-presents-stories-explored-commission-segregation

Mesaritou, E., Coleman, S., & Eade, J. (2016). Introduction: Guiding the pilgrim. *Tourist Studies, 16*(1), 3–22. doi:10.1177/1468797615588425

Ming, L. (2007). *Research in Campus Tourism: Actuality, Problems and Advice - a Literature Review of 10 Years Research in Campus Tourism in China*. doi: CNKI:SUN:ZTKB.0.2007-06-030

O'Mara, M. P. (2012). Beyond town and gown: University economic engagement and the legacy of the urban crisis. *The Journal of Technology Transfer*, *37*(2), 234–250. doi:10.100710961-010-9185-4

Orbasli, A. (2008). *Architectural Conservation. Principles and Practice*. Blackwell.

Oxford City Council. (2017). *Listed buildings - overview*. Retrieved 7 24, 2017, from https://www.oxford.gov.uk/info/20192/listed_buildings/867/listed_buildings_-_overview

Pereira Brando Albino, S. (2015). *Tourism in university cities: the role of universities in place branding* (Unpublished Doctoral Dissertation). Instituto de Geografia e Ordenamento do Território. Lisbon: Universidade De Lisboa.

Pevsner, N., & Williamson, E. (1985). *The Buildings of England. County Durham* (2nd ed.). Yale University Press.

Pocock, D. (2013). *The Story of Durham*. The History Press.

Popescu, A. I. (2012). Branding Cities as Educational Centres: The Role of Higher Education Institutions. *Management and Marketing. Challenges for the Knowledge Society*, *7*(3), 493–510.

Roberts, M. (2003). *Durham. 1000 years of history*. Tempus Publishing Ltd.

Salazar, N. B. (2005). Tourism and glocalization "local" tour guiding. *Annals of Tourism Research*, *32*(3), 628–646. doi:10.1016/j.annals.2004.10.012

Salazar, N. B. (2006). Touristifying Tanzania: Local Guides, Global Discourse. *Annals of Tourism Research*, *33*(3), 833–852. doi:10.1016/j.annals.2006.03.017

Schofield, P., & Fallon, P. (2012). Assessing the Viability of University Alumni as a Repeat Visitor Market. *Tourism Management*, *33*(6), 1373–1384. doi:10.1016/j.tourman.2011.12.021

Shaw, K., & Fincher, R. (2010). University students and the creative city. *Journal of Policy Research in Tourism, Leisure & Events*, 2(3), 199–220. doi:10.1080/19407963.2010.512203

Smith, L. (2006). *Uses of Heritage*. Routledge. doi:10.4324/9780203602263

Stephenson, A. L., & Yerger, D. B. (2014). Does brand identification transform alumni into university advocates? *International Review on Public and Nonprofit Marketing*, *11*(3), 243–262. doi:10.100712208-014-0119-y

Tucker, H. (2016). Empathy and tourism: Limits and Possibilities. *Annals of Tourism Research*, *57*, 31–43. doi:10.1016/j.annals.2015.12.001

Tucker, H., & Carnegie, E. (2014). World heritage and the contradictions of 'universal value'. *Annals of Tourism Research*, *47*(July), 63–76. doi:10.1016/j.annals.2014.04.003

Urry, J. (1990). The consumption of tourism. *Sociology*, *24*(1), 23–35. doi:10.1177/0038038590024001004

Winter, T. (2013). Clarifying the critical in critical heritage studies. *International Journal of Heritage Studies, 19*(6), 532–545. doi:10.1080/13527258.2012.720997

Woodward, S. (2013). Campus tourism, universities and destination development. In M. Smith & G. Richards (Eds.), *Routledge Handbook of Cultural Tourism* (pp. 265–272). Routledge.

ENDNOTES

1 https://www.dur.ac.uk/about/shaped/dates/ accessed 8/1/17
2 http://www.virginia.edu/factsaccessed8/1/17
3 https://www.dur.ac.uk/about/facts/ accessed 8/1/17
4 http://www.virginia.edu/factsaccessed8/1/17
5 https://www.dur.ac.uk/about/facts/ accessed 8/1/17
6 https://www.dur.ac.uk/about/facts/ accessed 8/1/17
7 http://whc.unesco.org/en/list/370accessed12/1/17
8 http://whc.unesco.org/en/list/442accessed12/1/17
9 http://whc.unesco.org/en/list/370accessed12/1/17
10 http://whc.unesco.org/en/list/442accessed12/1/17
11 http://durobs.durham.gov.uk/ia/ia1008/ accessed 12/1/17
12 https://www.census.gov/quickfacts/table/PST045216/5114968,51accessed12/1/17
13 https://www.durhamworldheritagesite.com/management-plan/5-audiencesaccessed12/1/17
14 http://www.countydurhampartnership.co.uk/media/12826/Index-of-Deprivation-2015-Factsheet/pdf/IndexDeprivation2015Factsheet.pdf Accessed 24/7/17
15 http://www.durham.gov.uk/media/1020/Regeneration-Statement-2012-2022/pdf/RegenerationStatement.pdf Accessed 24/7/17
16 https://www.census.gov/quickfacts/table/PST045216/5114968,51accessed24/7/17
17 http://www.charlottesville.org/departments-and-services/departments-a-g/department-of-finance/financial-reports/cafr-2016accessed24/7/17

Chapter 5
Tourist Guidance in Sport Tourism

Orhan Icoz
https://orcid.org/0000-0002-3077-8476
Yasar University, Turkey

Onur Icoz
https://orcid.org/0000-0002-1778-9744
Aydin Adnan Menderes University, Turkey

EXECUTIVE SUMMARY

Sport tourism can be viewed as the amalgam of specific sporting activities, being undertaken by people in various forms of competitive interplay or other interaction, that occur in the places where might be instrumental to the sport and/or tourist experience. It is one of the fastest growing market segments in tourism and travel industry and is receiving increased attention for its social, environmental, and economic effects on destinations. It is estimated that there will be a larger market potential than ever for sport-related tours and tourism. On the other hand, it is well known that tour guides are tourism professionals that lead their guests through the many parts of any destination and they are responsible for planning, organizing, and execution of the tours as a very critical component of the tourism value chain. In this chapter, the basic functions and services of tour guides for sport activities and sport tourism will be discussed and the basic objective is to provide detailed information about this enhancing field of tourism for young generation tour guides.

INTRODUCTION

Tourism and sports are two popular human activities that create a great deal of social interactions, economic transactions, travels to other destination etc. Both industries are considerably growing as societies and economies develop worldwide. Depending on this growth, the sport related travels have been becoming one of the fast growing market segments of tourism, and the concept of "*sport tourism*" has widely been recognized in recent years in order to portray sport-related travels (Gibson, 1998a) and it is considered

DOI: 10.4018/978-1-7998-3725-1.ch005

as a favorable alternative to overcome traditional seasonality problems of tourism destinations (Higham and Hinch, 2002) and as an effective tool for sustainability.

Tourism guidance, on the other hand, is a noteworthy component of travel and tourism industry, since tour guides are the key actors of the tours to unknown or less known places. Traditionally, tourism guidance is extremely significant mission for those who are traveling abroad, unexperienced, having foreign language difficulties, but curious about historical/geographical places and local cultures.

As tourism industry is growing all over the world, new and diverse types of tourism are emerging or some existing tourism activities are gaining more importance than ever before. Although, sport tourism is not a new phenomenon, but it is flourishing over the years and getting more and more popularity. Moreover, among various tourism types, sport tourism is one of the most active and multifarious type of tourism, especially for individual sport participation. Therefore, sport tourists may need more technical, practical and geographical assistance than classical tour participants may have. Their key assistants are the guides who are the specialists in their particular fields.

On the other hand, sport event tourism has their own significance for both sports and tourism sectors. These organizations are the best occasions in order to watch directly popular athletes, teams and players on the ground. For international organizations, thousands of people as spectators from all over the world travel to destinations where the sports events are held. These spectators also need some assistance by tour guides in order to facilitate their travels to the destinations and to learn more about the places they visited.

A brief literature review shows that there are number of studies about sport tourism, whereas no specific study on sport tourism guidance. The well-known authors who studied on sports tourism are Gibson (1998b), Stevenden and De Knop (1999), and Deery et al. (2004). There are also some other studies focusing on various dimensions of this type of tourism such as; sport events tourism (Chalip, 1988; Fairley 2003; Deery, 2004; De Aragao, 2015; Achu, 2019;), sport tourism markets (Gibson et al., 1998) and economic benefits of sport tourism (Hurley, 2018; Mc Bride (2018), active sport tourism (Gibson, 1998a; Cousquer and Beames, 2013; Gibson et al., 2018; Heather et al., 2018), sport tourist behaviors (Hengpoom and Weeraspoon, 2019), sport tourism destinations (Higham 2007), sport tourism development (Hinch and Higham (2019), and active sport tourism motivations (Hodeck and Hovemann, 2016; Kim, 2013; Kirkup and Sutherland, 2015; Neirotti et al., 2001; Rinaldi, 2011). On the contrary to the availability of rich literature on sport tourism, there are quite a few numbers of studies referring the different aspects of tourism guidance. The noticeable authors who have studied in this field could be mentioned as Cohen (1985), Mancini (2000), Ahipasaoğlu (2001), Huang (2010), Oter (2019), Overend (2012), and Ababneh (2017).

Accordingly, this chapter aims to generate an understanding about the significance of sport tourism guidance in this sector. For this purpose, the relevance between tourism and sport in a historical perspective, the types of sport related tourism, its economic and social aspects, motivations for sports travel, the importance of guidance for sport tourism and the future of sport tourism are discussed as the main topics in this chapter.

Sport, Tourism, and Sport Tourism

Sports and tourism activities have been acting together from the ancient times to date that sports tourism has become one of the most significant industries characterized by ever increasing economic contributions. Contemporary tourists are many in numbers, curious about new things, want to explore new destinations, they are the new type of tourists-adventurers who are interested in extreme sports during their travel.

In a deeper analysis, sport and tourism are two complementary and counterpart fields of the leisure industry where most of the individuals actively or passively participate and these social experiences have become very significant economic activities worldwide (Standeven and De Knop, 1999).

At the beginning of the 1990s, "sport tourism" was briefly defined by Kurtzman (*quoted in* Van Rheenen et al., 2017) as "*the use of sport as a touristic activity*" in his opening speech to the *Sports Tourism International Council* where the participants was mainly consist of the organizers of big sports events and the representatives of attractions, resorts, tours and cruises. As a result of the developments in this phenomena all over the world, somewhat more comprehensive and technical definition was suggested by Gibson (1998b) as "*leisure-based travel that takes individuals temporarily outside of their home communities to play, watch physical activities or attractions associated with these activities*". Standeven and De Knop, (1999: 12) also defined sport tourism as "*All forms of involvement in sporting activity, participated in casually or in an organized way for non-commercial or business/commercial reasons, that necessitate travel away from home and work locality*".

There are number of factors influencing the rise of sport tourism, while the most important one is the physical wellness and fitness of individuals. Along with the economic development of societies and the enhancing living standards, people begin to care more about the protection of their own health. Thence, as an effective tool for health protection, the sports tourism emerged as a new type of tourism combining tourism with the sports, (Xinhua et al, 2016). Therefore, sport tourism, as a specific tourism category, has also become an academic involvement and governmental concern. It is also perceived to be significant owing to its contribution to economy as well as other effects such as the positive impact on social welfare and the sense of honor that sport tourism may produce (Deery et al., 2004). Therefore, sport tourism is classified among the sub-categories of tourism, which is also considered as alternative tourism (Yildiz and Cekic, 2015; Perna et al., 2019).

Historical Background and Development of Sport Tourism

The link between tourism and sport dates back to the early beginnings of the development of sport and travel. Even in ancient Rome and Greece, sport was the trigger of traveling people, and visiting the major sport events was very popular among some layers of communities. Since the ancient times, the people have travelled to other places in search of sports competition and the history of travel for sports purposes dates back to 776 B.C. (Yildiz and Cekic, 2015). The ancient Olympic Games which were started in that year and continued for 1200 years is commonly cited as one of the earliest forms of sports competition that stimulated considerable travel flows with the motivation to participate or to simply watch the games. Historically, colossal events during Roman Empire were the well-known example of sports spectatorship and the most popular sport among Byzantine peoples was known as chariot racing. Likewise, during the Middle Ages, a group of chevaliers from Great Britain travelled across Europe and, in many cases, participated in games lasting a couple of days, around 1200 A.D. (Zuahar, 2004; 50).

Travels for sports have been developed through the ages and reached today's understanding. The sport events that were not organized for centuries due to the storms of history and have been regularly held again since the late 19th century. Ten millions of people travel overseas for sports purposes, but surveys focusing on the impacts of sport competitions on travel have a little history, as opposed to the researches on traditional 3S tourism (Sotiriadis, 2016). As one of the mega sports events, the first contemporary Olympics were organized in 1896 in Athens. Following this successful organization, National Olympics Committee gave the next Olympics of 1900 to Paris (Yildiz and Cekic, 2015). For the later stages of history, sport tourism gained a broader role in societies and different relationships have been built between sport and tourism (Jakovlev et al. 2017).

According to Higham (2018: 166), "an important part of the dynamic expansion of sport over the past 30 years has been the extension of traditional sporting seasons. The reasons for this growth include a variety of technological progress and innovations, changing social and economic conditions and the effect of globalization, while one of the most important factors has been the professionalization of many sports at the high level of competition". In addition to above mentioned factors, Hinch and Higham (2019: 5) argued that the growth of sport tourism in recent years stands at the cross-section of following major trends;

- the changing and varying demographics of sport participants,
- growing interests for fitness and well-being in western societies,
- more demand for active participation in recreational activities during holidays,
- growth of interest in the major roles played by sports and sports events in urban renewal and urban imagery, and the potential to leverage tourism opportunities associated with sports events

Nowadays, most of the destinations and countries have many different sport opportunities that some of them professionally organized and managed, while some others are designed for individual participants. For the latter category, there are still less number of visitors who have pre-experience in the activity involved and it is well known that there is an increasing interest for different sport activities among young travelers.

Sport tourism, as a vigorous part of tourism industry has gained importance in tourism destinations and it becomes an objective of many studies owing to its growing importance and significant contribution to tourism revenues. As a consequence of this growth, such studies as the effects of major sport events on economic growth of tourism are expanding in tourism literature. These studies were started in the 1980s, about the different aspects of sport tourism consisting of not only big sport events related travels but also other travels for individual sports. These individual sports also create significant effects on local economies. The studies broadly assume sport tourism as a niche and specific market in tourism and the studies on sport tourism were significantly increased in the 2000s.

Sport as a notable special interest of tourism, is expressed under three related tourism domains, including *hallmark events* [the Olympic Games], *outdoor recreation* [adventure tourism - canoeing, skiing and surfing] and *tourism associated with health and fitness* [from therapeutic spas to tennis in resorts] (Hinch and Higham, 2019: 22). Sport and tourism industries have been accompanied to one another in terms of their development phases over the past 5 decades. They both have been democratized and increased freedom, become easier to get involved, become a social experience worldwide and have been open to the effects of globalization. Furthermore, sport events, sport teams, sport clubs and sport facilities have become major tourism appeals by themselves. Similarly, popular individual sports players are also seen as interesting pull factors *(attractions)* for tourists. The development of sports as a new type of tourism attraction generates new chances for sport teams, competitors, and managers to expand their market potential and number of supporters. The objective of investments for sport tourism in a region or country is not only projected for the development of local residents in sports, but also to attract visitors and enhance the image of the destination and turning into a trustworthy brand. This is why, sport as an attraction is widely used by the policy makers all over the world, and the increasing demand for mega sports events organizations such as World Soccer Cup and the Olympic Games by the competitive countries are the most important evidence of this involvement.

Classification of Sport Tourism

The types of sport tourism display a number of variety. This variety primarily depends on the type of sport tourist as an active or passive participant, and then the hardness and structure of sports in question. The length of sports event and the activity itself may also be a determinant. Three generic categories of sport tourism have been acknowledged as *active sport participation, visiting sports related attractions* and *watching sporting events* (Gibson, 1998a). Among these, the first and second categories imply only *passive participation* to sports related events or attractions, but the third requires *active participation.* The main distinction between these two generic classification could be seen in the behavior of physically taking part as a *'competitor'* or watching a sport event as *'supporter'* (Heather et al, 2018).

Table 1 shows the classification of sport tourism. On this table, as the types of sport tourism under the category of "*Attractions, Tours and Events*" display a characteristic as passive participation or sport events, the other types under the category of "*Resorts, Adventure and Cruises*" are the examples of active sport tourism.

Economic and Social Importance of Sport Tourism

It is well known that major sports events substantially contribute to the economic expansion and tourist flow in a destination and these events are the critical components of tourism products for destinations. Sport tourism, like many other kinds of tourism, is also considered as an effective tool for economic progress in metropolitan areas. In recent years, it has been witnessed that the dimensions and magnitude of sport-related travel have considerably grown in these urban areas.

Recent data has shown that sport related activities contributes considerably to developed western economies. This contribution is measured as between 1% - 2% of GDP and the sport sector accounts for 1.76% of the European Gross Value Added (GVA) that means that it contributes more to the economies than any other sectors such as agriculture, forestry and fisheries combined. Converted to Euro that is about 175 billion. If also added all the sectors that are influenced by sport, the share of sports at the European GVA is 2.98% or almost €300 billion, which actually is a lot (Sportyjob, 2015). Similarly, the contribution of tourism to the World Gross Domestic Product(GDP) is counted between 4% and 6% (UNWTO, 2018). On the other hand, the World Travel and Tourism Council's (WTTC) research reveals that the sector accounted for 3.2% *(Direct Contribution)* and 10.4% *(Total Contribution)* of global GDP and 319 million jobs, or 10% of total employment in 2018 (WTTC, 2019). Sports tourism is a ever growing part of the global tourism industry which is regarded that generate profits near to $800 billion [in 2016] a year and about 32% of the revenues of the tourism industry in the world are committed to sports tourism (Askarian et al., 2015). As a result of the development of these two fields, both tourism and sport have gained considerable amount of interest from industry practitioners and academics over recent years (Ritchie and Adair, 2010).

Authors claim that the mega sport events do boost tourism. In South Africa [in 2010] for example, the total increase in international visitors was only about 307,000 total spending $444 million, mainly in shopping, food and accommodation during the World Soccer Cup. Likewise, the World Cup [in 2014] generated $3.0 billion in the overall Brazilian economy, by the 3.7 million national and international tourists during the event (De Aragao, 2015: 8). 2008 Summer Olympics in Beijing generated $3.6 billion in revenue, and 2012 Summer Games in London generated $5.2 billion (McBride, 2018). The number of tickets issued for 2012 London Olympics was 8.8 million (BBC, 2012), and in 2016 Rio Olympics it was over 7 million (Timeout, 2019) as well. Sochi 2014 Winter Olympics was the most expensive Olympics at $21.9 billion from 1964 to 2016, according to a study reported by Humankinetics (2019).

Table 1. The scope of sport tourism

Sport tourism categories	*Nature based or man-made activities*
Attractions	§ Mountains § Canyons § Sport heritage sites § Sports museums § Stadium § Sports theme parks § Unique sports facilities
Resorts	§ Ski resorts *(Winter/Summer)* § Integrated resorts § Health and fitness clubs § Sport training facilities
Cruises	§ Transportation for participation in sports tourism § On-board sports facilities § Hosting of sports personalities/conferences
Tours	§ Sport related tours
Events	§ Football/Soccer games § Mega sporting events *(Super Bowl)* § the Olympic Games § Major tennis tournaments *(Wimbledon/Grand Slam)* § Marathons
Adventure/Active Sport Tourism *(Including extreme sports)*	§ Camping § Canoeing/kayaking § Cave diving § Cliff hanging § Cycling tours § Equestrianism *(Horse riding)* § Golf tourism § Hiking § Mountain bike § Mountain/Rock climbing § River tourism § Running/Jogging § Scuba diving § Snow cycling § Snow rafting § Snowboarding § Trekking § Walking § Whitewater rafting § Windsurfing/surfing/kite surfing

Source: Adapted by the authors from Ritchie, B. and Adair, D. (2002).

Followings are some of the statistical data about sport tourism worldwide (WST, 2020);

- Sports related tourism activities creates 12 -15 million international arrivals annually and estimated short term yearly growth is measured as around 6%.
- The leading countries for sport tourists are the Germany, Italy, UK, Sweden, Spain, Denmark and Finland.
- In some countries, the revenues obtained by sports tourism are more than 25% of the total, while the share of Australia recorded as 55%.
- In 2007, winter sports identified about 49% of travels while the share fell to around 42% in 2013.
- More than 6.5 million amateur and over 10.000 professional athletes have been determined worldwide.
- Majority of destinations were determined through the web (35%) and secondly word of mouth (32.3%) by travelers.

Researchers have claimed that sport tourism and sport related events might also affect destinations in social perspectives as well. These destinations mostly generate sport events, which may launch the participation of amateur or professional sport-tourists, promote regional services, create an extra demand for the consumption of local products and support involvement in local activities (Snelgrove and Wood, 2010). Authors also recognize the social functions of sports and sport tourism as the followings (Zuahar, 2004: 7);

1. The socio-emotional function: the socio-psychological stability of the individuals.
2. The socialization function: the individual involvement to cultural values and the development of personality characteristics.
3. The integrative function: the level of collectivity.
4. The political function: the role of sport as a political tool.
5. The social mobility function: the movement of individuals.

Consequently, sports activities and events establish potential association between guests and local people, with the interaction between tourism and sports playing a significant role in the tourism industry (Perna et al., 2019).

Factors Affecting the Development of Sport Tourism

Sport tourism is growing all around the world and there are a number of important indicators representing this growth. Event sport tourism as a specific type of mass tourism is more visible and obvious to some extent and it can be measured easily by the number of recorded spectators in the stadiums and gymnasiums. For example, the number of people travelling for a soccer match to a certain destination or foreign visitors of an Olympic Game can easily be measured by the recorded spectator numbers, and these figures can be found in the international official statistics representing the number of sport tourism travelers is steadily increasing for the recent decades.

Although it is very hard to count the number of active sport tourism travelers, it is obvious that the participation of active sport tourism is still growing all around the world. Global forecasts and the studies on this subject also recognize this growth and a variety of conditions have assisted in the enhanced popularity of sport-tourism.

A number of factors affecting the growth of sport tourism have been determined by the authors. Some of these factors are very traditional ones that have an effect on the development of any kind of tourism, while others are particularly effective on sport tourism as listed below (*adapted from* Jakovlev et al., 2017; Higham 2018, 148; Hengpoom and Weerasophon, 2019);

I - Factors affecting demand for sport recreation services.
§ Economic forces
§ Urbanization and installation *(institutionalization of sport activities)*
§ Increase in disposable income of the population
§ Increase of free time of individuals and paid holidays
§ Education and training
§ Globalization and international openness
§ Sport equipment
§ Sharing economy
§ Subjective and irrational factors *(fashion, behavior, habits, prestige)*
II - Intermediary factors
§ Tourist sport agencies and specialized intermediaries who are the experts on certain sports
§ Electronic intermediaries/agencies
§ Fan/supporter associations/unions
§ Sport organizations
§ Type of sport/event
§ Level of competition
III – Factors of sport recreation quotes
§ Natural-sources
§ Professionalism *(trainers, instructors, guides)*
§ Sport-recreation and references
IV – Information/media
§ TV programs and documentaries
§ Web sources/Sport specific web pages
§ Other information sources *(publications on specific sports)*
§ Blogs *(travel blogs)*
§ Social media interaction
§ e-Word of mouth
§ Existence of information services
V – Technological innovations, devices and effective communication
§ GPS Systems
§ Mobile/Smart phones
§ Satellite connected phones *(Iridium)*
§ Portable computers/tablets
§ Internet
§ Newly emerged sports *(kitesurfing, paragliding)*
VI – Rational/objective factors

§ Health protection
§ Well-being
§ Fitness
§ Special interest
§ Adventure
§ Exploration/discovery
§ Curiosity
§ Achievement
§ Self-actualization
§ Socio-demographic trends
VII – Irrational/Subjectives
§ Emulation
§ Prestige
§ Competition experiences
§ Snobism
§ Affection
§ Attitudinal/value changes

While above mentioned factors have a positive effects on the development of sport tourism, there have been identified some factors that may restrict individual travel behavior in order to participate in a sport tourism activity. For instance, event-based tourism is often concerned with the problems of overcrowding and high prices, while active sport tourists may be constrained by their fitness and health concerns. External barriers can be listed as geographical distance, a lack of facilities and a lack of time and money, whereas personal interests, abilities, capacities and knowledge are referred to as inner constraints (Humankinetics, 2019). In addition to these sectoral constraints, is well known that this industry faces some of global challenges that can be listed as the followings (Sport Management Degrees, 2020);

- Climate Change - This makes the sports tourists to shun away from attending the sports event. For example extreme cold and rains that occur during the time taken.
- Security Concerns - Tribal conflicts, terrorism and civil war in some territories also create a danger for sports tourism creating safety concerns and insecure environment.
- Natural disasters - These are unforeseen natural occurrences that create a danger for sports tourists.
- Lack of political will - This is a situation where the government of the place involved in hosting a sport event does not co-operate or provide support and motivation to the organization of the sports event.
- Competition *(neighboring communities)* - This is a situation whereby neighboring communities or places have got major sports events/activities happening there.
- Inadequate government support - The government may not give sufficient support to sport tourism industry especially in terms of allocation of funds.

Sport Tourists

Sport tourists are defined as "the individuals and/or groups of people who actively or inactively participate in competitive or recreational sport, while travelling to and/or staying in places outside their usual environment" (Hinch and Higham, 2019: 19). Apart from traditional leisure tourists and visitors, sport tourists display some specific life-style and demographic characteristics and they might be defined in a different way. Schreiber (1976: 86) and Gee (1988: 498) support this approach describing the sports travellers as '*better off, well educated, more active and more affluent than the average traveller in general*'. They are classified into three major groups (Humankinetics, 2019);

- Sport tourists as participants who travel to destinations to be an active part of sport.
- Event-based sport tourists who travel to a destination to watch players or athletes.
- Celebratory sport tourists who travel to destinations in order to visit sport museums, stadiums and other memorable places.

Among these groups, first group is described as "active" sport tourists, while others are "passive" sport tourists. Active sport tourists can also be seen as examples of "*allocentrics*" among the tourist typologies based on the theories by Stanley Plog (2001) as one of the most distinguished author in this field. As a matter of fact, the people who like sports and participate to various sport activities are usually known as extroverts and very active in their daily life. Moreover, if these activities are considered as a part of extreme sports which may also contain some degree of risk *(i.e. rock climbing, rafting, and surfing),* then participants are also categorized as *explorer, adventurer* and *novelty seeker* who are ranked at the extreme ends of the typological scale.

Inherently, the relative share of active sport tourists in total visitor figures is of little value, thence their economic contribution to the destinations is also rather negligible for most cases. However, since the education and income level of these tourists are higher than average, their per-person expenditures are higher than those visitors who have traditional motivations for travel. Likewise, these visitors may stay longer at the destination and they might be in touch with local people more, depending on the sports category they participated.

Another characteristic of nature sport participants is the fact that their contributions to the area they visited are very high with respect to sustainability and environmental protection concerns, because of their high level of environmental awareness and sensitivity to the nature issues. Similarly, active sport tourists are referred as ideal tourist profiles in the context of the sustainability of local social and cultural values as well. These behaviors of sport participants are also defined as "*green attitudes*" engaging in pro-environmental behaviors (Achu, 2019).

A profile for the typical active sport tourist described by Gibson et al. (1997) is shown on Table 2. This table highlights that the majority of active sport tourists are male, well-educated and long distance travelers.

Table 2. Profile of individuals likely to engage in active sport tourism.

Characteristics of the active sport tourist
§ More likely to be male § Wealthy individuals by a majority § University graduates § Prefer long distance journeys to participate in favorite sports § Likely to engage in active sport tourism well into retirement § Inclined to engage in repeat activity e.g. not a 'one-off' vacation

Passive sport tourists, on the other hand, display different characteristics from active participants in terms of their travel habits, attitudes and behaviors during their travels. Their basic travel purpose is participating to a sport event, activity or game as spectators only, not for playing any kind of sport game, nor participate in it. For most cases, this participation may occur as watching a sport event together with many other spectators or as a part of a tour group, instead of taking an independent and individual trip to the place where the sport activity is organized. Therefore, it might be accepted as an extension of mass tourism, since sport events have a variety of sport activities and games, from wrestling which has a few numbers of participants, say a hundreds of spectators to mega sport events where the number of spectators may reach more than a hundred thousand audiences for per event. The participants of these events usually organize their trips as a group, holding a travel itinerary, flying by charters, staying just one overnight at the destination accommodation, even for most cases no overnight at all.

Although per person expenditure of this group of tourists is lower than those active sport tourists and other traditional holiday tourists, due to the large number of participants, total number of local/regional trips, accommodations and other total expenditures are also very high in volume and visitors spend great amount of money in the location they visit. Even though passive sports tourists vary on their personalities depending on the sport event they participate, their communication and interaction with local people is so limited that even zero for most cases. Likewise, their education levels, social belongings, consumption behaviors during holiday are also close to mass averages.

However, it is well known that participant and spectator profiles display important distinctions based on the type of sport activity they participated in. For instance, the socio-cultural profile of the spectators of an elite sport like tennis is undoubtedly higher than those spectators for an *American Wrestling*. Likewise, researches have shown that the education level of spectators of the Olympic Games and basketball matches is higher than those of the soccer games (Chalip et al. 1998).

On the other hand, passive sport tourists not only consist of the audiences of any sport event, but also there are more people who are somewhat the part of this business, listed as the followings (WST, 2020);

- All visitors who may practice a specific sport at the destinations.
- Amateur level practitioners *(e.g. sailing clubs)* to competitions, tournaments, etc.
- Professionals who travel specific sport purposes *(e.g. medical people, trainers).*
- Employees of sports clubs.
- Referees and juries prepared for the checks, evaluations etc.
- Journalists, television and radio crews to report news about the events
- The "spectators" who travels to watch sports events *(e.g. group or individual sports enthusiasts).*
- The travelers who move to learn how to practice the sport or to see what goes "in fashion".

Motivations for Sport Tourism

The theory of motivation comprises of some internal and external factors that provoke and raise a person's behavior and hence "*tourist motivation*" defined as "a meaningful state of mind which adequately disposes an actor or a group of actors to travel" (Kirkup and Sutherland, 2015).

The focal point and essential motivation for sport tourism is naturally sport activity itself and motivations for sport tourism are regarded as relatively upper level motivations in comparison to other generic tourism motivations. An examination of study of travel and tourism literature reveals that motivations for travel are in accordance with the Maslow's model for '*hierarchy of needs*'. According to Maslow's theory, the motivation for traditional holiday tourism is oriented to meet a basic human need such as "*having rest and relaxation*", but active sport tourism motivations are mostly about such high-level needs as "*achievement*" and "*self-actualization*" that can be found on the uppermost layer of Maslow's pyramid. On the other hand, with the exclusion of safety needs, other four generic needs *(physiological, self-actualization, social and esteem)* of the pyramid have been suggested in the sport literature as motivating factors for sport participation and spectatorship, Maslow's hierarchy seems to be an appropriate base upon which build a theory of sport activity (McDonald et al. 2002).

The authors have identified some of the motivations for sport tourism as well. Crompton (1979) categorizes the sport tourism motivations as *escape, self-exploration, social interaction, prestige, regression, relaxation* and *improvement* in kinship relations, while Fairley (2003) in his research, found that *nostalgia (visitation to sport museums, heritage events etc.)* as one of the most important motivational factor for sport tourism. For a deeper approach, McDonald et al. (2002) developed the *Motivations of Sport Consumers Scale* and they have determined thirteen broad sport motivations, listed as *achievement, stress reduction, skill mastery, self-esteem, risk taking, affiliation, aggression, physical fitness, aesthetics, social facilitation, competition, value development* and *self-actualization.*

With regard to the motivations for "*sport events tourism*", these tourists mainly prefer to travel for sport events, since it will be much better to watch their favorite team/player or event live on the stage rather than that of a screen. Some other authors have identified the motivational factors for sport events tourism participants as social interaction, drama, achievement, acquisition of knowledge, escape, interest in sport, interaction, rivalries, love games, socialization, viewing outstanding players, historical significance, experiencing firsthand the excellence of athletic competition, the cultural (including sport subculture) experience, and game entertainments (Neirotti et al., 2001; Green, 2001; Rinaldi, 2011: 70; Nezakati et al., 2013; Kim, 2013).

Above mentioned motivations for active sport tourism are generally considered as inner*(push)* factors that motivate sport tourists. However, some external factors that lead active sport tourists to particular destinations are described as "*pull factors*", and the destinations that have a number of amenities and sport attractions such as favorable natural resources required by the sports involved will be the best places to visit for sport tourists, if they are supported by convenient supply conditions. External or pull factors are the most powerful motivations for "*sport events*" tourists. Because, the existence or availability of these sports events creates a direct reason for travel.

While the tourist types and motivations in both categories differ from one tourist to another, sport tourists also differ in their own scale based on the type of travel and activity of sport. Just as individual features of spectators for athletics is different from those of soccer game, personal features and expectations of participants of trekking, equestrian tourism or canoeing will also be different from one another.

For this reason, the leadership styles, roles and functions of sport tourism guides will vary depending on the expectation of participants, and the type of sport or activity.

Authors identified that sport event participant and spectator behavior consists of social and psychological needs, acquisition of knowledge, escape, aesthetic qualities of the sport, vicarious achievement, physical attractiveness, social interaction and appreciation of physical skills (Trail and James, 2001; Trail et al., 2003). In addition to these motivations, '*identification*' was also identified as a key motive emerging from the sport spectator in the literature meaning that a self-awareness with respect to other objects, identification results in feelings (Kirkup and Sutherland, 2015).

Researches reveal that there is almost no significant difference between different active sport tourists regarding the importance of motivational factors. Some of the studies concentrate on surfers, golf players, climbers, ski resort customers, and the results show that the most common motivations are listed as experience a lot, move around, pleasure, being free, being active, doing something for health, refueling, doing sport actively, enjoying healthy climate, experience nature and landscape and family togetherness (Hodeck and Hovemann, 2016).

FUTURE OF SPORT TOURISM

Sport tourism is subject to specific niche-market demand and sport tourists may easily change their preferences and purchase patterns swiftly follow, if their expectations cannot be met satisfactorily by the service providers (Coakley, 2004, 566). It is also one instrument of social interaction that may produce a platform for the development of sports (Hinch and Higham, 2004).

For future prospects, Coakley (2004: 566) argues that power and performance sports will probably be the most obvious sports that likely affect the forms of active sports tourism. This trend has a potential to affect the expansion of travel flows involving major sports events. These sports are expected to continue as the principal forces in sport spectatorship and passive sport tourism participation rates, namely performance sports might be a driving force for both active and passive participation. The increasing importance of involvement and pleasure sports represents two important directions: that '*there are a number of ways to do sports*' and '*power and performance sports may not be consistent with personal values and experiences*' (Coakley, 2004: 567). He also estimates that some current factors and trends may cause the growth of sport participation such as, sustained interest in health and fitness, advancing age, more female sports than ever, increase in alternative sports.

Gibson et al. (2018) also proposed that "new circumstances ought to be investigated as little is known about large and growing segments of the field including *eSports, obstacle course racing, youth sports, themed events, and fitness events such as CrossFit competitions"*. On the other hand, as for the supply side of sport tourism, authors estimated that "the sports resources that can be transported and reproduced will be developed in response to changing technologies, urbanization and demands for immediate accessibility" (Higham, 2007: 288). The development of some sport categories and services such as *skateboard facilities, sports theme parks, artificial climbing walls, golf-driving ranges,* and *a wide range of water sports* has been a response to demand for participation and pleasure sport in urban centres. Likewise, the fast progress of new technologies such as internet, virtual and augmented reality, interactive television, and new generation electronic games display the convergence of interests in sport and tourism (Higham, 2007: 288). Enhancing the activity of sports tourism in any destination allows to following additional benefits (WST, 2020);

- Spreading out the classical tourism season.
- Appealing to new tourist segments.
- Building and increasing a positive destination image.
- Extending visitor overnights owing to high spending patterns of sport tourists.
- High level of word of mouth promotion through recommendation of recent visitors to their colleagues.
- More long distance visitors who travel to watch sports events.

The Roles and Functions of Tourist Guides for Sport Tourism

Tourist guides play a principal role in insuring repeat visitation to a destination or country through creating a positive image. Their major task is to introduce tourism products to visitors and interpret the sights that they are visiting. They are not only responsible for their guests, but they also need to consider other stakeholders who are directly or indirectly influenced by their activities and roles. Tour guides also act as active representatives of the country, they are the first to encounter and greet visitors and usually the last ones to take leave of them when they depart the country.

There are many field of expertise for tourist guides from traditional *culture, archeology, history,* to *specific events, special interest, nature, adventure,* and so on. *Sport tourism* is also one of these specific expert areas for tour guides. There are a number of different sport activities that visitors want to participate during their travel in various destinations or sport itself may be the main purpose of travel. Some of these occurrences are organized for only spectators *(i.e. tennis tournaments),* while others need active participations of visitors *(i.e. mountain climbing, horse riding, windsurfing aerotourism-gliding etc.)* (Ristova et al, 2019). The visitors who participate to any kind of sport activity may need a professional assistance in order to be fully satisfied with their travel and to overcome unforeseen difficulties. Accordingly, the role and importance of expert tour guides for sport events and sport tourism are better understood when considering some of the troubles for especially nature based sports and for those who need professional assistance.

Before identifying their role and functions for sport tourism, it is necessary to define their profession and identity in tourism industry. *Oxford English Dictionary* literally defines *the guide* as "*a person who exhibits the way to others, especially one employed to prove tourists around places of interest*" and as a worldwide representative body of this profession. More technical definition of tourist guides on the web page of World Federation of Tourist Guide Associations (WFTGA) is that "*a person who guides visitors in the language of their choice and interprets the cultural and natural heritage of an area which person normally possesses an area-specific qualification usually issued and/or recognized by the appropriate authority*". WFTGA also mentions about their basic function as the representatives of the host country is to meet tourists and act in such a way as to bring credit to the country visited and publicize it as a tourist destination (WFTGA, 2018).

In addition to these official definitions, there are some other definitions suggested by authors. Mancini (2000, 5) for example, shortly defines them as "*somebody who takes people on sight-seeing tours for a limited time*", while Overend (2012, 53) describes as "*interpreters and performers*" of the tours. As one of the recent approach, Öter (2019) define the tour guide as "*a person who provides broad information and presentation of a destination for both domestic and foreign tourists within the context of tour program*". Actually, they are the frontline staff in tourism sector who play a significant role in forming the tourist experience during their travel. Hence, tour-guiding service as a travel business function is also the essential component of various tourism services provided by travel intermediaries (Huang et al., 2010). Cohen (1985) on the other hand, emphasizes the vanguard role of tour guides as a pathfinder who leads the way through the environment and to unknown places for most cases. He also indicates that the definitions for tourist guides shape their contemporary roles in tourism as *mentors, leaders* and *effective information sources*.

In addition to their professional qualifications, the tour guides are also expected to have some individual traits as listed on the Table 3 (Sahin, 2018);

Table 3. Individual characteristics of tourism guides

· open minded	· respectful
· enthusiastic	· trustworthy/reliable
· effective speaker	· confident
· honesty	· responsibility
· tolerant	· helpful
· friendly	· sociable
· polite	· fair
· warm hearted	· punctual
· decisive	· flexible

The term "*tourist guide*" often creates an image as a person walking on the street holding a flag or an umbrella on his/her hand in front of a group of people, and telling them something about the historical places visited, which is frequently categorized as "*heritage guide*". Although, heritage guidance is the most common and well-known type of tour guiding, but there are also many other types of guidance specified in such areas as local cuisine, religion, mountaineering, trekking, skiing, safari, city and some special interest tours. However, tourist guides are responsible not only for telling stories to tourists, but also responsible for the introduction of local culture, satisfaction of participants, effective time management etc. Their role on the foreign visitors is so important for a country to create a good image that the guides are required to be omniscient representatives in their field. Therefore, a tour guide plays numerous roles during the tours he or she undertakes such as a *mediator, interpreter, translator, culture broker, leader, teacher, host conduit* and *ambassador* (Ababneh, 2017).

Cohen (1985) listed the roles of tourist guides as; leading the way, directing the trip, getting access to non-public regions, controlling the group and ensuring security, tension management, integration, making people enjoyed their tour, animation, representation, organization, selection of objects of interest, information, and interpretation. However, it is well known that the roles of tourist guide may also differ based on the tourism types they engage. For instance, if the purpose of the tour is to taste wine and visiting wineries and/or vineyards, the guide must be good at wine-tasting and has depth knowledge about wines and grapes. If the tour is organized for birdwatching, he/she has to know about the names and sounds of the different type of birds, how they make their nests when they breed etc. Since a number of tourism types can be listed depending on the variety of tourist interests so the expertise areas of tourism guides can be listed as well. Each type of tourism naturally requires different skills, abilities and capabilities of tour guides involved. Consequently, whereas sport tourism is a specific category of tourism as explained earlier, it also requires some area specific capabilities and qualifications of professional tourism guides.

Sport Tourism Guidance

There is a wide consensus among the authors that the three major reasons for tourism and other temporary travels have long been *trade, sport and religion* in the course of history (Ahipaşaoğlu, 2001: 5) and sport as a tourist motivation is still a prevalent factor affecting the movements of tourists today. As mentioned previously, active sport tourism participants inherently need some degree of assistance, additional support and leadership unlike from traditional tourist guides. Although there is no particular definition for sport tourism guidance in the literature, but it is well known that there are specialist tour guides serving activity and sport specific travels that can be categorized under the heading of sport tourism guidance.

Before discussing the basic requirements for sport tourism guidance, it should be mentioned that the nature of sport tours that might be self-guided or organized, based on access, location and kind of the activity. The followings are the typical tour categories sport tourism guides might be involved (Zuahar, 2004: 13);

1. Particular tours to sport attractions over a certain number of days *(sport museums etc.),*
2. Compound visits to sports attractions and significant sports events *(sport games etc.),*
3. Participation in a number of important sports events (team competitions),
4. Attendance at congresses and prominent sports events *(i.e. Academic Meeting before the World Cup Tournament),*
5. Trips involving the indigenous characteristics of a territory tracked by visitors for physical and/or aesthetic concerns *(extreme individual sports etc.).*

Nowadays, passive sports tourism has nothing to do with guides since there is almost no need for specific guidance for these events and organized tours as a service of travel agents or tour operators cover all the activities during their travels. However, expert tour guides might be an integral part of active sport trips due to the possible risks and difficulties they may encounter during their trips to less known places. Hence, sport tourism guidance requires more job specifications than that of any traditional ones.

Cohen (1985) explained that the initial function of tourism guidance was the pathfinding for the trips to unknown places where the travel reasons were hunting, safari, mountaineering etc. Specialist mountain climbing guides, for example, lead the participants around major mountain territories of the world. Cohen (1985) also indicated that such tour guides were the pioneers of active sport tourism guidance.

Travelers who are interested in mountaineering still want to participate to the tours to the Alps or the Himalayas. There are number of trekking or climbing tour routes to the highest point of the Alps. While these routes can be accomplished by self-guided travelers, it also can be escorted by a specialist tour guide who is well qualified and certified in climbing and/or other winter sports like skiing. Mountaineering guides should have some practical qualifications other than heritage guides such as crisis management, risk management, equanimity, first aid knowledge and they are responsible for the security and safety of the group as well.

As another sport tourism category, historical *Lycian Road (Route)* along the southwestern coasts of Anatolia (Turkey) is one of the most famous trekking routes and this route is used by many domestic and foreign visitors who are curious about the nature, trekking, camping in an historical environment, archaeology, and the ruins of ancient civilizations. This route takes more than 30 days by hiking as self-guided for a single trip. There are also shorter guided tours on the route and unexperienced visitors can trek on the roads where camping breaks are possible as well. On this type of route, it might be better to trek accompanied by an expert tourist guide in order to have historical information. Guides for *Lycian Road* know the tracks, camping areas, the history of the region, ruins and ancient cities around the area, archaeology, flora and fauna forms, first aid when necessary and any other information required by the participants on the route.

Undoubtedly, there are differences between traditional culture tour guidance and sport tourism guidance, because of the nature of sport activities. According to Cousquer and Beames (2013), the major difference between mountain guidance and traditional tourist guidance is that of the main responsibility of mountain sport guides is to concentrate on the safety concerns and physical danger of the mountain climbing or hiking trips. Keeping the group atmosphere positive is more important on sport and adventure tours than culture tours and cultural tour guides are not expected to check the status of group members as frequently as sport tourism guides do.

Sport tourism guides should also be aware of the following distinctions that segregate active sport tourists from traditional culture and holiday tourists;

- They are more mobile than holiday tourists,
- They prefer to travel remote and less known destinations,
- They are known as "*wanderlust*" tourists while holiday tourists are "*sunlusts*",
- They may visit more than one destination per travel,
- They may need variety of technical equipment,
- Their per person expenditure is reasonably high on countryside,
- They need moderate level accommodation, they may stay overnight even in the primitive cottages or tents,
- They spend less money on shopping and eating out, they may cook their meals on a natural environment instead,
- They represent explorer and novelty seeker character.

With regard to passive sport tourism, the role and responsibility of tour guides for major sport events focuses on the capability of manipulating and managing the large number of people, since they are required to manage and guide the group members. In summary, sport tourism guidance requires the following knowledge and skills depending on specific areas of sport tour guides;

- dealing with the physical requirements of the position,
- to be able to become an effective member of teamwork,
- to be able to overcome the problems they may encounter,
- to have a professional knowledge about his/her field of expertise,
- to speak foreign languages fluently for foreign participants,
- to have a good knowledge about the geographic locations,
- to have good knowledge about local culture,
- to have a good knowledge about first aid and other health issues,
- to have a good knowledge about contagious diseases and local epidemics,
- to have a good knowledge about local flora, fauna and geological structure *(rocks, cliffs, deep scarps etc.),*
- to have good verbal and non-verbal communication skills,
- to have good communication skills in an intercultural environment,

- to have organizational and leadership skills, including effective coordination,
- to take responsibility for leading the group,
- assembling necessary equipment and supplies,
- effective use of navigation systems,
- to warn the tour participants about potential risks of the sport,
- instruction and demonstration of related skills and techniques.

Sport tourism guidance is very active and multi-dimensional profession. This profession also requires somewhat different motivations from traditional guidance in order to execute this business and to accompany sport tourists deservedly. In addition to an extrovert personality, the basic motivations of sport tourist guidance might be considered as; challenging the hassles, seeking novelty, liking natural environment/wildlife, being a part of sport activity and desire to help others.

Consequently, sport tourism guidance encompasses all kind of guiding services during the tours that they are responsible for, including a variety of sports activities as mentioned earlier as well as their extreme sports skills and knowledge.

CONCLUSION

There are variety of studies that examine sport tourism and tourism guidance separately in their specific domain, and both fields have a substantially long history. However, a literature review shows that there has been no specific study or research on sport tourism guidance and its distinctive features. Therefore, the basic objective of this study is to fill the present gap in this field where almost no study has been made so far. This study is designed as a preliminary theoretical approach in order to reach this objective.

In this paper, the historical and actual trends/dimensions of sport tourism were studied respectively, and brief information about the literature review on this topic was given. Later section covers the information about the roles, missions and functions of tourist guides, by examining tourism guidance as a professional field in tourism. The final section of this paper is all about the basic qualifications, competencies and skills of sport tourism guides they should have in terms of active and passive participants of sport tourism, particularly for the nature sports and other individual sports that degree of difficulty is relatively high.

Sport tourism is expected to continue its growth and development in terms of diversity, quality and number of participants all over the world like many other types of tourism. The major factors for this development might be listed as the contribution of technological progress to participation of individuals for nature sports, opening more destinations to sport tourism, increasing disposable income and education levels of the individuals, and quick dissemination of the information regarding sport tourism which may increase the awareness of people to sport tourism through information and telecommunication technologies.

As previously mentioned in the text, lack of direct studies regarding sport tourism guidance creates a need for area specific researches and studies on sport tourism guidance. Sport tourism guidance can be an academic interest and study field in the near future, if attention can sufficiently be drawn. In this way, more guided sport tours to different parts of the world can be organized, more sport tourism related investments and facilities can be developed, and above all, this specific area of tourism can be conveyed to further levels through the specialization of more tourist guides on sport tourism.

REFERENCES

Ababneh, A. (2017). Tourist Guides and Job Related Problems, Analysis and Evidence from Jordan. *Journal of Quality Assurance in Hospitality & Tourism, 18*(2), 200–217. doi:10.1080/1528008X.2016.1189867

Achu, F. N. (2019). Pro-Environmental Behaviour of Attendees at a Major Sport Event in Cameroon. *Geo Journal of Tourism and Geosites, 27*(4), 1307–1320. doi:10.30892/gtg.27416-435

Ahipaşaoğlu, H. S. (2001). *Guidance in Tourism*. Detay Publishing. (in Turkish)

Askarian, F., Shojaei, M. K., & Yavari, U. (2015). Challenges in the Development of Sports Tourism in Aras Free Zone According to Managers and Experts Opinions. *Arabian Journal of Business and Management Review., 3*(7), 59–69. doi:10.12816/0014521

BBC. (2012). *Olympics in Numbers.* Retrieved November 10, 2019, from https://www.bbc.com/turkce/haberler/2012/07/120723_olympics_innumbers

Chalip, L., Green, B. C., & Velden, L. V. (1998). Sources of interest in travel to the Olympic Games. *Journal of Vacation Marketing, 4*(1), 7–22. doi:10.1177/135676679800400102

Coakley, J. (2004). *Sports in Society: Issues and Controversies* (8th ed.). Mc Graw Hill Higher Education.

Cohen, E. (1985). The tourist guide: The origins, structure and dynamics of a role. *Annals of Tourism Research, 12*(1), 5–29. doi:10.1016/0160-7383(85)90037-4

Cousquer, G. O., & Beames, S. (2013). Professionalism in mountain tourism and the claims to professional status of the International Mountain Leader. *Journal of Sport & Tourism, 18*(3), 185–215. doi:10.1080/14775085.2014.904246

Crompton, J. (1979). Motivations for pleasure vacation. *Annals of Tourism Research, 6*(4), 408–424. doi:10.1016/0160-7383(79)90004-5

De Aragao, M. M. (2015). *Economic Impacts of the FIFA World Cup in Developing Countries. Honors Theses. Paper 2609. Western Michigan University*. Lee Honors College.

Deery, M., Jago, L., & Fredline, L. (2004). Sport tourism or event tourism: Are they one and the same? *Journal of Sport & Tourism, 9*(3), 235–245. doi:10.1080/1477508042000320250

Fairley, S. (2003). In search of relived social experience: Group-based nostalgia sport tourism. *Journal of Sport Management, 17*(3), 284–304. doi:10.1123/jsm.17.3.284

Gee, C. (1988). *Resort Development and Management.* Educational Institute of the American Hotel and Motel Association.

Gibson, H., Attle, S., & Yiannakis, A. (1997). *Segmenting the sport tourist market: a lifespan perspective.* Paper presented at Teaming for Success: A Forum on Sport Tourism, Arlington, VA.

Gibson, H. J. (1998a). Active sport tourism: Who participates? *Leisure Studies, 17*(2), 155–170. doi:10.1080/026143698375213

Gibson, H. J. (1998b). Sport tourism: A critical analysis of research. *Sport Management Review, 1*(1), 45–76. doi:10.1016/S1441-3523(98)70099-3

Gibson, H. J., Lamont, M., Kennelly, M., & Buning, R. J. (2018). Introduction to the Special Issue Active Sport Tourism. *Journal of Sport & Tourism, 22*(2), 83–91. doi:10.1080/14775085.2018.1466350

Green, B. C. (2001). Leveraging subculture and identity to promote sport events. *Sport Management Review, 4*(1), 1–19. doi:10.1016/S1441-3523(01)70067-8

Heather, J. G., Lamont, M., Kennelly, M., & Buning, J. R. (2018). Introduction to the Special Issue Active Sport Tourism. *Journal of Sport & Tourism, 22*(2), 83–91. doi:10.1080/14775085.2018.1466350

Hengpoom, P., & Weerasophon, W. (2019). Sport Tourist Spending Behavior in Thailand. *The 2019 International Academic Research Conference*. https://www.icbtsconference.com/17096732/e-proceedings-amsterdam-2019

Higham, J. (2007). *Sport tourism destinations: issues, opportunities and analysis*. Elsevier, Butterworth Heinemann. doi:10.4324/9780080474434

Higham, J., & Hinch, T. (2002). Tourism, sport and seasons: The challenges and potential of overcoming seasonality in the sport and tourism sectors. *Tourism Management, 23*(2), 175–185. doi:10.1016/S0261-5177(01)00046-2

Hinch, T., & Higham, J. (2004). *Sport Tourism Development–Aspects of Tourism*. Channel View Publications.

Hinch, T., & Higham, J. (2019). *Sport Tourism Development*. Channel View Publications, Cromwell Press.

Hodeck, A., & Hovemann, G. (2016). Motivation of active sport tourists in a German highland destination – a cross-seasonal comparison. *Journal of Sport & Tourism, 20*(3–4), 335–348. doi:10.1080/14775085.2016.1235988

Huang, S., Hsu, C. H. C., & Chan, A. (2010). Tour guide performance and tourist satisfaction: A study of the package tours in Shanghai. *Journal of Hospitality & Tourism Research (Washington, D.C.), 34*(1), 3–33. doi:10.1177/1096348009349815

Humankinetics. (2019). *The Remarkable Growth of Tourism*. https://humankinetics.me/2019/04/04/the-remarkable-growth-of-sport-tourism/

Hurley, I. (2018). *Is there any real economic benefit to hosting a world cup?* https://psmag.com/economics/is-there-any-real-economic-benefit-to-hosting-a-world-cup

Jakovlev, Z., Dimitrov, N. V., Koteski, C., & Serafimova, M. (2017). Sport Tourism as the Most Important Alternative Form of Tourism - Strategies and Challenge, *Economic Development/Ekonomiski Razvoj, 19*(3), 197-207.

Kim, Y. H. (2013). An Assessment of Sport Event Tourists' Motivation with a Framework: A Case Study at a Southeast Conference Football Game. *Journal of Tourism Research & Hospitality, 2*(2). Advance online publication. doi:10.4172/2324-8807.1000115

Kirkup, N., & Sutherland, M. (2015). Exploring the relationships between motivation, attachment and loyalty within sport event tourism. *Current Issues in Tourism*, *20*(1), 7–14. doi:10.1080/13683500.2015.1046819

Mancini, M. (2000). *Conducting tours: A practical guide* (3rd ed.). Nelson Education.

McBride, J. (2018). *The Economics of Hosting the Olympic Games*. Retrieved July 25, 2019, from https://www.cfr.org/backgrounder/economics-hosting-olympic-games

McDonald, M. A., Milne, G. R., & Hong, J. B. (2002). Motivational factors for evaluating sport spectator and participant market. *Sport Marketing Quarterly*, *11*(2), 100–113.

Neirotti, D. L., Bosetti, H. A., & Teed, K. C. (2001). Motivation to attend the 1996 Summer Olympic Games. *Journal of Travel Research*, *39*(3), 327–331. doi:10.1177/004728750103900315

Nezakati, H., Chin, M. C., & Akhoundi, M. (2013). Factors Influencing People to Travel Abroad for Sport Tourism (Malaysia Evidence*). Journal of Applied Sciences (Faisalabad)*, *13*(24), 5708–5715. doi:10.3923/jas.2013.5708.5715

Öter, Z. (2019). Guided Tour. In Encyclopedia of Tourism–Turkey: The Basic Concepts of Tourism and Hospitality Industry. Detay Publishing. (in Turkish)

Overend, D. (2012). Performing Sites: Illusion and Authenticity in the Spatial Stories of the Guided Tour. *Scandinavian Journal of Hospitality and Tourism*, *12*(1), 44–54. doi:10.1080/15022250.2012.678070

Perna, F., Custódio, M. J., & Oliveira, V. (2019). Local Communities and Sport Activities Expenditures and Image: Residents' Role in Sustainable Tourism and Recreation. *European Journal of Tourism. Hospitality and Recreation.*, *9*(1), 49–59. doi:10.2478/ejthr-2019-0006

Plog, S. (2001). Why destination areas rise and fall in popularity, An update of a Cornell Quarterly Classic. *The Cornell Hotel and Restaurant Administration Quarterly*, *42*(3), 13–24. doi:10.1016/S0010-8804(01)81020-X

Rinaldi, G. (2011). *Sport tourism: An exploration of the motivations and tourism experiences of Australian Football League interstate sport tourists travelling to Western Australia* (Thesis). Edith Covan University. Retrieved September 30, 2019, from https://ro.ecu.edu.au/theses_hons/34

Ristova, C., Popeska, B., & Dimitrov, N. (2019). Possibilities for development of sport-recreational tourism of Kozuf and Plackovica Mountains. *The 11th International Scientific Conference "Modern Trends of Physical Education and Sport"*.

Ritchie, B., & Adair, D. (2002). The Growing Recognition of Sport Tourism. *Current Issues in Tourism*, *5*(1), 1–6. doi:10.1080/13683500208667903

Sahin, S. (2018). Competencies of Tourist Guides. In S. Eser, S. Sahin, & C. Cakici (Eds.), *Tourism Guidance* (pp. 47–74). Detay Publishing. (in Turkish)

Schreiber, R. (1976). Sports interest. *A travel definition. The Travel Research Association 7th Annual Conference Proceedings*, 85–97.

Snelgrove, R., & Wood, L. (2010). Attracting and leveraging visitors at a charity cycling event. *Journal of Sport & Tourism*, *15*(4), 269–285. doi:10.1080/14775085.2010.533918

Sotiriadis, M. (2016). Events sponsorship as a business partnership: Suggesting the critical success factors, Special issue on Special Interest Tourism. *TOURISMOS: An International Multidisciplinary Journal of Tourism.*, *11*(5), 1–25.

Sport Management Degrees. (2020). *Five Challenges facing the sports industry*. Retrieved January 24, 2019, from https://www.sports-management-degrees.com/lists/five-challenges-facing-the-sports-industry/

Sportyjob. (2015). *The contribution of sport to economic growth and employment in Europe*. Retrieved November 10, 2019, from https://www.sportyjob.com/blog/the-contribution-of-sport-to-economic-growth-and-employment-in-europe/

Standeven, J., & De Knop, P. (1999). *Sport Tourism*. Human Kinetics.

Timeout. (2019). *Contribution of Sport to Economic Growth*. https://www.timeout.com/istanbul/tr/spor-ve-fitness/2016-yaz-olimpiyat-oyunlari

Trail, G., & James, J. (2001). The motivation scale for sport consumption: Assessment of the scale's psychometric properties. *Journal of Sport Behavior*, *24*(1), 108–127.

Trail, G. T., Robinson, M. J., Dick, R. J., & Gillentine, A. J. (2003). Motives and points of attachment: Fans versus spectators in intercollegiate athletics. *Sport Marketing Quarterly*, *12*(4), 217–227.

Van Rheenen, D., Cernaianu, S., & Sobry, C. (2017). Defining sport tourism: A content analysis of an evolving epistemology. *Journal of Sport & Tourism*, *21*(2), 75–93. doi:10.1080/14775085.2016.1229212

Weed, M. (2014). After 20 years, what are the big questions for sports tourism research? [Editorial]. *Journal of Sport & Tourism*, *19*(1), 1–4. doi:10.1080/14775085.2015.1032505

WFTGA. (2018). *What is a Tourist Guide?* Retrieved December 05, 2019, from www.wftga.org/tourist-guiding/what-touris-guide

WST. (2020). *Sport Tourism*. Retrieved January 24, 2019, from https://www.wst-show.com/en/sport-tourism

WTTC. (2019). *Economic Impact Research*. Retrieved November 10, 2019, from https://www.wttc.org/-/media/files/reports/economic-impact-research/regions-2019/world2019.pdf

Xinhua, L., Juntao, C., Jinmei, Z., & Lin, L. (2016). A Study on Sports Tourism Competitiveness Based on Factor Analysis Method. *12th International Conference on Computational Intelligence and Security*, 16-19.

Yildiz, Z., & Cekic, S. (2015). Sport Tourism and its History and Contribution of Olympic Games to Touristic Promotion. *International Journal of Science Culture and Sport,* (SI4), 326-337.

Zuahar, J. (2004). Historical perspectives of sports tourism. *Journal of Sport & Tourism*, *9*(1), 5–101. doi:10.1080/1477508042000179348

Chapter 6
Best Practices for Tour Guides in the Northern Region of Jordan:
"Land of Olive Oil and Indigenous Food"

Sawsan Khreis
Yarmouk University, Irbid, Jordan

Foued Benghadbane
Institute of Management of the Urban Techniques, University of Oum El Bouaghi, Algeria

EXECUTIVE SUMMARY

Tourism investments are concentrated in the southern region of Jordan (Petra and Aqaba). This chapter argues that if food tourism delivers its sustainable benefits, it could be a vehicle for regional tourism development by strengthening the regional food on offer so that deprived areas are regenerated. Villages in Ajloun are selected for the investigation as they are well-known for food production. These villagers have inherited the talent for preparing traditional Jordanian food. This study is the first scientific study conducted to compare two villages in Ajloun, namely Orjan village and Ibbin village for logic justifications: 1) Provide the best example illustrating the best practices of tour guides in Jordan who unconsciously apply the principles of alternative tourism that lead to a sustainable tourism development. 2) This chapter aims to identify many problematic issues such as whether tour guides apply the best practices that are the concrete reflection of sustainable tourism principles.

INTRODUCTION

In the post-globalization era, many changes have touched all aspects of life commencing with values and ending by lifestyle. Tourism as a global phenomenon has been affected, in that it becomes the sole solution for some countries to confront crisis, especially in areas with tensions, where employment opportunities have decreased. Urban tourism witnesses a growth in developed countries while rural

DOI: 10.4018/978-1-7998-3725-1.ch006

tourism becomes the 'dernier recours' for revitalizing the communities in developing countries, which have nothing to do with urban tourism.

Much research has been conducted on rural tourism from the 1990s until now. It started in Europe with bed and breakfast especially in the UK, France and Switzerland. Britain promoted this type of tourism by launching many slogans, 'Feel as you are at your home' and 'Breathe fresh and clean air.' France and Ireland imposed some conditions for rural dwellings. Spain has a competitive advantage with its wide offer that visitors could sleep in historic homes, towers and historic palaces. Italy was also the pioneer in the adoption of this type of tourism as the government took the initiative of constructing homes and pensions for vacations. Hungary is a distinct model as the government gives investors fiscal exemptions, besides organizing an equestrian festival. In sum, European governments support these types of tourism and contribute by promoting it online. However, other types of sustainable tourism such as agritourism and ecotourism are equated with rural tourism. As there are a lot of definitions, defining rural tourism is problematic.

Even in regions not suitable for tourism, tourism can function as a vehicle for rural development (Neumeier & Pollermann, 2014). Above all, some studies consider that not all types of tourism in rural areas are listed under the general umbrella of rural tourism. In some countries, rural tourism takes a limited seasonal aspect, such as Jordan that has a peak season in summer in Ajloun.

Rural tourism provides travelers personal contact with the local people, which is considered a unique insight into their lives (Canadian Tourism Commission, 2004). Rural tourism has the advantage over urban tourism, in that it is cheaper as it needs primitive services. This is a rule in developed countries, meanwhile in developing countries; a limited number of visitors frequent rural areas, as they are high cost destinations. For example, an overnight stay in Ajloun costs 50 euros.

The future plan for the economic sector, including the tourism sector insists on the necessity of developing many types of tourism, such as cultural, medical, environmental religious, adventure etc. However, it does not mention rural tourism, which is a niche market (Jordan Economic Growth Plan, 2018). Studies were conducted on a wide range of themes; however, they put emphasis on economic implications. Tourism can diversify local rural economies and increase their visibility (Lun, Penclaner & Volgger, 2016). The research by Neumeier and Pollermann (2014) conducted in five eastern German rural regions shows that tourism induced small economic impacts, while it caused important non-economic implications by enhancing the social and cultural aspects (Neumeier & Pollermann, 2014).

Developing sustainable tourism is a big challenge for planners and policy makers. The tourism industry faces a multitude of significant sustainability-related challenges including energy and greenhouse gas, water consumption, waste management, loss of biological diversity, and planning (UNWTO, 2012). Food has been considered a useful instrument of destination development (Henderson, 2009). Others stress the impressive role of food as *the key to understand homeland ...it is central to our perception of home and identity. In sum it is a tool for building a nation* (Raviv, 2015).

Anthropologists started studying gastronomy since the 1970s. Gastronomy was first defined as *the intelligent examination of whatever concerns man's nourishment* (Arnott, 1975). These days, our modern world witnesses the suppression of traditional cooking and its replacement by fast food. It serves the interests of traders and fills the stomachs of consumers with junk food found all over the world. It is known for its dangerous impacts on health, putting away the recipes of ancestors who invented special recipes from the gift of their lands. These cuisines are not just simple recipes that reflect the lifestyle, religion, traditions, ideology, beliefs, and intelligence to prepare meals using ingredients well conserved even during the harsh times of the year, but also their interactions with other civilizations. For example, if a recipe is imported from another country, it means at that time they had good relations with them.

To become a major tourist attraction, cuisines must create a complementary relationship with the local tourism image and tradition or history, as is the example in France (Horng and Tsai, 2012). Food is a feature that can add value to the destination (Handszuh, 2000). Food can also provide a sustainable competitive advantage of the destination. Tourism has the potential to provide opportunities to augment the economic benefits of local foods through the selling of food and meals to tourists, which will at the same time, enhance and revive the use of local indigenous food (Mak et al, 2012).

Food is a major source of profit in tourism. *Tourists' spending on food can constitute up to one third of the total tourist expenditure* (Long, 2004). Local food markets serve the needs of a wide range of local stakeholders, including local inhabitants, retailers, farmers, producers, restaurants and, eventually, the entire community. Contemporary consumers are increasingly searching for locally-produced ingredients and food, highlighting a business opportunity of interest for both local food producers and service providers.

The study by Al Obeidat (2016) insists on the significance of food as an effective tool for preserving identity. The researcher investigates the importance of the traditional Jordanian dish "Al-Mansaf." The study concluded that Al-Mansaf *has changed over past decades* while it is still *a tie to local heritage and part of local and national identities.* Sidali, Kastenholz, and Bianchi (2013) confirmed the importance of local food in reinforcing personal identity. Their study identifies seven dimensions that transform food products to an attractive culinary niche: anti-capitalistic attitude, coherence, struggle against extinction, mutual disclosure, personal signature, rituals of spatial and physical proximity, and sustainability-related practices.

Local food could create a sense of place. Haven-Tang and Jones (2005) concluded that *gastronomic tourism and food festivals can be used as vehicle by which rural food could enhance the local identity of a destination.* A study conducted by Tsai (2013) in Taiwan found that Taiwanese dishes are the best way to promote Taiwanese culture and that Taiwanese food improved the comprehension and appreciation of Taiwanese culture.

Tourists usually travel for the sake of recreation, entertainment, newness and strangeness. However, to what extent are they ready to be familiar with the traditional food in strange destinations? (Cohen, 1972)

There is no previous study that studied the tourism situation and the role of tour guides together in any destination in Jordan. This study will be the first to highlight the primordial role of local guides who could with their individual efforts, confront a number of obstacles such as total refusal from the local community, the absence of any financial support, the poor infrastructure, the absence of ancillary services, the absence of facilities, the non-existence of any type of accommodation, the negligence of these regions by the Ministry of Tourism and Antiquities (MOTA), the absence of training for local tour guides, the absence of pre-defined and traced trails that made the beginning very difficult.

All previous studies conducted in Jordan were focused on religious sites, cultural tourism and ecotourism in Ajloun Nature Reserve, such as the study by Rawashdeh (2012). The study aims to investigate whether the characteristics of ecotourism can be applied to Ajloun Forest. The study shows that age, income, education of the sample, and the nature of the tourist movement to the area match the ecotourism characteristics. The study recommends increasing night activities and events in order to attract more visitors and increase their length of stay. It also recommends improving the language and skills of the staff (Rawashdeh, 2012).

The study by Gharaibeh, (2012) aligns with the study by Rawashdeh, (2012) in that it was conducted in Ajloun Nature Reserve. This study aimed to highlight its role in biodiversity and conservation of Ajloun Nature Reserve, which is characterized by the mountains of Ajloun. It also aims to demonstrate its role in the sustainable development of the rural population in the middle of the surrounding environment, through training in traditional industries, famous in the region. Gharaibeh (2012) has a set of results and recommendations that will contribute to improving the environmental performance of reserved areas, and improve the quality of life of the rural population around them.

Our study, as mentioned earlier, is the first scientific study conducted to compare two villages in Ajloun, namely Orjan village and Ibbin village for the following reasons:

1. Provide the best example to illustrate the best practices of tour guides in Jordan who unconsciously apply the principles of alternative tourism leading to sustainable tourism development, by their individual efforts. Their first priority is the promotion of their region that they like, and believe that tourism is the key to improving their income, preserving their natural environment and safeguarding their authentic culture. The best and sole example is Orjan village where the authors met all the local guides who agreed to be interviewed.
2. To demonstrate the uniqueness of the best example where tour guides fulfill their role, the authors selected another village, which is located in Ajloun and has the same natural assets. The local community here did not take any initiative to attract visitors. On the contrary, the residents of Ibbin took the worst type of tourism and applied it. Twenty families from the local Ibbin community constructed twenty well-equipped apartments to host Arab tourists from Arab Gulf countries, especially Saudi tourists. These tourists come for fun and leisure with their male friends; the majority consumes alcohol and brings their girlfriends. Their behavior is considered a shock for the residents of Ibbin as it contradicts local culture and religion. Therefore, the dominant type of tourism applied in Ibbin is mass tourism.

BACKGROUND

Jordan

Jordan not only has unique cultures, customs and traditions but also a range of distinctive local food as part of its gastronomic products. Jordan considers its food and cuisine as a niche market (an emerging tourism market) as well as pull factor in the promotion of the northern region of the Kingdom of Jordan. Jordanian cuisine represents a core element of the destination's intangible heritage, demonstrating the national cultural characteristics, which can be developed into an international brand. This will enhance local economic development, if cuisine becomes a major tourist attraction in this region. The methodology primarily involved in-depth interviews and document analyses.

In the past decade, the tourism sector in Jordan suffered from a sudden drop in tourist numbers. This severe situation can be explained by many reasons. First of all, the over dependency on international tourists as the MOTA Tourism Board concentrated their marketing and development efforts to satisfy international tourists. Secondly, tourism in Jordan is dominated by cultural tourism, which means that medical, religious and other types of tourism are niche markets that did not have any priority, at least until now. Food tourism in Jordan is considered to be in its infant stage.

The authors of this study argue that food tourism, if well planned and developed, will save Jordan from the dilemma of a concentration of tourists in the southern region of the country, represented by the golden triangle (Wadi Rum, Petra and Aqaba) that deprives the northern region from a fair distribution of tourism benefits. This type of tourism will have a direct impact on the region's economy. Moreover, this type of tourism will empower women who are responsible for preparing and cooking traditional food. Figure 1 is a map of Jordan with its tourist attractions

Figure 1. Map of Jordan (Ministry of Tourism and Antiquities (MOTA, 2020).

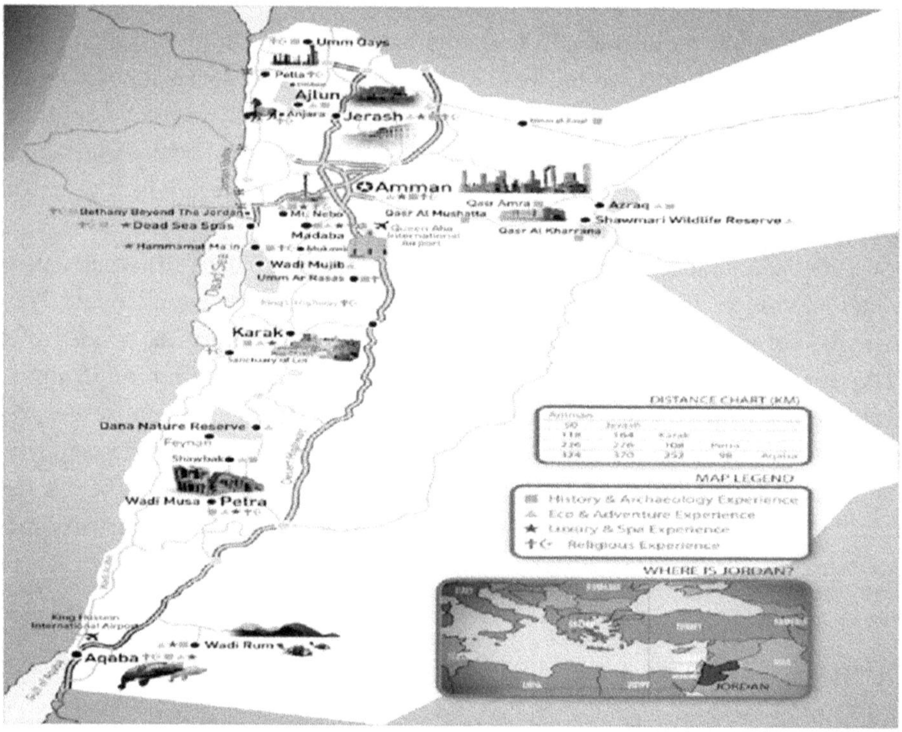

This chapter combines food tourism with the multiple roles of tour guides to demonstrate the decisive role of tour guides in the revival and sustainability of food tourism in the northern region of Jordan. If some sort of tourism is adopted in this region of Jordan, food service offered to tourists will play an important role in the differentiation and promotion of villages. Food tourism can become the tourism capital through which the local resources of food can be turned into marketable attractions. Food tourism will be a tool for regional re-equilibrium.

The Northern Region (Study Area)

The Northern Sector

This area has the governorates of Jerash, Ajloun, Irbid, Mafraq and Balqa. The sights and topological characteristics of this sector can be divided into three regions:

1. The cultural zone: contains the archaeological city of Jerash, Ajloun and Irbid.
2. The coastal region: includes the eastern part of the Dead Sea beaches to the north of the governorate of Balqa.
3. The desert region: containing the desert area east of the Mafraq governorate.

Food tourism in recent years tends to become the main reason for tourists traveling to a destination other than for basic residence. Experiences of living in different regions of the world lead to the development of culinary tourism. In Jordan, there is potential for the development of food tourism, there are still old recipes and dishes that are well preserved (WTO, 2019).

Case Study

Identification of tourism destinations in Ajloun

Ajloun is just a short journey from Jerash, through pine forests and olive groves. It boasts scores of ancient sites, including watermills, forts and villages, all in the beautiful hills and valleys of northern Jordan (JTB, 2020). Ajloun is located to the northeast of the capital, Amman. It is the smallest governorate in Jordan as it has an area of 320 km^2. Despite being the smallest governorate, it has the highest density in green cover that occupies 75% of the total area, which means that 240 km^2 of its land is green. According to the department of statistics, the population of Ajloun was estimated to be 194,700 in 2019. The most notable towns and villages in Ajloun (other than Ajloun itself) are Ibbin, Sakhra, and Mrajjam.

The village of Orjan is located 12 km to the northeast of Ajloun, specifically in the Orjan District in the northern part of the Ajloun Governorate, which enjoys a strategic tourist location. Meanwhile it is difficult to infer its ancient history through the current location of the village. The village is built on caves surrounded by valleys and springs. Orjan is situated on the route to the Ajloun Natural Reserve, although the absence of official and organized tourism programs prevents its local community from tourism benefits (USAID, 2009). According to a study conducted by USAID, the infrastructure and services provided in Orjan village are weak in that there is a need to organize walkways in the forests, natural valleys and agricultural roads as well as providing general tourist and health facilities on farms located in Wadi Orjan.

In addition to the absence of lighting for the road leading to the village, there is also no internal lighting. The village also lacks sanitation services and tourist services such as restaurants, parks, accommodation, and handicraft stores. The village is located in one of the most beautiful areas of northern Jordan, covered in forest and fruit trees. The village has many values that are unique, such as aesthetics. The beauty of this village is due to the marvelous geology of the region as it is surrounded by mountains in the north and south.

Figure 2. Location of case studies (Orjan and Ibbin) Personal elaboration by authors (2020)

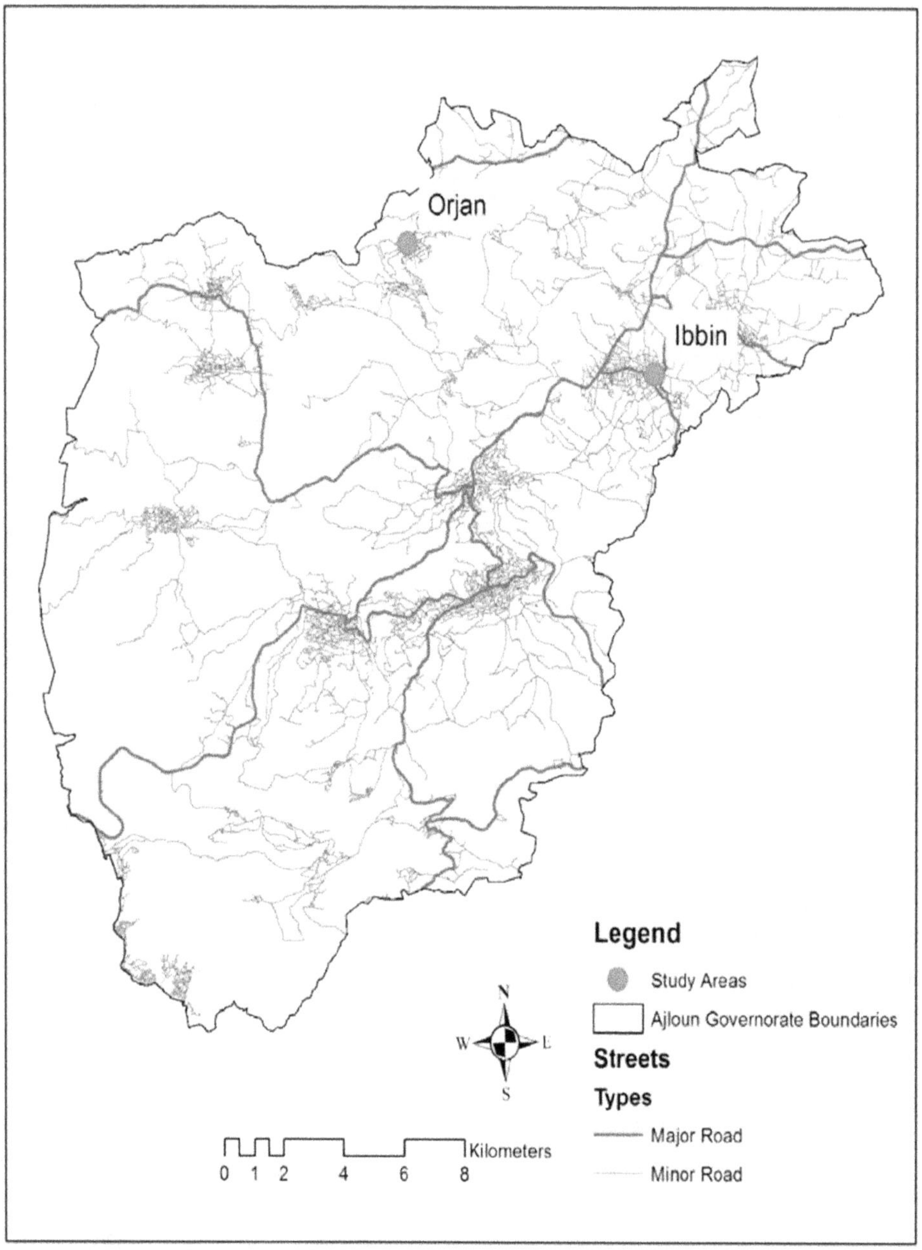

It has many caves and springs that transformed the village to a beautiful green landscape covered by grapes, figs and ancient Greek olive trees that distinguish it from other parts of the country. It also has the most ancient olive oil mills, which produce the finest and best olive oil in the country. The valleys of Orjan are decorated with enchanting pomegranate and fig trees, and the streams are lined with fragrant wild mint. (MOTA, 2019).

The Soap House

At the Soap House located in Orjan village, local women use a variety of local ingredients, including lavender, geranium, mint and pomegranate, to create high-quality, handmade, natural olive oil soaps called Orjan soap. Visitors are invited to tour the workshop and view the soap-making process.

The Biscuit House

The Biscuit House located in Orjan village consists the kitchen, where 'Tasali Jordanian Delights' are created. These products are made by local women employed in the workshop. All the products are handmade and consist of natural ingredients, some of which are produced by the local community (RSNC, 2019).

The historic importance of the village is worth mentioning. The village of Orjan gained importance as part of the Yabish Gilad valley, and it constitutes a point of economic continuity and a focal point for the Kingdom of Moab, due to the abundance and fertility of its lands, abundance of water and springs, and its extension to the northern regions of Jordan. The area was inhabited about 5,000 B.C. There is much historical evidence that we are unaware of, its historical extension, including the Mahrama and Qableh region, Al-Asif, Alden, Iraq, the monks, the lake and the area of ammunition, which is spread along the Wadi Al-Rayyan (Wadi Orjan), an area more than 6 km in length (USAID, 2009).

Jordan's department of statistics estimates the total residents of the Orjan subdistrict to be 28,727. There is no urban density in the urban planning of houses and the style of construction of houses is rural, which means that residents use the most modest materials in house construction. The majority of houses have no surrounding walls; all the houses are spread between the farms and foothills.

Infrastructure

Poor public transit services (mainly buses and taxis) add to the accessibility problem in Ajloun. All of these facts put negative remarks on tourism development in Ajloun. A detailed assessment of the infrastructure-related problems in the Ajloun tourism area was studied by Al-Kheder, Al-Malabeh, and Al-Momani (2016). The road network and transport system in Ajloun, given all the recent developments mainly represented by the third tourism project sponsored by USAID, still carries many threats and challenges that affect local community safety and tourism sustainability.

Roles of Tour Guides

According to the study by Khreis (2018), professional tour guides should ensure the success of an authentic tourist experience and they could modify tourist behaviors. Tour guides should possess excellent verbal and non-verbal communication skills and they should organize tourists in the place. Moreover, tour guides should integrate the tourists in interesting activities according to their tastes and preferences. Above all, tour guides should be neutral and have a prompt response to problems. Khreis' study was conducted in Petra. The authors considered these findings when they prepared the survey questions for this study (Appendix 1).

Chronological History of Rural Tourism Development in Orjan

In the semi-structured interviews with stakeholders in the region, they all agreed that the beginning of interest in receiving tourists was initiated by a pioneer, Eissa Dweikat. He says, "I began by designing the tour. It started by contacting Jordanian travel agents to suggest a tour of one or two days, then after years I designed a program for one week … the tour is about hiking in precise trails that differ in terms of length, difficulty and activities according to the visitors' ages, preferences, health, seasons, and local permissions."

It was this pioneer, Mr. Eissa Dweikat who started trails including hiking as the dominant activity and motivation of the visit. He continues, "Before 2004, foreign visitors used to come according to individual arrangements. It wasn't before 2004 that we could talk about tourist packages. Tourists used to come according to coordination between me and local travel agents. Later on, a larger number of visitors such as academics from Manchester used to visit Orjan, this continued until 2008. In 2008, the ex-Minister for MOTA, Her Excellency Maha Al-Khatib visited the region. She declared the first tourist trail that extends for 3 km, around Rasoun Mountain. The mayor of the region, Mohammad Sawalmeh accepted the idea and he tried to persuade the locals who refused tourism. In the same year, Abraham Path began with the support of American and international organizations, they tried to establish the 24 Project" In 2009, Her Excellency tried to promote this tourist trail; however, residents of Orjan did not accept the idea of receiving tourists. The tour guide, Ali intervened to explain the residents' rejection of tourism, "The local people were afraid of the impact of tourism on their culture as they have a prejudgment that the visitors have no religion, so they may diffuse their ideas and their western culture that they completely reject."

The tour guide, Mo'ath explained how local guides in the region organize the tours. "When we have a tour group of ten people, it is sufficient to have one local guide who could manage everything. When the number of visitors becomes twenty, we need two local guides. One leads the group, he takes the position of leader and the other guide takes the position of the observer, who makes sure that all the visitors follow the same trail. Rarely do we have thirty visitors who need three local guides to manage everything." All the visitors engage in a hiking activity that varies, the local guides told us. According to the local guide Ali, "All the visitors come here to walk for long distances that vary from 10 km at least to 16 km maximum without any rest!"

"Despite the continuous rejection by the locals, local guides continued to sell their programs to travel agents so that more and more visitors came to the region. Gradually, locals observed the economic benefit they could gain, if they got involved in the tourism process, so they opened their homes to receive visitors in their homes," the third local guide, Abdullah added. "In 2012, six homes opened their doors to let visitors stay and sleep in their homes, they offered bed and breakfast … this enabled these families to earn money … over the years, they became dependent on tourism!"

In 2016, among women who helped in the development of rural tourism in the area is Madame Mona Haddad who built (Bait Al-Baraka) that aims to train women from the region to master skills of cooking, welcoming visitors, and to accompany female visitors along trails. Her initiative helped women in the village to have a better standard of living.

Tourism Development Stages in Orjan

The authors conclude that tourism development in the region is conducted and started by the following stages:

1) Growth stage: Individual initiatives that struggled to make their region a tourist destination despite lack of acceptance or support from the local community. Tourism passed through the growth stage. After eight years (2004-2012) locals started to accept the idea as they realized two things:
   A) Visitors who come to their region are respectful; they respect their environment as well as their dress code, culture and lifestyle.
   B) The visitors' expenditure has a direct economic impact that enables the local community to earn an extra income.
2) Development stage: During this stage, the locals started to receive visitors in their homes; six local families receive visitors all through the year. They come for many activities. The dominant activity is hiking, some come for climbing mountains, and all of them are interested in the local culture, history and food rituals. They all like the traditional cuisine, which distinguishes this village. The ingredients of local traditional dishes are composed mainly of wheat and olive oil, which is produced from their olive trees. The government plays a minor role as the MOTA has recently declared that touristic trails should be given more attention, to support rural regions located in the northern region of Jordan. In 2019, MOTA began to improve infrastructure. In 2019, the government also began to put Orjan on the tourist trail of Darb Al-Oyoon. One could find a description of the Orjan trail in a brochure edited by the Royal Society for Nature Conservation.

The details of the Orjan trail described in the brochure said: Guided only, Grade: Moderate, Distance: 12 km, Time: 6 hours, Group size: 4-20 people, Trail Description: Continuing from Rasoun village, this tour takes you through wooded valleys to the lush green orchards of Orjan village, giving you the chance to discover the rural lifestyle of the region. A poplar-lined spring provides a wonderful contrast to the arid mountains of the previous stages of the trail. As you hike, you will notice the dramatic rock walls, old water mills once used by local people to grind seeds and the hollow walnut trees. Enjoy a delicious traditional brunch in an orchard of fig, pomegranate, apple and apricot trees. At the end of the trail, you will pass through old olive trees dating back to the days of the Roman Empire.

The Current Situation of Tourism in Ibbin

The department of statistics estimated the total number of Ibbin residents to be 15,286 in 2019. The authors met local people and the mayor of this rural town. There are no local guides in this region that have the same natural assets as Ajloun. In Ibbin, the response to the tourism process is completely different as they have another approach, which is considered sustainably negative. The tourist offer in Ibbin comprises twenty apartments owned by local people. These are well equipped for the tourists' preferences and needs. The authors observed that this region is distinguished by its target market, which is Arab tourists from Gulf countries, especially Saudi Arabia. However, according to the results of our survey, these Arab tourists have low incomes so they do not spend a lot of money in this region.

Figure 3. Tourist apartment in Ibbin, (Photo by authors).

The tourists' behavior in Ibbin and their presence does not coincide with sustainability as the majority of them (98%) come alone or with a group of male friends. They come for a fun time. The owners of these apartments look for this segment of tourists as they look for money, which is their preoccupation. The authors contacted all the owners and they all had a sole preoccupation, which was money!

The owner Youssef said: "Our region is isolated; it is a target for those who look for serenity and escape from routine ...we host Arab tourists who don't even respect the rules of hygiene in my apartment. They throw their cigarettes on the carpet, they spoil furniture ... I don't have a choice as I took loans from the bank to construct it and I need clients. Sometimes I receive Jordanians who are a little bit more aware about hygiene and cleanliness issues." The authors were surprised by an offer from the owner Youssef who asked them to market his apartment among clients and he promised to give a commission if the authors did this for him, as he does not know how to bring in tourists regularly. He continues: "I knew a student who promoted my apartment ...I was paying her 20 Jordanian dinar for each group."

When the authors asked another owner called Bisher, to cooperate with local guides to take tourists on tours in the region, he replied, "Do you want me to pay from my money to those local guides? I don't mind, if tourists enjoy the region or not … I offer tourists all what they need from security, by installing cameras for surveillance, I offer them breakfast in the morning but other meals such as lunch and dinner aren't provided. If they want these meals they should pay more."

When asked about the traditional cuisine and if owners cared about providing traditional food for tourists, the third owner, Abdullah answered, "It doesn't matter if I provide tourists with traditional and local dishes or not. Saudi tourists ask for their own dishes such as kasbah, mandi and biryani. Sometimes grilled meat." The fourth owner, Moheeb was asked about environmental issues. He was asked if he had a chance to follow the ethical type of tourism that preserves nature and does not have any negative impact on the environment, was he prepared to follow this principle? He answered, "If this so-called type brings me more money, why not? I built this apartment as an accommodation for tourists now; to earn more money …I have seven children who need education, health expenses and security for their future. ...Tourism is a chance for us to have a better life." The fifth owner Ghaleb was asked the same question, his answer was, "No, I don't want to change my business! My tourist apartment is my only source of living. I sold my father's land to construct this apartment." The sixth, seventh and eighth owners replied with answers that focused on money without any concern or intention to change this type of tourism.

It is evident through these open question interviews with owners of the apartments that:

1. All of them have a dominant concern, which is material gain.
2. All the owners of the apartments are unaware of environmental issues.
3. All the owners have a commercial mentality and they do not care about increasing the value of tourists' experiences.
4. All the owners do not care about tourist behavior as long as they earn money.
5. No owners know how to promote their apartments through mass media.
6. No one is ready to prepare a tour for tourists that are organized by local guides.

Table 1 shows a comparison of tourism facilities available in Orjan and Ibbin.

Table 1. A brief comparison between Orjan and Ibbin. (Source: Author's personal elaboration, 2019).

Name of destination	Orjan	Ibbin
Number of tourist campuses	2	0
Number of home stay	6	0
Type of tourism	Alternative tourism: food tourism, volunteer tourism, ecotourism, soft adventure tourism	Mass tourism
Participation of locals	Yes	No
Guided tours	Yes	No
Existence of food tourism	Yes	No
Presence of multiplier impact	Yes	No
Nationalities of visitors	Italian, French, Spanish and American	Arab tourists from the Gulf countries
Motivation of visit	Tasting of local food, hiking, local culture	Leisure

Method

For this study, the authors used the descriptive approach, which uses available scientific references from Arabic and foreign languages, scientific studies and tourism brochures that are related to this study. Secondary data were collected from literature related to food tourism and tour guiding. The authors also prepared a questionnaire to distribute among travelers and used participant observation by staying for one week in Orjan. In addition to this, the authors conducted a semi-structured interview with the local guides who assume the responsibility of preparing the tour programs, coordinating with tour operators and local travel agents, and guiding visitors. The authors visited villages in the northern region of Jordan namely, Irbid, Jerash and Ajloun to conduct semi-structured interviews with tour guides, women and visitors.

Statistical Treatment

After collecting data from the returned responses, the researchers used the Statistical Package for the Social Sciences SPSS v23 to analyze the data. Suitable statistical treatments from the following tests were used for each question and hypothesis:

1. Cronbach's alpha (α) to test reliability. Cronbach's alpha value is acceptable if it is >0.60
2. Percentage and frequency to describe the sample.
3. Arithmetic mean and standard deviation to answer the study questions.
4. Simple linear regression analysis with the F test.
5. 5. Relative importance, that is assigned using:

$$\text{Class Interval} = \frac{\textit{maximum class - minmum class}}{\textit{number of classes}}$$

$$\text{Class Interval} = \frac{5-1}{3} = 1.33$$

Low degree <2.33, median degree from 2.34–3.66 and high degree ^{3}3.67.

FINDINGS

This study used a descriptive approach in addition to questionnaires, semi-structured interviews, and participant observation to find out the best practices of tour guides and how they fulfill their roles. A total of 243 questionnaires were distributed, 200 were returned completed and only 43 questionnaires were excluded, as they were incomplete. Table 2 shows the Cronbach's alpha values calculated for the reliability of each variable.

Table 2. Cronbach's alpha value for each variable

Cronbach's alpha value	Variables
0.86	Managerial competence
0.83	Tour guide qualification (professionalism)
0.79	The tour finds success
0.81	Further recommendations that will add value to the tour
0.89	Total

Table 3. Means and standard deviations for each item of the 'Managerial competence' level

Item No.	Statement	Mean	Standard Deviation	Order	Importance
1	Rural tour is well organized in terms of time.	4.25	0.494	1	High
7	Tour guide pays attention to security issues.	3.95	0.450	2	High
6	Overcrowding of visitors mitigates the value of the touristic purpose.	3.90	0.405	3	High
2	Rural tour is well organized in terms of program planning and implementation.	3.77	0.496	4	High
3	Tour guide assumes his ethical responsibility in that he respects group's preferences and beliefs.	3.75	0.453	5	High
5	Time reserved to stay on rural site is sufficient to consume the place and enjoy your experience.	3.70	0.526	6	High
4	Time reserved to stay on rural destination, is sufficient to do the religious rituals.	3.66	0.480	7	Moderate
8	Tour guide knows how to maximize visitor's satisfaction as he understands your preferences and needs.	3.40	0.452	8	Moderate
Total		3.79	0.469	---	High

Table 4. Independent sample t-test to test the mean difference for the 'Managerial competence' level by places

Place	N	M	SD	T	P.
Orjan	100	3.97	.595	3.12	0.03*
Ibbin	100	2.23	.597		

N: number, M: mean, SD: standard deviation, F: F value, P: P-value

Table 5. Means and standard deviations for each item of the 'tour guide qualification (professionalism)' level

Item No.	Statement	Mean	Standard Deviation	Order	Importance
3	Tour guide encouraged visitors to taste food while insisting on its healthy ingredients.	3.99	0.899	1	High
2	Tour guide motivates visitors to become involved in preparing local food.	3.90	0.741	2	High
1	Tour guide is polite and respectful to visitor's culture and preferences.	3.85	0.880	3	High
4	Tour guide's interpretations of the history of traditional food are exciting in that it persuades tourists to enjoy eating it.	3.82	0.586	4	High
5	Tour guide exposes visitors to new experiences contradicting your expectations.	3.77	0.857	5	High
6	Tour guide shifts opinions on behalf of the local community's hospitality.	3.70	0.830	6	High
12	Tour guide encourages visitors to use all their senses to have a unique unforgettable experience recalling the history of this region represented by visiting local people who keep the same traditional lifestyle.	3.65	0.741	7	Moderate
8	Tour guide answers all questions posed by visitors.	3.60	0.742	8	Moderate
9	Tour guide talks all the time thus he hinders us from enjoying the splendor of the landscape.	3.59	0.666	9	Moderate
10	Tour guide's behavior differs from other tour guide's behavior in rural destinations in other parts of the world (according to past experiences).	3.55	0.843	10	Moderate
11	Tour guide has excellent communication skills.	3.50	0.708	11	Moderate
7	Tour guide is a good interpreter who masters well, the history of this rural destination.	3.43	0.720	12	Moderate
Total		3.69	1.030	---	Moderate

Question One: Does the Tour Guide Have Management Competences?

Table 3 shows the computed means and standard deviations for each item of 'the managerial competence level.'

For the managerial competence level, the range of means for items in this domain is 3.40–4.25. The highest mean is for item 'Rural tour is well-organized in terms of time' and the lowest mean is for the item 'Tour guide knows how to maximize visitor's satisfaction as he understands your preferences and needs.' The total mean is 3.79 by high agreement.

Difference Between Places

Table 4 shows the independent sample t-test to the test mean differences for the 'managerial competence' level by places. The result shows a statistically significant difference between places for managerial competence ($T=3.12$, $P<0.05$) in favor of Orjan. The reason visitors estimate that local guides do not understand visitors' preferences and needs is declared in the last question in the questionnaire related to visitor's recommendations. The authors found that according to 56 visitors, local guides do not let visitors wear what they want, for example short skirts and local guides do not allow visitors to drink alcohol.

Table 6. Independent sample t-test to test mean differences for the 'Tour guide qualification (professionalism) level by places

Place	N	M	SD	T	P.
Orjan	100	3.72	0.542	2.98	0.042
Ibbin	100	3.19	0.431		

N: number, M: mean, SD: standard deviation, F: F value, P: P-value

Question Two: Is the Tour Guide Qualified (Professional) in That He Fulfills all His Duties Towards the Visitors?

To answer this question, we computed mean and standard deviations for each item of 'Tour guide qualification (professionalism),' as shown in Table 5.

For the 'tour guide qualification (professionalism)' level (Table 5), the range of means for items in this domain is 3.43–3.99. The highest mean is for the item 'Tour guide encourages visitors to taste food while insisting on its healthy ingredients' and the lowest mean is for item 'Tour guide is a good interpreter who masters well the history of this rural destination.' The total mean is 3.69 by high agreement. This is explained by the fact that local guides are not well trained to shift from the role of a tour guide who presents to a tour guide who interprets. Academics realize well that interpretation exceeds the presentation of a destination. This result coincides with the recommendation of visitors who find that local guides should follow intensive lectures to master the skill of interpretation that enhances the value of the tourism experience.

Table 7. Means and standard deviations for each item of the 'Tour finds success' level

Item No.	Statement	Mean	Standard Deviation	Order	Importance
1	Tour increases positive perception about the uniqueness of this visit.	3.89	0.813	1	High
6	Tour boosts a sense of nostalgia to past times.	3.85	0.772	2	High
4	Tourism is a magic tool towards peaceful tourism.	3.83	0.754	3	High
3	Tour enhances mutual understanding between cultures.	3.80	0.759	4	High
2	Tour changes perceptions towards Jordanians in that they respect the visitor's cultures.	3.15	0.771	5	Moderate
5	I don't feel that there is any bias to a particular culture.	3.00	0.733	6	Moderate
Total		3.58	0.767	---	Moderate

Difference Between Places

Table 8. One-Way ANOVA to test the mean difference for the 'Tour finds success' level by places

Place	N	M	SD	T	P.
Orjan	100	3.00	0.874	3.55	0.03
Ibbin	100	2.56	0.753		

N: number, M: mean, SD: standard deviation, F: F value, P: P-value

Question Three: Did the Tour Find Success?

To answer this question, we computed the mean and standard deviation for each item of the 'Tour finds success' level as shown in Table 7.

Table 7 shows the 'Tour finds success' level. The range of means for items in this domain is 3.00–3.89. The highest mean is for the item 'Tour increases positive perception about the uniqueness of this visit' and the lowest mean is for item 'I don't feel that there is any bias to a particular culture.' The total mean is 3.58 by moderate agreement.

Difference Between Places

Table 8 shows the independent sample t-test to test mean differences for the 'Tour finds success' level by places. The result shows a statistically significant difference between places for 'the tour finds success' (T=3.55, P<0.05) in favor of Orjan. Visitors found a bias towards local culture and a somewhat non-acceptance of foreign cultures. Authors attribute this phenomenon to the newness of tourism, which exposes locals to new cultures. The openness to new cultures will take place gradually. It is predicted that locals will allow visitors freedom in terms of the dressing code and drinking, for example.

Table 9. Means and standard deviations for each item of the 'Further recommendations that will add value to the tour' level

Item No.	Statement	Mean	Standard Deviation	Order	Importance
5	Tour guide should take the Italian experience applied, in the food tours.	3.90	1.36	1	High
2	Duration of stay should be extended to enjoy the visit more!	3.79	1.41	2	High
3	How much time do you think should be dedicated to this visit?	3.76	1.32	3	High
4	Tour guide shouldn't expose us to a narrative related to politics or religious issues.	3.65	1.34	4	Moderate
1	How long do you spend on the current trip?	3.19	1.42	5	Moderate
Total		3.67	1.37	---	High

Question Four: Do These Recommendations Add Value to the Tour?

To answer this question, we computed the mean and standard deviations for each item of the 'Further recommendations that will add a spiritual value to the religious tour' as shown in Table 9.

Table 9 shows the 'Further recommendations that will add a value to the tour' level. The range of means for items in this domain is 3.19–3.90. The highest mean is for the item 'Tour guide should take the Italian experience applied to the food tours' and the lowest mean is for the item 'How long do you spend on the current trip?' The total mean is 3.67 by high agreement. Visitors do not feel that their stay should be more than five days; they think their visit needs a maximum of four days. They also feel that the tour guides are vigilant to all issues of sustainability. However, the guides should permit them to drink alcohol and wear clothes that suit them, not the local community.

Difference Between Places

Table 10. One–Way ANOVA to test mean differences for the 'Further recommendations that will add value to the tour' level by places

Place	N	M	SD	T	P.
Orjan	100	3.33	0.768	2.13	0.04
Ibbin	100	3.01	0.681		

N: number, M: mean, SD: standard deviation, F: F value, P: P-value

Table 10 shows the one-way ANOVA used to test mean differences for 'Further recommendations that will add value to the tour' level by places. The result shows a statistically significant difference between places for further recommendations that will add value to the tour ($T=2.13$, $P<0.05$) in favor of Orjan.

RECOMMENDATIONS

1. Give intensive courses for local guides in Orjan to be perfect tour guides as they should respect the rights of visitors. For example, they should permit the visitors to drink alcohol as currently they are not permitted to drink wine.
2. Offer a rest to the visitors after a certain distance and take into consideration the physical ability of the visitors.
3. Stop the exploitation of visitors by bus drivers who ask for tips, which is illegal and unethical.
4. Regulate and monitor the tour guiding in Orjan by the Ministry of Tourism and Antiquities.
5. Put signs along the trails that lead visitors, in case they are lost or exposed to any unexpected accident.
6. Organize musical nights by the local community to enhance the visitor's experience.
7. Encourage the local community to follow training courses to produce local clothes and souvenirs that recall the history and lifestyle of Orjan. These could be sold to visitors through stands installed along the trail.

8. Install caravans that provide snacks and drinks for visitors.
9. Stop any tourism activity in Ibbin as it does not take into consideration any sustainability issues and take the best practices applied by the local guides of Orjan.

FUTURE RESEARCH DIRECTIONS

Further studies should be conducted in the future, using the same methodology to obtain a follow-up of the tour guides performance in other rural areas in Jordan. Comparative studies should be conducted to position Jordan as a niche market for rural tourism in comparison with similar competitor destinations such as Lebanon and Turkey.

CONCLUSION

The results indicate that the respondents, who scored higher on the estimation of the tour guides performance in the two villages (Orjan and Ibbin) in the Ajloun governorate, show a bias towards tour guides working in Orjan village as they are qualified in terms of managerial competences, while they think that tour guides should shift to the role of interpreters. In addition, the tourists recommend that tour guides should take the Italian experience and apply it to food tours.

In summary, international visitors are satisfied with the tours in Orjan village except for some restrictions by tour guides as they do not allow them to drink alcohol or wear what they want. International tourists were dissatisfied with the touristic experience in Ibbin. The authors conclude that the tour guides apply the principles of sustainability and the global code of ethics in Orjan village, while the tour guides in Ibbin are not aware of the principles of sustainable tourism.

The authors predict the continuity and prosperity of rural tourism in Orjan village due to the exceptional efforts of tour guides and that the bad example given by the touristic experience in Ibbin village will lead the region to discourage visitors to return so the region will not continue to be a tourism destination for a long period.

ACKNOWLEDGMENT

We would like to express our gratitude to the editors who gave us time and pertinent remarks to finish our chapter. We would like to express our gratitude to Mr. Awni Kawar, Mr. Wael Janineh and Mr. Eissa Dweikat who facilitated our mission.

REFERENCES

Al-Kheder, S. A., Al-Malabeh, A. A., & Al-Momani, R. (2016). Spatial Analysis and Transportation System Review for Tourism Areas in Jordan: Ajloun City as a Case Study. *Jordan Journal of Civil Engineering*, *10*(4).

Al Obeidat, A. A. A. (2016). The sociocultural and economic evolution of Mansaf in Hartha, Northern Jordan. *Humanities (Washington)*, *5*(22). Advance online publication. doi:10.3390/h5020022

Arnott, M. L. (1975). *Gastronomy: The anthropology of food and food habits*. Mouton Publishers.

Cohen, E. (1972). Toward a Sociology of International Tourism. *Social Research*, *39*, 163–182.

Gharaibeh, K. M. (2012). The Developmental Role of Ajloun Forest Reserve in Jordan. *Journal of the Planner and Development, 25*.

Handszuh. (2000). *Culinary Tourism Strategic Development: an Asia-Pacific Perspective Library*. DOI: doi:10.1002/jtr.834

Haven-Tang, C., & Jones, E. (2005). Using local food and drink to differentiate tourism destinations through a sense of place: A story from Wales– dining at Monmouthshire's great table. *Journal of Culinary Science & Technology*, *4*(4), 69–86. doi:10.1300/J385v04n04_07

Henderson, J. C. (2009, April 18). (20029). Food tourism reviewed. *British Food Journal*, *111*(4), 317–326. doi:10.1108/00070700910951470

Horng, J. S., & Tsai, C. T. (2012). Culinary tourism strategic development: An Asia-Pacific perspective. *International Journal of Tourism Research*, *14*(1), 40–55. doi:10.1002/jtr.834

Long, L. M. (1998). Culinary tourism: A folkloristic perspective on eating and otherness. *Southern Folklore*, *55*(3), 181.

Mak, A. H. N., Lumbers, M., & Eves, A. (2012). Globalization and Food Consumption in Tourism. *Annals of Tourism Research*, *39*(1), 171–196. doi:10.1016/j.annals.2011.05.010

Neumeier, S. & Pollermann, K. (2014). Rural Tourism as Promoter of Rural Development – Prospects and Limitations: Case Study Findings from a Pilot Project promoting Village Tourism. *European Countryside, 6*(4).

Raviv, Y. (2015). *Falafel Nation: Cuisine and the making of national identity in Israel* (H. Hartman, Ed.). Nebraska Press. doi:10.2307/j.ctt1d989k8

Rawashdeh, A. A. (2012). Eco-tourism in Ajloun Forest Reserve: An Investigative Study. *Jordanian Journal for Social Science*, *5*, 151–170.

Sidali, K. L., Kastenholz, E., & Bianchi, R. (2015). Food tourism, niche markets and products in rural tourism: Combining the intimacy model and the experience economy as a rural development strategy. *Journal of Sustainable Tourism*, *23*(8-9), 1179–1197. Advance online publication. doi:10.1080/09669582.2013.836210

Tsai, C. T. L. (2013). Culinary tourism and night markets in Taiwan. *International Journal of Business and Information*, *8*(2).

ADDITIONAL READING

Demirović, D., Košić, K., Surd, V., Žunić, L., & Syromiatnikova, Y. A. (2017). Application of Tourism Destination Competitiveness Model on Rural Destinations. *Journal of the Geographical Institute. Jovan Cvijic.*, *67*(3), 279–295. doi:10.2298/IJGI1703279D

Ildiko, I. O. A. N, Radulescu, C.V. & Florina, B.R. A. N. (2014). Romanian rural tourism: status and prospects by innovative organizational approaches. *Journal of Tourism–Studies and Research in Tourism.*

Mbaiwa, J. E., & Stronza, A. L. (2010). The effects of tourism development on rural livelihoods in the Okavango Delta, Botswana. *Journal of Sustainable Tourism*, *18*(5), 635–656. doi:10.1080/09669581003653500

Murphy, P. E. (2013). Tourism: A community Approach (RLE Tourism). Abingdon-on-Thames, UK: Routledge.

Roberts, L., Hall, D., Copus, A., MacLeod, M., & Boyne, S. (2001). *Rural tourism and recreation: principles to practice.* CABI., doi:10.1079/9780851995403.0000

Ryglova, K., Vajcnerova, I., Sacha, J., & Strojarova, S. (2015). The Quality as a Competitive Factor of the Destination. *Procedia Economics and Finance*, *34*, 550–556. doi:10.1016/S2212-5671(15)01667-6

APPENDIX

Dear Guest,
You are part of a touristic trip in the northern region of Jordan that includes the visit of rural destinations in Ajloun. This survey is conducted for scientific purposes in order to develop a doctoral research. Your evaluation of the tour in rural destinations is really valuable for this research.
Please enjoy your visit to your second home "Jordan".

V) **Socio-demographic variables**

What is your age? ______________

Your Gender?

a) Female
b) Male

Education:

a) primary education
b) secondary education
c) College degree
d) Master degree
e) PhD degree

Monthly Income in USD?

a) Less than 1000
b) 1000- 1499
c) 1500-1999
d) 2000-2499

e)3000 3499

f) 3500- 3999
g) More than 4000

Your Nationality? ______________

Your Religion? ______________

VI) Please give your personal comments related to this tour explaining reasons of satisfaction or disappointment.

Table 11.

No	Item	Totally Agree	Agree	Neutral	Disagree	Totally Disagree
I)	**Managerial competence**					
1	Rural tour is well organized in terms of time.					
2	Rural tour is well organized in terms of program planning and implementation.					
3	Tour guide assumes his ethical responsibility in that he respects group's preferences and beliefs.					
4	Time reserved to stay on rural destination, is sufficient to do the religious rituals.					
5	Time reserved to stay on rural site is sufficient to consume the place and enjoy your experience.					
6	Overcrowding of visitors mitigates the value of touristic purpose.					
7	Tour guide pay attention to security issues.					
8	Tour guide knows how to maximize visitor's satisfaction as he understands your preferences and needs.					
II)	**Tour guide qualification(professionalism)**					
1	Tour guide is polite and respectful to visitor's culture and preferences.					
2	Tour guide motivates visitors to become involved in preparing local food.					
3	Tour guide enhances visitors to taste food while insisting on its healthy ingredients.					
4	Tour guide's interpretations of the history of traditional food are exciting in that it persuade tourists to enjoy eating it.					
5	Tour guide exposes visitors to new experience contradicting to your expectations.					
6	Tour guide shifts opinions on behalf of local community's hospitality.					
7	Tour guide is a good interpreter who masters well the history of this rural destination.					
8	Tour guide answers all questions posed by visitors.					
9	Tour guide talks all the time thus he hinders us from enjoying the splendor of the landscape.					
10	Tour guide's behavior differs from other tour guide's behavior in rural destinations in other parts of the world. (according to past experiences)					
11	Tour guide has excellent communication skills.					
12	Tour guide encourage visitors to use all their senses to have a unique unforgettable experience recalling the history of this region represented by visiting local people who keeps the same traditional life style.					
III)	**The tour finds success in that:**					
1	Tour increases positive perception about the uniqueness of this visit.					
2	Tour changes perceptions towards Jordanian in that they respect visitor's cultures.					
3	Tour enhances mutual understanding between cultures.					
4	Tourism is a magic tool towards a peaceful tourism.					
5	I don't feel that there is any bias to a particular culture.					
6	Tour boosts a sense of the nostology to past times.					
IV)	**Further recommendations that will add a value to the tour:**					
1	Duration of tours should be extended					
2	Duration of stay should be extended to enjoy more the visit!					
3	Time dedicated to visit rural areas And interact with locals is sufficient					
4	Tour guide shouldn't expose us to a narrative related to politics or religious issues.					
5	Tour guide should take the Italian experience applied, in the food tours.					

Chapter 7
Tea Tourism and the Importance of Tea Tourists' Guidance in India

Birsen Bulut Solak
https://orcid.org/0000-0003-1583-1504
Selcuk University, Turkey

Sakib Bin Amin
https://orcid.org/0000-0003-2363-9045
North South University, Bangladesh

EXECUTIVE SUMMARY

The importance of tea tourism in India is immense as it has a dual impact by creating a regional tea market and securing jobs for tea labor. Therefore, proper guidance for the tea tourists is essential regarding the tea tourism destinations by addressing the issues, challenges, and opportunities in promoting local culture. Through the public-private partnership, tea tourism should be included in the mainstream tour packages, and local tea customs can be conveyed through a variety of activities promoting sightseeing and other forms of entertainment and tourism experiences. A proper planning is required for making tea tourism more attractive and developed in India. Development of tea tourism sustaining the environment and preserving the heritage and culture will benefit the Indian regions by creating employment opportunities and boosting the rural economy and thereby alleviate the insurgency and other socio-economic problems. It is expected to contribute to the literature on tourist guiding and the promotion of tea tourism and the tourist guides within tea garden boundaries in India.

INTRODUCTION

Tourism has changed from a small-scale industry into one of the world's largest and fastest-growing industries of the world economy since 1960 onwards (Amin, 2010; Amin, Kabir, Khan & Rahman, 2019). The tourism and travel industry has been ranked as the 4th largest industry in the world after fuels,

DOI: 10.4018/978-1-7998-3725-1.ch007

chemicals, and automotive products. Tourism's total contribution of the global economy in 2017 was 8.7 trillion USD, which equates to 10.4% of total global GDP. The tourism sector supported 313,221,000 jobs in 2017, which is equal to 9.9% of the total employment. This sector also generates 1.5 trillion USD in the export sector in 2017, which is 6.5% of the total exports. International tourist arrivals reached a new record figure of 1.3 billion in 2017, which was only 25 million in 1950. It is forecasted that international tourist arrivals are to reach 2.09 billion by 2028 (Amin, Kabir, Khan & Rahman, 2019).

Tea tourism, a form of nature tourism, is a significant tourism concept associated with tea gardens, and also a kind of unique tour experience connected to nature [1]. The concept of tea tourism gains the momentum since the beginning of the 21st century and one of the most researched topics nowadays (Koththagoda & Thushara, 2016; Sanjiv & Suvamay, 2016). At present, tea plantations, tea culture and tea production have also been interacted by the tea tourism industry in the world (Fernando, 2015). Hence, the tea gardens, the process of tea plucking, tea producing, undertaking a tea-testing session, trying one's hand at plucking tea leaves, trekking in the natural beauty of a lush tea garden, visiting a modern tea factory, cultural festivals of the tea tribes and staying at the tea bungalows are part of tea tourism, and within the idea of tea tourist as well. Tea culture is conveyed through a variety of activities promoting sightseeing, learning, shopping and other forms of entertainment and tourism experiences (Zhang, 2004).

Tea is farmed in over 30 countries of the world including, China, India, Sri Lanka, Bangladesh, Kenya, Indonesia, Japan, Thailand, Turkey, etc. India is the second-largest producer after China also increased to reach 1.2 million tonnes in 2013, according to the Food and Agriculture Organization of the United Nations (FAO) (FAO, 2015). Because of this reason, India is known for the best place for tea production, which has also seen a great jump of 304%, while the cultivated tea area increased by 160% in the last five decades (Shah & Patel, 2016). Tea production in India is growing in 16 states, of which North-East India accounts for about three-fourths of total tea production because of suitable climatic conditions as well as socioeconomic conditions suitable for tea production. Moreover, climatic and geographic conditions of India have formed an appropriate possibility for tea cultivation as tea tree has a long history dated back to the late 18th century (Shah, 2013). Therefore, India holds the immense potential for developing a tea tourism industry for future export diversification.

India has lots of tourist potentialities because of its natural beauty, hilly regions, and historical destinations. The economy of many Indian districts is also based on tea, agriculture, and tourism. Their significant tourism economic activity has also been cantered on the tea industry. Therefore, tea tourism has a tremendous scope for expansion in India, which can encourage as niche tourism segment as the number of tea plantations. Among the teas cultivated in India, the most popular tea comes from Darjeeling Himalaya, where the tea has been cultivated for the last 150 years. Darjeeling tea is well known as one of the finest tea in the world (Khawas, 2006). The bulk (75 to 80%) of tea production comes from Northern India, Assam and West Bengal being the major contributors (Kumar, Badal, Singh, & Singh, 2008). However, it is not only growing in north-eastern states, but the commercial cultivation of tea in India is spreading in 16 states. More than 50% of the Indian tea is produced in Assam. The remaining tea growing states are West Bengal, Tamil Nadu, and Kerala, which are producing around 25.8%, 14.5%, and 5.3% of the total tea production in India, respectively (Shah & Patel, 2016).

The famous tea producing areas like Darjeeling, Upper Assam, Cachar, Nilgiri, etc., could also be promoted for tea tourism. For example, Assam tea is renowned for its strong liquor; Nilgiri tea is well known for its delicate flavor, and the Darjeeling tea is famous for its unique color and aroma. Recently, the rising number of tourists have been rushing to the lush green tea estates of Darjeeling, the Nilgiris, and Assam. Moreover, Assam is also one of the best destinations in tourism, where tourists can also experience a great variety of wildlife, excluding tea tourism (Horen &Dhurbojyoti, 2011). Tea estates in India have also started accommodating guests in their bungalows, flagging off tea tourism (Sharma & Bhowmick, 2016).

The tea is enjoyed worldwide for their distinctive features and flavors. Cheng, Xu, Zhang, J. and Zhang, Y. (2010) find that tea tourists are tea lovers among the ages of 31 and 40, and tea-drinking habits have significant influences on people's attitude towards tea tourism. Jolliffe (2007) defines tea tourists as a tourist experiencing history, culture, and traditions related to tea consumption. Zhang (2004) indicates that tea tourism has advanced to discover tea planting areas with a beautiful natural environment or special historical heritage. Tea culture is conveyed through a variety of activities promoting sightseeing, learning, shopping, and other forms of entertainment and tourism experiences. Horen and Dhurbojyoti (2011) reveal that demographic factors are associated with tea tourism expectations. Thus, in planning market division and creating promotional strategies, tea tourism marketing should take into account tourist age, gender, and cultural background.[2]

From the comprehensive review of literature, it can be understood that though some studies have been conducted on different aspects of tea tourism in different regions in India and even in foreign countries, a study specifically for tea guiding in India is missing in the literature. Hence, this chapter highlights the tourist attitudes towards tea-based tourism in India and addresses some issues, challenges, and opportunities in India for tourists as well as the impact on the community from tea tourism. Another objective of the chapter is also to promote the potential tea tourism destinations in India in terms of tourist guiding in tea tourism. The chapter is further expected to contribute to the literature on tourist guiding and the promotion of tea tourism within tea garden boundaries in India.

BACKGROUND OF INDIAN TEA MARKET

The offering of a cup of tea is a universal sign of hospitality, in either a home or commercial hospitality setting, which was dated back 4500 years ago in China. For hospitality, tea consequently serves as both a symbol and a resource with considerable potential for the provision of commercial hospitality as the tea culture in the world revolves around creating a comfortable and positive experience for guests (Fernando, Kumari, & Rajapaksha, 2017). Tea was discovered in Assam for the first time in the year 1828 by two British travelers Robert and Charles Bruce (Sarma, 2014). In Assam, tea plantation started with the arrival of Europeans, and commercial cultivation was started from 1835 onwards (Shah, 2013). While tea was grown on an experimental basis from 1840, the industry was established as a commercial enterprise in 1856. Moreover, the first commercial tea gardens were planted in 1852. The introduction of the narrow gauge railway in 1881 increased the accessibility of Darjeeling in the colonial period, and the tea industry was the pivot of growth and development. By 1874, there were 113 tea gardens, and by 1905 there were 148 tea gardens (O'Malley, 1907). The former British government also took the initiative to develop tourism activities indirectly through offering restricted accommodation for the tourists in Government Bungalow/Dak Bungalow at many places of tourist interest. The end of the British era in 1947 left behind its footprints in the Darjeeling Himalaya, where colonial bungalows on plantations are preserved to this day, and the hill schools still retain British traditions. There are several gardens in Darjeeling that are over 150 years old and were established in the1800s during the British period. These Bungalows were also typically the residence of British tea planter built during the old colonial time. Moreover, the major economic activity in Darjeeling was centred on the tea industry and the seeds imported from China in the days of old.

Almost 80% of India's annual production could consume by its population (Hazarika, 2011). Depending on the oxidization and fermentation of tea leaves, different types of tea are traded in the market. Orthodox tea and CTC tea, two major forms, are most popular among the traditional tea lovers (Shah & Patel, 2016). Moreover, developing sustainable tourism in tea plantations would help develop market retention and stability of tea, which has gradually fallen in the face of competition, and the industry recognizes more in value addition rather than exporting in bulk (Athukorala, 2008). Major importers of Indian tea are Russia, UAE, United Kingdom, and Poland (Shah, 2013; Tea Board, 2015).

SOCIO-ECONOMIC IMPACT OF INDIAN TEA TOURISM

Tea plays a vital role in ensuring rural development, poverty reduction, and food security in the tea producing countries and is one of the most important cash crops in the World (Chang &Brattlof, 2015). Therefore, tea tourism can have immense potential for developing the local economy as well as for creating employment opportunities, earning foreign currency, reducing poverty, and social impact on the locality of India. The tea industry has been playing a significant role in the socio-economic development of India right from the historical period. Moreover, employment opportunity is one of the biggest socio-economic impacts of Tea tourism in India. The tea tourism industry has great significance in populous and developing countries like India (where unemployment and underemployment is one of the major problems), as it provides sufficient employment opportunities to a larger section of the rural people in marketing, plantation, processing, and ancillary industries.

A quite high number of semi-urban and urban people gets employment opportunities from the distribution and marketing activities of tea production, which attracts the attention of tourist around the globe. It is worth mentioning here that according to the recent statistics, the tea industry facilitates 3 million people in India (Talukdar & Hazarika, 2017).

It is quite evident that over the time period, the employment level in the tea tourism sector has an upward trend from 1 million employees in the 1990s to racing up to almost 1.3 million employment in 2018. This sector is not only creating employment opportunities but also having a positive impact on developing rural area and their livelihood.

Tourism has been a center of attraction for the past few decades. With globalization and open market economy trend around the World, new dimensions have been added to the tourism sector, such as tea tourism. This tea tourism sector has been new hope for bringing up economic prosperity in rural India by channeling foreign currency to the local economy. For example, tourists from outside of the locality come to spend their leisure period enjoying the wonderful tea garden, which gives them historical colonial flavor as the tourism industry was established during the colonial period. With this, the tourists come to know about different products produced by the local Tea industry. As a result, the demand for local tea increased globally.

Besides, India is the second-largest nation with respect to the total area allotted for the production of tea and volume of production in the world. Its contribution in those areas is 20% and 25% respectively. It is also found that India maintains its predominant place in the global business arena in tea production/export over the past 150 years (Nagoor, 2009).

Figure 1. Trend of total employment from Tea Tourism Industry
Source: Tea Board of India

Figure 2. Trend of the total export of tea
Source: Tea Board of India

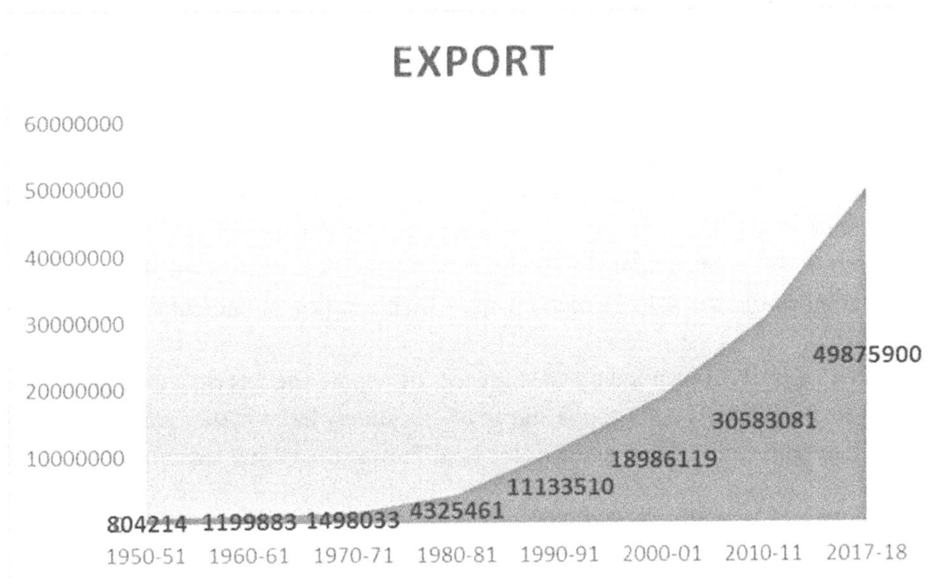

Moreover, India has been earning foreign currency by exporting tea with steady and positive growth over seven decades. This can be related to the tea tourism promotion by which tourists have information about the products and quality tea. Therefore, the contribution of tea tourism in earning foreign currency has immense potential. This has an important economic impact not only channeling foreign currency to the local economy but may have the potential to draw foreign direct investment in the country.

The most potential workers in the tea tourism industry are women who contribute to 70% of total industry workers (Sarma, 2014). Women are an integral part of this tea tourism sector which assures the women participation in labor forces in India. This allows women to engage in economic activity and earning a livelihood, which gives them better socio-economic conditions reducing poverty. By promoting women's employment, the tea industry in India can also achieve many SDG Goals such as no poverty, zero hunger, good health and wealth being, gender equality, decent work, and economic growth, reduced inequality, etc. [3] Therefore, the tea tourism industry will give a platform to local people to change their socio-economic conditions by engaging in economic activities.

PROSPECTS AND OPPORTUNITIES OF TEA TOURISM IN INDIA

Tea tourism is operated in the tea growing regions in the world, and moreover, India has the utmost potentiality to advance the region into a strong force through tea tourism (Horen & Dhurbojyoti 2011). Hence, tea tourism ultimately improves the livelihood of the community. This section evaluates these potential prospects in India for the development of the segment.

The diverse scenery, perpetual weather, and the hills with tea estates make up most of the tourism in Darjeeling. There is a diversity in charming magnetism that tourists would come across here where has a stronghold on the map of India for being a place having the utmost tea exports. Thus, tea tourism in Darjeeling calls for thousands during the plucking season when women in the fields are plucking the leaves (Bhuta, 2015). Darjeeling tea, with its unique Muscatel flavor and exquisite bouquet, is the World's most exclusive tea, fetching the highest prices across the World, and is also known as the Champagne of teas in West Bengal (Kumar, Badal, Singh, & Singh, 2008). Assam is also the World's largest distinct tea growing area and produces the instantly recognizable, rich, malty, full-bodied, bright teas that have established themselves as favourites around the World.

Some tea estate workers and villagers have converted their houses to low budget accommodation and offer them homemade food and a nice smile into the tea garden life as well as areas (Sharma & Bhowmick, 2016). Moreover, tea estates bungalows are often well-preserved vestige of the colonial era, with large courtyards, century-old trees, and green lawns, which capture an ancient world attraction (Horen & Dhurbojyoti, 2011).

In India, tea is extensively consumed by all segments of people such as rich and poor, young and old, men and women-across all faiths and groups and in all the states. India is the largest user of the tea, but has more potential, as per capita consumption of tea is still lower than other tea consuming countries (van der Wal, 2008). The tea industry should exploit a vast local market that yearly consumes 1,000 million kg or 80% of tea produced in India, the largest producer of black tea in the World (Shah & Patel, 2016). The FAO estimates that the Indian domestic market for black tea would grow by 2.2% per annum and attain 1.03 million tonnes of tea consumption annually by the year 2021 (Shah & Patel, 2016).

ISSUES AND CHALLENGES OF TEA TOURISM IN INDIA

discussed below under different heads. There are also constraints which are coming into the way of its development are self-explanatory which can be identified as (a) Lack of infrastructure, (b) Communication bottlenecks, (c) Geographical isolation, (d) Ethnic clashes leading to political instability, (e) Growth of insurgent activities (f) Lack of fund (g) Absence of leadership and entrepreneurship ability, (h) Indifferent attitude of the Government both at State and Central level (i) Defective socio-economic and political set up (j) Lack of a proper tourism development policy of the Government (k) Lack of people's co-operation and consciousness, (l) Lack of proper friendly investment atmosphere (m) Absence of congenial atmosphere to attract the tourists and such many other factors (Das, 2012).

Horen and Dhurbojyoti (2011) reveal that SWOT analysis and pros and cons of the tea tourism sector in Assam based on the interface with the guests, relevant officials, tour organizers, hotels, and other private stakeholders in the state. It is revealed that 40% of both domestic and foreign tourists are motivated by their friends and relatives, whereas 10% of them are motivated by travel agencies. It is observed that the operative of the tourist information bureaus and travel agencies has left much to be desired. For the expansion of the tea tourism sector, travel agencies should use distillate on motivating both overseas and local tourists. About little awareness (26%) among the tourist to visit tea garden, the respondents are asked to choose the main causal factors for it, no hotel and restaurants in tea garden campus is identified by 46% respondents followed by poor road connection (31%) and lack of security (23%) respectively.

Unplanned development of tourism has led to the worsening of environment and ecology, which has in turn been affecting the tourist movement in the area (Khawas, 2002). Even though there are certain challenges for tea production in India like low labor productivity, high cost of production, climate change, seasonal effect, injudicious nutrient management, pest and disease, and consistent quality aspects of tea.

Low Labor Productivity

In India, the labor shortage is a common challenge. Temporary workers are also engaged during the peak plucking season, which affects the quality of tea adversely. Some growers harvest tea leaves by using sickles due to a shortage of workers. One of the reasons for the poor quality of the green leaf is harvested of tea leaves with sickles. However, small tea growers are many times ignored the quality of tea life they pluck (Hazarika, 2011; Shah & Patel, 2016).

Lack of Facilities

Residents are of the opinion that certain tourism activities like ecotourism, tea tourism, and flora-tourism contribute to the preservation of the natural environment and that tourism has also contributed in many ways to the conservation and protection of the local flora and fauna within the communities. The situation of electricity supply varies across tea estates. During the British period, there was no electricity supply in the region (Darjenning). The supply of electricity is a relatively recent phenomenon in the region. Even at present, less than 50% of the region is electrified. Across the tea estates, although all the tea gardens have the facility of electricity connection, there are cases where only the offices, factories, and the houses of white color officials are provided with electricity depriving the laborers (Khawas, 2006). Tea estates have very poor or no provisions for drinking water facilities, housing, latrines, health care, electricity, and education, even for permanent workers. Tea estates still operate in the colonial master-slave mode.

Lack of Government Support

According Horen and Dhurbojyoti (2011), tourism services are characterized by many issues like lack of maintenance of road and infrastructure, poor quality of transport service and tourist staff, lack of security, lack of coordination among several agencies like Department of Tourism and Department of Archaeology in handling the demands of the tourists in places of both historic and religious importance. Lack of affordable and reliable communication network, clean and hygienic food and accommodation, availability of water sports equipment also hamper the development of the tourism industry in India.

Absence of Untrained Tourist Guides

In many areas in India, there is clear evidence of a lack of trained tourist guides in promoting the tourist attractions and facilitating the tourist destinations among the tourists. Therefore, the tea planters have to play a dual role as a planter as well as a host to the tourist without proper training (Horen &Dhurbojyoti, 2011). Overall, tea tourism is still not an organized sector in India, as well. Tour operators or agencies are not active in tea tourism, mainly because the tea estate owners are very selective about their guests. Prices in the main retreats are kept far higher to keep that out of reach from the vast majority. Home-stays, of course, are low scale alternatives, which tend to be attracted mostly to foreign tourists on low budgets, but due to improper promotional activity, these are still out of reach areas (Das, 2017; Sharma and Bhowmick, 2016).

INITIATIVES TO PROMOTE TEA TOURISM IN INDIA

Developing a realistic tea tourism policy is an urgent requirement for the promotion of tea tourism in most of the tea destinations in India. The scenic beauty and wildlife attract a large number of domestic and foreign tourists to India every year. Harmonization with the management of the tea gardens can efficiently do a lot in promoting tea tourism in the state. Many of these tea gardens have polo fields and golf courses. There are as many as 30 airstrips and helipads maintained by the tea garden management. These facilities can form into an attractive package for tourism. The road communication to most of the tea gardens is fairly well maintained, and the rest houses and bungalows with modern facilities located there are generally kept ready for visitors and guests (Horen & Dhurbojyoti, 2011). Strategies need to be articulated for sharing income from tea tourism among estate owners and the state or local government, and procedures laid down for the development of tourism infrastructures. The existing lease contracts between the state government and tea producers need to be amended to allow tea estates to build the necessary constructions to promote tea tourism.

A proposed model for the development of the Indian tea tourism industry is given below. As discussed above, a long term plan is needed for the development of tea tourism in India. There should be proper coordination between the ministry of tourism, the foreign ministry, and the tea board of India. Travelers across the world who are visiting India do not include the tea gardens in their main destinations. Therefore, a constant effort is needed to popularize tea tourism among foreign tourists by creating awareness among them. A few of the tourist operators are offering individual tea tour packages for the tourists; however, the tea tour packages should be included in the mainstream tour packages to attract more tourists in the future. This will certainly help the foreign tourists to include the visits of tea tourism in their plans. There should be proper coordination among the public and private tour operators in creating awareness, designing, promoting, and introducing the tea tourism policies. The accommodations and communications of tea tourism destinations also need to be developed, where required. It is also necessary to prepare skilled tea tourist guides for the tea tourism industry in the future. [4] Finally, a proper marketing and promotional tools will enhance the development of the tea tourism industry in India.

Figure 3. A model for promoting tea tourism in India

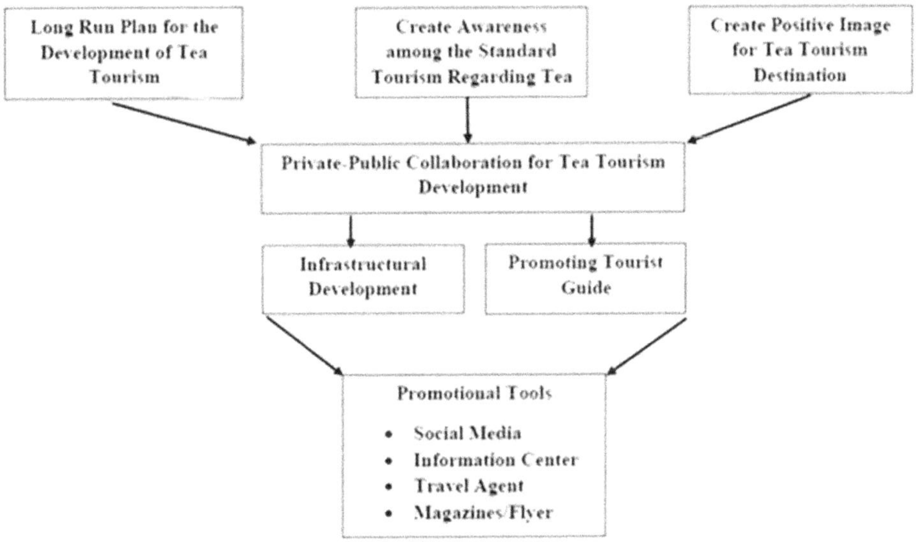

IMPORTANCE OF TOURIST GUIDES IN INDIAN TEA TOURISM

Tourist guides play an essential role in a tourist experience at a destination to leave a long last expression in the tourist mind. Tour guides are frontline staff who provide the 'moment of truth' for tourists and can make or break the trip (Zhang & Chow, 2004). As a frontline line service provider, they leave a lasting impression. Since the tourism industry is combined with multiple stakeholders, each has different expectations from a tour guide. Therefore, tourist guides play a different role in a different place and act, thereby. They facilitate tourists and help them build experience for themselves.

In India, tour guiding is considered a somewhat fragmented professional work. Individuals must be eligible to obtain a license to work as tour guides. Three different types of tour guiding systems, such as local level guides, state-level guides, and regional level guides, are evident across the world nowadays. For obtaining a tour-guide license, it requires professional training. Such a license is issued by the relevant authority to allow an individual to guide tourists in a particular area (Prakash and Chowdhury, 2010).

Expressively, tea tourism is a segmented new concept in the tourism industry, which has immense potential for further growth and contributing to the existing industry. Many of its areas are undiscovered for which tourist guiding can play an important role in exploring the new window in tea tourism. It is suggested that tourist guides help to tourist to discover and create an expression with satisfaction. According to Ham (1992) and Moscardo (1998), detailed information on tourism destinations is necessary, which increases visitor satisfaction and contributes to the commercial acceptability of tourism operations. Their satisfactory performances can also increase the image of their tour companies and even the destinations themselves.

There are three primary ways in which tour guides can enhance the quality of the tourists' experience- (1) providing information for the tourists to make the optimal choices about their activities; (2) providing information to the tourists for ensuring safety and comfort to cope with and better manage encountered difficulties and understand messages given by the warning signs; and (3) providing the actual experience so that tourist can participate in activities such as heritage walks, adventure, visit art galleries, etc. (Moscardo, 1998). Therefore, the role of tourist guides in tea tourism can play a vital for the further flourishing of this segmented subsection of tourism.

In India, as in many other Asian destinations, culture and heritage are the main tourism products such as tea tourism, which has been very popular recently among tourists, and the above notion of a tour guide's role is largely accepted even now. Guides are expected to play an important role as information disseminators (Holloway, 1981) whose primary responsibility is to selecting, glossing, and interpreting sights. According to Mancini (2001), a tour guide is a person who takes people on sightseeing tours of limited duration. Cohen (1985) insists that tour guides should display a high level of professional skill and intimate knowledge of the local culture.

Tea tourism in India has many aspects such as tourism destination visits, experiencing tea products, and spending time in the tea garden, which has been a long tradition since the colonial period. However, many other aspects are not yet known to tourists. For which, tourist guides come to play a vital role in tea tourism. They can disseminate information and lead the tourists to a better experience. Moreover, trained tourists' guides can ensure proper promotion of tea tourism, which will enhance further boost in the industry. Additionally, information and proper knowledge about tourist destination are very important to attract the tourist. In this case, Assam, West Bengal, Northern India as well as Southern India are not popular for tea tourism expectedly for their insurmountable nature. Therefore, tourist guides should come forward to the dissemination of information and leading the tourist to these areas.

However, proper policy regarding tourist guides should be implemented. According to Chowdhury and Prakash (2011), there are at least five broad categories of challenges identified in Indian Tourism destiny. These include general tourism environment, working conditions of tour guides; their relationships with local authorities; their relationship with trade intermediaries; and problems in handling tourists/ customers. To overcome these problems, proper training institutions should be established to make professional tourist guides, which can further generate service-related income in this sector.

CONCLUSION

Tea is one of the main resources that can promote as a niche tourism segment in India because of its major economic activity. Tea gardens have a lot of tourist potentialities because of their natural beauty and hilly regions etc. Tea tourism in India also helps to improve the economic conditions of the local community who engaged with the tourism business and generates good income and benefits than losses for the local community. Tourism industry also encourages both consumption and the development of relationships with potential customers as not harming the Tea estates, but provides double income (Fernando, Kumari, & Rajapaksha, 2017). In this chapter, we highlight the key issues, challenges, and opportunities of tea tourism in India with special reference to Assam, Darjeeling, West Bengal, Northern

India, and Southern India, as well as to whole India. We also propose a marketing model for Tea tourism in India as a tourism destination. We argue that the prevailing problems of tea tourism need to be addressed adequately in order to satisfy the inquisitiveness of the tourists.

Therefore, the governments of Indian states should take some initiative for the development of tea tourism in India's landscapes. For instance, they can open information centers at different popular tourist destinations at India to encourage both local and foreign tourists to visit the tea gardens and also provide all the logistics to motivate them. Private companies should also come forward to provide tea tourism venture in the state. Social media should also be involved in the promotion of tea tourism by tourist guiding. Proper planning and design are required for improving tea-tourism more attractive, and tourist destination centers associated with tourist guides The Department of tourism should graduate more trained tourist guides from tourist guide training programs. The tourism department and the tea board should coordinate together for developing tourism infrastructure, tour guides, and skilled professionals for the tourism sector. Local people should also need to be engaged in the decision making process of tea tourism to make the more industry tourist and management friendly (Datta, 2018).

The micro and macro climate affect the quality and quantity of tea production. Small tea growers are ill-equipped with the scientific management practices because of ignoring the quality of tea life they pluck. Therefore, it is important to train small tea growers for scientific tea management practices. Since the tea plants are exposed to different pests and diseases, it is recommended to grow several varieties for planting. Mechanical plucking types of machinery are a better alternative in the areas where there are labor shortages. Mechanical plucking has the potential to reduce one-third dependency on laborers. This will further strengthen the Indian tourism industry.

REFERENCES

Amin, S. B. (2010). Tourism development and economic growth in Bangladesh: An application of the ARDL testing approach. *Bank Parikrama*, *34*, 60–82.

Amin, S. B., Kabir, F. A., Khan, F., & Rahman, S. Z. (2019). Impact of seaside tourism on host community in Bangladesh: The case of cox's bazar, north south business review. *NSBR*, *102*(10), 12.

Athukorala, P. C. (2008). Export performance in the reform era: Has India regained the lost ground. *ASARC Working Papers*. Doi: doi:10.1057/9780230228337_6

Bhutia, S. (2015). Sustainable tourism development in Darjeeling Hills of West Bengal, India: Issues & challenges. *Global Journal of Human-Social Science: B*, *15*(3), 1-10.

Chang, K., & Brattlof, M. (2015). *Socio-economic implications of climate change for tea producing countries*. Food and Agricultural Organisation of the United Nation Report.

Cheng, S., Xu, F., Zhang, J., & Zhang, Y. (2010). Tourists' attitudes toward tea tourism: A case study in Xinyang, China. *Journal of Travel & Tourism Marketing*, *27*(2), 2010. doi:10.1080/10548401003590526

Cohen, E. (1985). The tourist guide: The origins, structure and dynamics of a role. *Annals of Tourism Research*, *12*(1), 5–29. doi:10.1016/0160-7383(85)90037-4

Das, D. (2012). Tourism industry in North-East Indian States: Prospects and problems. *Global Research Methodology Journal, 2*(7).

Das, R. (2017). Prospects and problems of tourism in Assam. *International Journal of Applied Research, 3*(8), 32–35.

Datta, C. (2018). Future prospective of tea-tourism along with existing forest-tourism in Duars, West Bengal, India. *Asian Review of Social Sciences, 7*(2), 33–36.

Fernando, I. N. (2015). What competitive strategies way forward the regional competitiveness? A comparative economic approach to Sri Lankan tourism. *International Journal of Business and Management, 10*(4). Advance online publication. doi:10.5539/ijbm.v10n4p178

Fernando, P. I. N., Kumari, K. W. S. N., & Rajapaksha, R. M. P. D. K. (2017). Destination marketing to promote tea tourism socio-economic approach on community development. *International Review of Management and Business Research, 6*(1), 68–75.

Food and Agriculture Organization of the United Nations. (2015). *Global Forest Resources Assessment 2015*. Rome: Author.

Ham, S. (1992). *Environmental interpretation: A practical guide for people with big ideas and small budgets*. Fulcrum Publishing/ North American Press.

Hazarika, K. (2011). Changing market scenaria for Indian Tea. *International Journal of Trade. Economics and Finance, 2*(4), 285–287.

Holloway, J. (1981). The guided tour: A sociological approach. *Annals of Tourism Research, 8*(3), 377–402. doi:10.1016/0160-7383(81)90005-0

Horen, G., & Dhurbojyoti, N. (2011). Problem and prospect of tea tourism in Assam - A Swot Analysis. In *International Conference on Advancements in Information Technology with Workshop of ICBMG 2011 IPCSIT* (vol. 20). IACSIT Press.

Jolliffe, L. (2007). *Tea and tourism: tourists, traditions and transformations*. Channel View Publications. doi:10.21832/9781845410582

Khawas, V. (2002). *Environment and rural development in Darjeeling Himalaya: Issues and concerns.* Retrieved from https://www.academia.edu/2648241/Environment_and_rural_development_(2019, September 1)

Khawas, V. (2006). Status of tea garden labourers in eastern Himalaya: A case of Darjeeling tea industry. Council for Social Development, Sangha Rachana, New Delhi, India.

Koththagoda, K. C., & Thushara, S. C. (2016). Customer based brand equity for empowering Sri Lanka as a tourism destination: With special reference to tea tourism. *Kelaniya Journal of Management, 5*(1) January-June 2016.

Kumar, P., Badal, P. S., Singh, N. P., & Singh, R. P. (2008, January-March). Tea industry in India: Problems and prospects. *Indian Journal of Agricultural Economics, 63*(1).

Mancini, M. (2001). *Conducing tours* (3rd ed.). Delmar Thomson Learning.

Moscarado, G. (1998). Interpretation and sustainable tourism: Functions, examples and principles. *Journal of Tourism Studies*, *9*(1), 2–13.

Nagoor, C., & Kumar, N. (2010). Assessing the impact of the ASEAN-India FTA on the Tea Industry. *Economic and Political Weekly*, *30*(44), 112–116.

Prakash, M., & Chowdhury, N. (2010). Tour guides: Roles, challenges and desired competences. A review of literature. *International Journal of Hospitality and Tourism Systems*, *3*(1).

Sanjiv, K., & Suvamay, B. (2016). Tea tourism in Darjeeling. *International Journal of Advance Research and Innovative Ideas in Education*, *2*(4).

Sarma S. (2014). Tea Tourism in North East India. *Employment News, 39*(4).

Shah, S. K. (2013). Prospects of Indian Tea Industry. *International Journal of Agricultural Sciences and Veterinary Medicine*, *1*(1), 18–25.

Shah, S. K., & Patel, V. A. (2016). Tea production in India: Challenges and opportunities. *Journal of Tea Science Research*, *6*(5), 1–6.

Sharma, S. K., & Bhowmick, S. (2016). Tea tourism in Darjeeling. *International Journal of Advance Research and Innovative Ideas in Education*, *2*(4), 1120–1122.

Talukdar, U., & Hazarika, C. (2017). Production and export of value added tea in India and its global competitiveness. *Economic Affairs*, *62*(4), 705–710. doi:10.5958/0976-4666.2017.00086.9

Tea Board. (2015). Retrieved from http://www.teaboard.gov.in/pdf/stat/Monthly _Export_ of_ Tea.pdf

van der Wal, S. (2008). *Sustainability issues in the tea sector a comparative analysis of six leading producing countries* (J. Turner, Ed.).

Zhang, H. Q., & Chow, I. (2004). Application of importance-performance model in tour guides performance: Evidence from mainland Chinese outbound visitors in Hong Kong. *Tourism Management*, *25*(1), 81–91. doi:10.1016/S0261-5177(03)00064-5

Zhang, L. J. (2004). (Master's). Zhejiang University. Retrieved from Wangfang Database (Y600349).

ENDNOTES

1 It is worth noting that tea is the most important and consumed beverages, excluding water around the world.

2 For more detail about the literature of tea tourism, see: Sultana, S., and Khan, R.S., 2019, "Factors Determining Tourism: A Framework to Promote Tea Tourism Destination in Chittagong", *Global Journal of Management and Business Research: F Real Estate Event and Tourism Management*, 18(1), 21-30.

3 For more details about the SDGs, see https://www.un.org/development/desa/disabilities/envision2030.html

4 The importance of tourist guiding in Indian tea tourism is discussed in the next section.

Chapter 8
Analysing the Tour Guides' Perceptions About Gastronomy Tourism in Himachal Pradesh

Debasis Sahoo
Central University of Himachal Pradesh, India

EXECUTIVE SUMMARY

Gastronomic activities like tasting of local cuisines, visiting of apple orchards, tea gardens, fruit marts, and attending cookery classes have always been an integral part of tour itinerary in Himachal Pradesh. Tourists rejoice it as a means of entertainment as well as cultural learning. The growing trend of gastronomy tourism and development of specific tourism policy under the theme of 'Agro/Organic Tourism' is certainly going to increase the demand for 'gastronomic tour guides' in coming years. Hence the present research tries to analyse the tour guides' perception about Gastronomy Tourism in Himachal Pradesh, their background knowledge, and opinion about the branding of Himachali Cuisine. Responses from 120 tour guides were collected through a survey instrument via convenience sampling. The analysis depicts the requirement of specialized food guide training programs, attracting young and women tour guides into the profession and emphasis on branding of Himachali Cuisine through listing of ethnic restaurants, preparation of gastronomic map, and organisation of culinary tours and events.

INTRODUCTION

Tourism has always been a key driver of socio-cultural and economic development due to its unique potential for creating infrastructures, generating employment, earning foreign exchange and in overall contributing to World's GDP (UNWTO, 2017). In this regard, the cultural attractions around the globe have always played a substantial role in the development of tourism industry from very ancient times. Association of tourism and culture can be traced back to the era of Herodotus (of Ancient Greece) who wrote about the "7 Miracles" of the ancient World in 440 BCE. This association got a major boost during the 19th century as culture was an essential part of education for the wealthy scions. But the synergistic relationship between tourism & culture brought about a paradigm shift in the tourism industry during

DOI: 10.4018/978-1-7998-3725-1.ch008

the 2nd half of the 20th century, which led to a much developed segment of the tourism sector, termed as "Cultural Tourism" (OCHTP, 2009). But with the rising interest among the tourists regarding various aspects of culture like history, art, religion, food, costume, festival, etc, it has lead to the development of various segments under cultural tourism like Heritage tourism, Pilgrimage tourism, ethnic tourism, cultural city tourism, gastronomy tourism etc (Csapó, 2012). Gastronomy tourism, alternatively termed as food tourism or culinary tourism refers to the "travelling of tourists to new destinations specifically for enjoying the local delicacies and learning new culture through these food preparations". Hence a gastronomic tour in total comprises several activities like tasting the local delicacies, tour of vineyards & wineries, attending cookery classes, participating in food /wine events and cooking competitions, etc (Hall & Sharples, 2003; Kivela & Crotts, 2005; Karimi, 2008). Though food tourism was initially a part of cultural tourism, over the last few decades it has created a niche market for itself across the Globe.

As per the Global statistics related to gastronomy tourism, one third of the total spending of the tourists is incurred towards food consumption during the tour. Also it was observed that food is one of the key factors for deciding the travelling destination in the case of 25% of leisure tourists (UNWTO, 2012). Apart from several researches that highlight the role of food as a travel motivator (Enright & Newton, 2004; Sharlene, 2012), it (food) has also been identified as a key factor in shaping the overall experience of the tourists (Timothy & Ron, 2013). Further, now-a-days 'gastronomic heritages' of a region are also being used as a tool for marketing, promotion and destination branding (Rand, Heath & Alberts, 2003; Lin, Pearson & Cai, 2011). Moving a step ahead, Cook & Crang (1996) stated that, as regional foods are associated with the geography & tradition of a particular place, hence they can be treated as a symbolic representation of the destination's identity. So in several countries they use their unique culinary heritages for building national identity such as Taiwan (Chuang, 2009), France (Frochot, 2003), Africa (Cusack, 2000) etc. In the above regard, Corigliano (2002) stated that 'Eno-gastronomic' elements (Indigenous food products and production processes) are the major reason behind the prosperity of food & wine tourism, whiskey tourism, olive oil tourism, *tapas* tourism etc. According to a recent survey by UNWTO (2017) it was found that, majority (87%) of the nations (from America, Europe, Asia & Pacific, Africa) have considered gastronomy as a strategic element for destination branding and identity creation. Apart from the fact that 46.5% of the Nations who have a food tourism strategy, as a part of their Tourism Action Plan, 24.6% of them have allocated a specific budget for the same.

The above facts clearly illustrate the potential of gastronomy tourism (as an alternative form of tourism) and the consistent increase in food travellers around the world (UNWTO, 2012) stands testimony to it. According to the UNWTO's report, 'food tourism' not only helps the rural communities to protect their heritage preparations (indigenous recipes) but also generates employment through ethnic restaurants, traditional chefs & local food guides. In this context, India being a culturally diversified country, is no way behind in capturing this food tourism segment and has capitalized on its culinary resources up to a great extent. This can be understood from the fact that the worth of 'food tourism sector' in India is estimated to be US$ 2 Billion each year (Mohanty, 2014). India with a rich cultural heritage of more than 5000 years and varied geographical conditions (from mighty Himalayas in the North to the Peninsular South), has numerous varieties of ethnic delicacies to offer. Ranging from the highly developed cuisines of the *Chettinan* community in Tamilnadu (South Indian State) to the customary '*Wazwan*' (Islamic Banquet) of 'Jammu & Kashmir' in north, each of them has been marketed as a part of Ethnic Food Tourism product (Laura, 2008; Sahoo, 2014). In a similar context, the simple, rustic and aboriginal style of preparations adopted in the North Indian state of Himachal Pradesh has the right mix of ethnicity and socio-cultural influence which is promoted as a part of the states' unique gastronomic affair. Apart from the hilly terrain and the geographical impact of the region, the cuisine of the state has also been influenced by four different cultures i.e. the ancient *Pahaadi* (Himalayas Mountain) culture, *Punjabi* Culture (nearby state of Punjab), *Tibetan* Culture (due to Tibetan Refugees) and the *Israeli* Culture (huge number of Israeli tourists). It can thus be stated that *'Himachali Cuisine'* has a perfect blend of ancient culinary practices and the richness of cultural diversity which attracts a large number of tourists to indulge in a variety of gastronomic activities during their visit to the state. It varies from 'tasting of local delicacies' to visiting of apple orchards, fruit marts & breweries. Therefore, in the recent tourism policy of Himachal Pradesh (2019) special provisions have been made for the promotion & development of gastronomic tourism.

The tour guides are often regarded as the face of the nation, as they are considered as the cultural ambassador of the country and provide the tourists with all the necessary information related to history, geography, culture, religion, food, costume, art form etc. Further, as they always remain in direct contact with the tourists, hence assist them in identifying the gastronomic specialities of the region and enhancing their overall culinary satisfaction (Köroğlu et al., 2007).

The present study tries to bank upon the growing 'food tourism segment' of the Indian & Himachal Tourism Sector and hypothesizes that, in the upcoming years the demand for 'Food Guides' is certainly growing to increase across the various states including that of Himachal Pradesh. It is primarily due to the fact that apart from assisting the tourists they will play a major role in promoting the ethnic delicacies of the region and development of the Gastronomic tourism. Hence the current research tries to analyse the tour guides' perception about Gastronomy Tourism in Himachal Pradesh by collecting their responses through a survey instrument (Questionnaire). Primarily it studies about four major aspects i.e. their knowledge of local delicacies, perception about food tourism in the state, opinion regarding the branding of Himachali cuisine and the requirement of specialized food guides. So the present research has the following four objectives.

OBJECTIVES

1. To identify the tour guides' knowledge about local cuisine of Himachal Pradesh.
2. To analyze the tour guides' perception about gastronomic tourism in the state.
3. To study their opinion regarding the branding of Himachali cuisine.
4. To analyze their opinion about the requirement of specialized food guides.

BACKGROUND OF THE STUDY / LITERATURE REVIEW

For the present research the literatures are reviewed under four major segments i.e. Gastronomy tourism, Gastronomy tourism and Indian perspective, 'Gastronomy & tourism' in Himachal Pradesh, and 'Gastronomy tourism & tourist guides'.

Gastronomy Tourism

Initially, food was an integral part of tourism as a basic necessity for the tourists; but with the passage of time and growing interest in the cultural aspects of the destination, local food of a region started being perceived as a cultural tourism product. This has resulted in the growth of a unique tourism segment i.e.

food tourism/ gastronomy tourism (Reynolds, 1993; Jones and Jenkins, 2002; Au & Law, 2002). Karimi (2008) defined gastronomy/culinary tourism as *"the travelling of people from their own geographical region to other parts of the country or may be outside the country with the aim of trying local cuisines of those areas for both enjoyment and learning purpose"*. In the above regard, a major role was played by ethnic food & beverages that were indigenous to a particular region and highly influenced by the cultural, natural and religious traditions of the place. These were known as Eno-gastronomic element (Corigliano, 2002). She also stated that as per the recent trends in consumer behaviour (in tourism), while selecting a destination the visitors give utmost importance to the "Eno-gastronomic" resources of the place, unlike its climate, heritage, landmarks & nature based activities. This has not only resulted in the creation of 'wine routes' & 'taste paths' but also given rise to various forms of gastronomic tourism like wine tourism and oleo/olive-oil tourism in Italy (Murgado, 2013), beer tourism in Canada (Plummer et al., 2005), whisky tourism in Scotland (Visit Scotland, 2016), food/wine tourism in France (Gonçalves & Maduro, 2016), '*tapas*' tourism in Spain (Jiménez-Beltrán et al., 2016), ethnic/local food tourism in Malaysia (Jalis et al, 2014) and religious food tourism in China, Japan and India (Son & Xu, 2013; Sahoo, 2019) etc. Later, depending upon the tourists' level of interest in food related activities and travel motivation, food tourism was segregated into different categories like (1) Gourmet tourism (tourists with highest interest & travels only for food), (2) Cuisine tourism (interest is high but also engaged in other activities), (3) Culinary tourism (moderate interest & enjoys food unlike other holiday activities) (4) Rural/Urban tourism (low interest but seeks something new to eat) (5) last category (no interest but travels to satisfy the basic need for food) (Hall & Sharples, 2003).

Gastronomy Tourism and Indian Perspectives

Indian cuisine is one of the highly diversified cuisines of the world as each & every state has something unique to offer as far as its gastronomic delights are concerned. Apart from the unique delicacies, they largely differ in terms of ingredients, cooking methods, utensils and serving procedure. In addition to the above, the local food culture also exhibits significant influence of Middle Eastern countries (Persia), South East Asia and European nations, which is primarily due to foreign invasions for over thousands of years and the religious diversity brought by them. Hence one single dish can never represent an entire country like India which is so diverse in terms of its regional culinary traditions (Thomas, 2019). As far as potential of gastronomic tourism in India is concerned several researchers have argued about the scope for its development, primarily based upon the culinary diversities of the regional states and union territories. The highly cited ones include studies about the food tourism potential of south Indian states like 'Karnataka & Tamilnadu' (Abraham and Ramraj, 2015), northern state of Punjab (Berry, 2014; Kumar & Rana, 2016), west Indian state of Maharashtra (Raut & Bhakay, 2012), eastern coastal state of West Bengal (Devi, 2017) and North-eastern states (Baruah, 2016) etc. Further, advancement in food tourism segment can be easily identified in several cities like Hyderabad, Kashmir, Lucknow, Goa, Rajasthan, Delhi and Mumbai, etc that are considered as the culinary hot spots of India and also witness a huge number of food tourists round the year (Sahoo, 2019). As far as the current development of gastronomy tourism in India is concerned it has shown a constant increase in the last two decades. Major development can be seen in the form of growing number of travel companies (like India Food Tour, Delhi Food Tour, Easy Culinary tour etc) offering "Organised food tours" in the above cities (Sahoo, 2014). Later while analysing the growing trend of culinary tourism in India, Vasudeva (2014) observed that the culinary tours which were initially targeted towards 'classical fine dining restaurants' (serving exotic Indian delicacies) have now started concentrating on the 'local ethnic cuisines' which are offered as a part of the cultural tour, both for enjoyment & learning purpose. Hence the 'formal cooking classes' as well as the daily traditional cooking practices at various Home-stays are among the favourite activities of foreign visitors in India. Apart from these, the other gastronomy tourism related activities include excursion tours to the Vineyards in Nasik by Wine Tour organisers like Black grape holidays, Unity wine tours, Wine tour India etc (Banerjee, 2015 and Kaddi, 2015), educational tour of spice gardens in Kerala & Goa by Periyar, Biovalley Herbal Gardens ltd. (Murugadoss et al., 2016) and organising of gastronomic festivals like 'Strawberry Festival' in Mahabaleshwar (Maharashtra), International Mango Festival in Delhi, Apple Festival in Shimla etc (Varsha, 2018; ITWD, 2019; ANI, 2019).

From an exclusive research by Duttagupta (2013) it was revealed that in the Asian Circuit India stands as a major food tourism destination parallel with Thailand, and the culinary enthusiast from various parts of the world (Europe, Africa, U.S.A, U.K. etc) patronize India to experience its unique food culture repeatedly. Further, a recent worldwide survey revealed that Indian cuisine is among the top ten 'list of best foods' preferred by the global tourists during their holidays (cited in Berry, 2014). Moreover, while analysing the culinary preferences of International tourists in India, Upadhyay & Sharma (2014) found that there are basically three types of culinary tourists i.e. taste seekers (taste and quality), localization seekers (locality of food) and experience seekers (culinary tradition, hygiene, nutrition etc).

Gastronomy and Tourism in Himachal Pradesh

Situated in the lap of the mighty Himalayas, the North Indian state of Himachal Pradesh possesses a varied food culture in spite of its geographical limitations. It can be observed in its cultural diversity which varies every 50 kms (Dube & Padalia, 2002). Among the major gastronomic affair of the state is an indigenous banquet arrangement called *'Dhaam'*. It comprises a series of unique dishes (from Rice, lentils, Curries, Gravies, sweets etc) that are cooked in brass vessels (*Charoti*) over 'rectangle wood fire hearths' and served on leaf plates in a particular order. The dishes are usually prepared by a specific category of cooks (*Boti*) and consumed by the guests squatting on the floor with the use of hand (Tanwar et al., 2018). This unique arrangement also has several versions as it varies from one region to the other not only in terms of the number or types of dishes but also their recipes and serving order. Apart from the above, being a neighbouring state to Punjab a lot of influence can also be seen in the staple diet of the local people (FSSAI, 2016). The majority of the delicacies which form the lifeline of *Pahaadi* (Mountain) people include Rice, *Roti/ Paratha* (flat Indian bread), *Rajma* (red kidney beans), *Maah Ki Daal* (Black gram), *Kadhi* (gram flour gravy) etc. Another jewel in the so called culinary diversity of the state is the influence of Tibetan refugees scattered throughout the state. Though Tibetans form only 1.15% of the state population (Census of H.P., 2011), their influence over the cuisine of the region is immense. It can be clearly seen from their traditional delicacies like *Momo, Chowmein, Thupka, Soups, Spring Rolls* which are now an indispensable element of local food culture (Onetikk, 2013).

As far as Tourism of Himachal Pradesh is concerned it is one of the most popular tourism destinations among the tourists as it is the home to some of the most beautiful hill stations of the country like Shimla, Kullu-Manali, Dalhousie, Khajjiar, Dharamshala-Macleod Gaunj etc. Apart from this it is also known as *'Dev Bhoomi'* (Land of Gods and Goddesses) and possesses numerous Temples, Monasteries, *Gurudwaras* & Churches, scattered throughout the state (Sharma et al., 2014). Hence it attracts millions of domestic and foreign tourists (approx. 16.4 million tourists in 2018) throughout the year pertaining to several categories like cultural, religious & leisure travellers (H.P. Tourism statistics, 2019). The tourists not only fall in love with the natural beauty and climate of the destination but also get indulged in the local culture, dance, music, food, costume and festivals of the place. It is the moment that leads to cultural exchange and certain amount of influence can be seen in the gastronomic aspect of the region as well.

The best example in the above regard can be seen in the beautiful villages of 'Kasol' (Kullu district) and 'Dharamkot' (Kangra district). The ever increasing number of Israeli tourists has exerted so much impact on the cultural aspect of the destination that presently they are named as 'Mini Israel' of India (Sehgal, 2017). Huge number of restaurants serving Israeli cuisine with signboards and menu cards printed in Hebrew language, stand testimony to the complimentary side of Gastronomy tourism in H.P. (Jain, 2018). The renowned Israeli delicacies that have made their way to the menus of these restaurants include *Hummus, Falafal, Pitta bread, Shakshouka, Lafa* etc. As far as importance of local cuisine as a gastronomic tourism product is concerned, it has also been capitalized in the form of 'traditional cookery classes' which is provided to the tourists seeking a whole new cultural experience. Other influences can be seen in the form of the native Cannabis-*Marijuana* (*Malana Cream*) which is quite popular among the Israelis and also a major motivation behind their visit to Himachal Pradesh (Sahu, 2018).

Apart from these, there are several "Regional specialities" that forms the basis of food tourism in Himachal Pradesh, as they are offered to the tourists during their visit to the state. In Kullu-Manali the major offerings include fresh 'River Trout', *Siddu* (local steamed bread), visit to apple orchard & fruit Market, and 'Ethnic Meals' (regional specialities prepared out of organic vegetables) etc (Bhalla & Bhalla, 2007). Similarly the Kangra valley offers 'Tea garden' visits, a sip of exotic 'Herbal Kangra Green Tea' and the delicious *Momo*/Dumplings at *Kirpal Momo Shop*-Dharamshala. Among the specialities of Kasauli are the varieties of *Achaar* i.e. Pickled Vegetable/Meat. It includes *Murgh ka Achaar* (pickled chicken), *Machhli ka Achaar* (pickled fish), *Mutton ka Achaar* (pickled goat), Pickled Mushroom, Ginger & Vegetables etc (Sapra, 2019). The summer capital Shimla is unique in various ways with the offering of *Paraati Kulcha* (flat Indian bread with chick pea curry) at Sharma Chat Shop, Tea & *Toshe* (snack) at Vijay sweet shop, Spoonie & chocolate cake at Trishool Bakers, *Mandyali Dhaam* at *Himachali Rasoi* restaurant (Sapra, 2019) and above all the signature 'apple orchard visit' in the upper Shimla area. In addition to the above specialities there is a huge variety of cultivations across the state like that of pomegranate, litchi, kiwi, plum, *Kinnu*/orange, mushroom, potato and sugarcane, honey bee etc, which are often visited by the tourists. Hence in the current Tourism Policy -2019 Himachal Government has classified the above resources under the theme of "Agro/Organic Tourism" for the necessary development and promotion in the upcoming years (H.P. Tourism Policy, 2019).

Gastronomy Tourism and Tourist guides

A 'tourist/tour guide' is a person who provides detailed information regarding various cultural, natural, historical as well as contemporary sites, to the tourists during an organized tour, who are completely or partially aware about the destination (Iriguler and Guler, 2017). Further according to H.P. Tourism Development & Reg. Act-2002, *"Guide means a person who engages himself as a paid guide for the tourists"*. Apart from being considered as a cultural mediator between the host and the guest, he is also regarded as the ambassador for the destination (Jensen, 2010). As he is always involved in face to face interaction with the tourists, his primary responsibility is to create a positive image about the destination in the tourists' mind through effective communication. It is basically carried out by assisting the guests in finding out best places to visit, standard eateries to dine, local cultures to be explored, ethnic cuisine to be relished etc. In a nutshell it can be stated that tour guides create an everlasting impression in the minds of the guests/tourists and enhances their overall satisfaction (Nguyen, 2015). Therefore a tour guide should posses certain level of education and skills like ability to respond to the guest query, zeal for knowledge up-gradation, higher standard of guiding skills and expertise in service delivery, particularly in cultural tourism (Tetik, as cited in Iriguler & Guler, 2016). Further they should be honest, disciplined & responsible and be able to use their presence of mind to tackle any kind of groups. From the above discussion it is clearly evident that the tourists' satisfaction is highly dependent on the four major aspects of the tour guide i.e. skill, knowledge, behaviour and integrity. Hence the "education and training" of tourist guides plays an important role in the above regard.

While tour guide training program in India is primarily conducted by the Ministry of Tourism (MoT), Govt of India (GOI), there are certain private institutions/Universities which provide similar type of courses. A list of tour guide training programs undertaken by the MoT in collaboration with Indian Institute of Tourism & Travel Management (IITTM) & Indian Institute of Skiing & Mountaineering (IISM) were presented below *(Table-1)*.

Table 1.Details of tour guiding training program conducted by MoT, GOI

Sl. No.	Name of the Training Programs	Eligibility / Duration (Th./Pr)	Course Contents
1	Heritage Tour Guide Program Capacity Building for Service Providers Scheme	10th Pass 18< Age>28 yrs 330/420 Hrs	Indian Heritages Indian Culture Tourism Industry Overview Behavioural Skills
2	Incredible India Tourist Facilitator (IITF) programme	Intermediate 7 / 7 days	Indian Society & Culture Tourism Products of India Handling Emergency
3	Regional Level Tourist Guide Training Programme (RLG): Stage:- General	Graduation Age > 20 years 20/ 6 weeks	Indian Culture Natural Tourism Products of India Historical Monuments of India Behavioural Training
4	Regional Level Tourist Guide Training Programme (RLG): Stage:-Expert Guide	Ph. D in Tourism / Hospitality 10 / 3 weeks	Indian Tourism/ Indian History/ Indian Architecture/ Indian Culture/ Indian Wild Life resources
5	Earn while You Learn Programme: 'Student Volunteer'	Intermediate/U.G. 18< Age>25 21 Days	India History, Culture & Tradition Art, Music & dance forms of India Indian Sculptures, Tourism Overview Behavioural skills
6	Paryatak Mitra	Intermediate/U.G. 18< Age>28 yrs 10 days	Overview of Indian Tourism Historical Monuments of India Dos & Don'ts at Monuments Transportation mediums Managerial & Behavioural skills
7	Adventure Trekking Training Course at Ladakh	10th Pass 18< Age>28 yrs 10/10 days	Intricacies of mountain trekking Practical experience

Source: (Website of MoT, IITTM-Gwalior & Bhubaneswar)

Analysis of the above information clearly depicts that, except the 'Heritage Tour Guide' & 'RLG-Expert Guide program', rest of the training programs are of general nature without any emphasis on producing specialized tour guides. Further there is strict variation in the duration of these courses i.e. 5 months (max.) to 10 days (min.) and all of them are only certification courses. Analysis of the course content reveals that there is not even a single course related to gastronomic practices of India or regional cuisines of any state. In present times, food & beverages is not merely considered as a basic necessity for the tourist rather as a complete tourism product for several gastronomic destinations of India. But there is hardly any training program which has been specifically designed for producing 'Specialized Food Guides' to cater to this segment of tourism. Hence there is a need for immediate reorganisation of the guide training courses as well as introduction of "Gastronomic Tour Guide Training Programs" at National as well as regional level, including Himachal Pradesh. Therefore the current research tries to analyse the tour guides' perception about gastronomic tourism in Himachal Pradesh as well as their opinion about the requirement of specialized food guides and branding of Himachali cuisine.

RESEARCH METHODS

This section describes various aspects of the research methodology like questionnaire design, sampling technique, method of data collection, data analysis and limitations of the study.

Questionnaire Design

In order to measure the tour guides' perception about 'gastronomy tourism in Himachal Pradesh' a structured questionnaire was designed for collecting their responses. The survey instrument developed by Köroğlu et al. (2007) and İriguler & Guler (2017) were studied, based upon which the present questionnaire was regenerated from the perspective of Himachal Pradesh. The questionnaire had five major segments i.e. (1) tour guides' knowledge of local cuisine, (2) perception about gastronomic tourism in the state, (3) opinion regarding the branding of Himachali cuisine, (4) opinion on the requirement of specialized food guides & (5) socio-demographic profile. While the initial segments comprised 7 close ended statements, the following two segments had 11 closed ended statements. The responses were recorded through a five point Likert scale varying between the lowest measuring unit of 1 (absolutely disagree) to highest unit of 5 (absolutely agree). The fourth segment had five close ended questions and the final segment recorded socio-demographic variables like age, gender, education, experience, region of operation, language and area of specialization.

Sampling Technique and Method of Data Collection

A total of 120 registered tour guides of Himachal Pradesh were surveyed across the four major tourism circuits of the state (30 each). They were (a) 'Manali-Kullu-Mandi' circuit, (b) 'Shimla-Naldehra-Sarahan' circuit, (c) 'Dharamshala-Palampur-Kangra' circuit and (d) Dalhousie-Chamba-Khajjiar Circuit. The data was collected via convenience sampling method, between September to December 2019. Primarily the responses were collected through personal interview (in person & via telephone) and e-mail.

Data Analysis

For the purpose of analysis, the collected data was initially loaded into statistical software package of SPSS-20. After checking the reliability of the instrument through the value of Cronbach's Alpha, the data was analyzed through percentage analysis, and the analysis of mean values and standard deviation.

Limitations of the Study

The major limitation of the study was the sample size. As most of the tourist guides were reluctant to respond to the questionnaire, only 120 samples could be collected. Further August to December being the peak season, the availability of the guides for interview was limited. As most of the guides contacted through mail didn't respond to the questionnaire, there is a need for higher number of personal interviews in order to increase the sample size and overcome the research limitations.

DATA ANALYSIS AND INTERPRETATION

The collected Data was analysed and interpreted under following five segments. They are as follows:

Socio-Demographic Characteristics of the Tour Guides

The analysis of socio-demographic characteristics of the tour guides was presented in (*Table-2*) which depicts that approximately 52% of the tour guides were between the age group of 21 to 40 years where as 27.5% of them belong to '41-50 years' age group and 14.2% were above 51 years of age. As far as gender is concerned it was observed that 92.5% of the tour guides were male while the rest 7.5% were female. From the education point of view it was found that 75.9% (i.e. 91 out of 120) of the tour guides were highly qualified with an education degree of Graduation and above. Among the 120 tour guides surveyed, 52 of them (43.3%) were having 5 to 10 years of experience where as 31.7% (38 nos.) had 11 to 15 years of tour guiding experience. Only 15.8% possessed more than 15 years of experience in this sector. When the respondents were asked about their language proficiency, it was revealed that 34.2% of the tour guides can interact only in local (*Pahaadi*) and Hindi language; where as 40.8% of them were comfortable to speak in Hindi & Punjabi along with their native language. In addition to this 19.2% of tour guides were found to have a good command over English apart from the above three languages. The majority (60%) of the tour guides' were also found to be efficient in using Punjabi language as Punjab is one of the bordering state of Himachal Pradesh and certain districts of H.P. were earlier part of Punjab state (prior to 1972). While interviewing the tour guides it came to light that 15% of them strictly adhere to the Shimla circuit and 19.2% deal with Manali region whereas 15.8% of the guides provide their service in both Manali and Shimla region. Similarly Dharamshala and Dalhousie circuits were guided by 12.5% and 18.3% of the tour guides respectively but 19.2% of them dealt with both these circuits.

Table 2. Socio-demographic characteristics of the participants

Socio-Demographic Variables		Frequency (N)	Percentage (%)	Cumulative (%)
Age	0-20 years	7	5.8	5.8
	21-30 years	26	21.7	27.5
	31-40 years	37	30.8	58.3
	41-50 years	33	27.5	85.8
	51 years and above	17	14.2	100
	Total	120	100	
Gender	Male	111	92.5	92.5
	Female	9	7.5	100.0
	Total	120	100	
Education	Matriculation (10th)	10	8.3	8.3
	Intermediate (+2)	19	15.8	24.1
	Graduate (+3)	58	48.4	72.5
	P.G. & above	33	27.5	100
	Total	120	100	
Experience	1 to 4 years	11	9.2	9.2
	5 to 10 years	52	43.3	52.5
	11 to 15 years	38	31.7	84.2
	Above 15 years	19	15.8	100
	Total	120	100	
Languages spoken	Local *Pahaadi* language	7	5.8	5.8
	Pahaadi& Hindi	41	34.2	40.0
	Pahaadi, Hindi, Punjabi	49	40.8	80.8
	Pahaadi, Hindi, Punjabi & English	23	19.2	100
	Total	120	100	
Regions mostly escorted by the Guide	Shimla circuit	18	15.0	15.0
	Manali circuit	23	19.2	34.2
	Both Shimla & Manali	19	15.8	50.0
	Dharamshala circuit	15	12.5	62.5
	Dalhousie circuit	22	18.3	80.8
	Dharamshala & Dalhousie	23	19.2	100
	Total	120	100	
Area of specialization	Religious tourism	37	30.8	30.8
	Culture, Heritage & Rural	20	16.7	47.5
	Adventure tourism	21	17.5	65.0
	Eco tourism, Health & wellness tourism	16	13.3	78.3
	Snow tourism	26	21.7	100
	Total	120	100	

The tour guides' adherence to their own region is primarily due to the geographical distance between these places whereas the practice of dealing with multiple regions is basically due to similar product offerings. Again, except the adventure (17.5%) & eco-tour guides (13.3%) all of them were of the opinion that they deal with variety of tourism products simultaneously and lack any kind of specialization but in terms of their knowledge and preference they identified themselves to certain specific categories like religious tourism (30.8%), Cultural, Heritage & Rural tourism (16.7%) and Snow tourism (21.7%).

OBJECTIVE-1

Tour Guides' Knowledge About the Local Cuisine of Himachal Pradesh

Before analysing the knowledge of the tour guides, the reliability of the instrument was checked through the value of Cronbach's Alpha, which was found to be 0.728. According to Tavakol & Dennick (2011) the value of Alpha anywhere between 0.70 and 0.95 is acceptable. Hence the data was considered suitable for analysis.

Table 3. Tour guides' knowledge about the local cuisine of Himachal Pradesh

Sl. no.	Statements...	Mean	Std. deviation
1.	I have general knowledge about local cuisines of Himachal Pradesh	4.212	0.833
2.	I have undertaken educational programs about the local cuisine	2.791	1.211
3.	I acquire information about local dishes from TV shows, cookery books, magazines and internet sources effectively	2.549	0.748
4.	I certainly inform the tourists about the food culture of the local sites visited by them	3.662	0.762
5.	I inform the tourists about the food culture of the place only if they ask regarding the same	3.250	1.291
6.	If I am escorting a group of tourists to a destination for the first time, I usually research about the local cuisine & keep myself informed before leaving for the tour	3.772	1.283
7.	I have sufficient knowledge about the local cuisine to serve as a gastronomic/food guide for the region	3.811	0.947

While studying the knowledge of the tour guides about local cuisine *(Table-3)* it was observed that, mostly they agreed to have general knowledge about the local food preparations (Mean-4.212) but they were indecisive about their responses to culinary courses undertaken by them (2.791). It might be due to the course curriculum of U.G and P.G. courses (in tourism) where very little information about the Indian cuisine is provided under the segment of 'Cultural heritages' of India. Similarly their response about acquiring information regarding local cuisine from various sources, was between disagree to undecided as the mean value was (2.549). While most of the tour guides agreed (3.662) to the fact that they usually inform the tourists about the regional food culture, certain guides were of the opinion that they provide such information only in case of an enquiry (3.250). Further it was found that though tour guides agreed to have sufficient knowledge about the local cuisine (3.811), still they feel the necessity of carrying hand on information about local cuisine, while escorting the guest to a new destination for the first time (3.772). This is primarily due to the difficulties of the unknown hilly terrains where the resources are very limited and knowledge of local resources (including food) is a must for all. Further while guiding religious tourists it is of utmost importance to have knowledge of *Satvik* (pure vegetarian) restaurants keeping the religious values in mind.

OBJECTIVE-2

Tour Guides' Perception About Gastronomic Tourism in the State

Unlike the above scale, the instrument for measuring the tour guides' perception about gastronomic tourism was also subjected to the test of reliability. The value of Cronbach's Alpha was calculated to be 0.811, hence proved the instrument a reliable one (Tavakol & Dennick, 2011). The collected responses were analyzed & presented in *(Table-4).*

Table 4. Tour guides' perception about gastronomic tourism in the state

Sl. no.	Statements...	Mean	Std. deviation
1.	I would say in general the group members like the flavours they try in Himachal Pradesh	3.423	1.313
2.	I think the food outlets placed in the tour itinerary offer the Menus which contain the Himachali food preparations	3.012	1.262
3.	I think the first time visitors are not much aware about Himachali cuisine	3.718	1.116
4.	I can say it positively that the group members are curious to know about the Local delicacies as well as the food culture	3.528	0.923
5.	I believe that I have an active role in the promotion of Himachali Cuisine	4.381	0.644
6.	I believe that food /gastronomic tourism will contribute significantly to Himachal Tourism	4.116	0.719
7.	I think Gastronomic/culinary tourism could be a healthy solution to the seasonality issue in Himachal tourism	3.873	0.698
8.	I believe Himachal Pradesh along with its culinary heritages is a highly potential destination for gastronomic tourism	3.827	1.314
9.	I believe that food tourism has already been developed in Himachal Pradesh	3.741	0.74
10.	I believe that Himachali Cuisine has an effective role in the overall promotion of the state	4.077	0.793
11.	I think culinary tourism might become the primary factor for visiting Himachal Pradesh	3.464	0.611

The current analysis revealed that, guides were not sure whether tourists liked the *Himachali* food preparations or not (3.423). This is mostly due to the tourists' regional diversity and their ethnic palates which are quite different from each other. While the North Indian tourists like the flavours of Himachal, others have mixed experience. The tour guides mentioned that food outlets placed in the itinerary, mostly offer local Himachali food in their menus (3.012) but their number & variety is quite limited. They also agreed that first time visitors are not much aware about the Himachali cuisine (3.718) as the ethnic delicacies are quite different in terms of their ingredients & preparations methods. Since the perception of guides' about the 'gastronomic curiosity of the tourists' was midway between 3-undecided to 4-Agree (i.e. 3.528), it can be understood that all the tourists were not having similar level of interest in local food culture. The two aspects regarding which tourist guides were very clear was the belief that gastronomic tourism will contribute significantly to Himachal tourism (4.116) and their active role in its promotion (4.381). Likewise the guides also agreed to the statements that food tourism could be a healthy solution to the seasonality issue (3.873) as Himachal Pradesh along with its culinary heritages is a highly potential destination for gastronomic tourism (3.827). The perception of the tour guides' about the development of food tourism in the state received a positive response (3.741) which may be due to the several gastronomic activities regularly undertaken by the tourists. For e.g. Tour of Apple orchards, Tea garden visit, Apple festival, relishing the ethnic *Dhaam* & regional specialities like *Siddu, River Trout,* Kangra Tea, Tibetan *Momo,* Israeli *Hummus, Falafal & Pitta bread* etc. Though tour guides had a strong belief that Himachali cuisine plays an effective role in the overall promotion of the state (4.077), they were not very sure (3.464) about the scope of gastronomy tourism becoming the primary factor for visiting Himachal Pradesh. It may be due to the pre-dominant image of Himachal Pradesh as a religious Destination (*Dev Bhoomi*) and vacationers paradise (home to world renowned Hill stations).

OBJECTIVE-3

Tour Guides' Opinion Regarding the Branding of Himachali Cuisine

Post reliability analysis i.e. value of Cronbach's Alpha giving significant result of 0.839, the opinion of tour guides about the branding of Himachali cuisine was analyzed and presented in *Table-5*. From the analysis it can be observed that most of the tour guides agreed (4.021) to the idea of organizing state level and regional 'Himachali Cuisine Days' by the hotels, on a regular basis in order to sell, preserve & promote the local delicacies. They also supported the use of organic fruits and vegetables for preparing indigenous dishes (4.228). Moreover a strong recommendation to carry out research work on Himachali cuisine was also received (4.519). As far as preparation of classical dishes of Himachal Pradesh is concerned, the guides were of the opinion that authentic recipes must always be followed. Further the respondents noted that a list of the 'brochure and books' should be prepared which contain detailed information about Himachali cuisine.

The tour guides participated in the survey also agreed (3.988) that restaurants offering special Himachali dishes on their menus must be registered and inspected at regular interval by tourism officials. Highlighting the close association between the gastronomic aspects of Himachal Pradesh and healthy eating practices was another aspect that was strongly agreed (4.713) by most of the respondents. They were also of the opinion that specialized training in Himachali cuisine should be provided to the qualified persons for conserving the heritage delicacies of the state. The guides also emphasized that tourism department should devise appropriate strategies for development of gastronomy tourism in the state as well as its promotion in the International market. Last but not the least, most of the tourist guides strongly agreed (4.782) for the mapping of important gastronomic zones and creation of supporting events & culinary tours.

Table 5. Opinion regarding the branding of Himachali cuisine

Sl. no.	Statements......	Mean	Std. deviation
1.	Hotels and other accommodation units should organize state level & regional Himachali Cuisine Days	4.021	0.910
2.	Food and beverages that are indigenous to Himachal Pradesh should be prepared by using organic fruits and vegetables	4.228	0.791
3.	Scientific researches should be carried out regarding the regional cuisine and be shared through print and audio-visual mediums	4.519	0.538
4.	The classical (special) regional delicacies of Himachal Pradesh should always be prepared as per their authentic recipes	4.612	0.819
5.	A list of the 'brochures and books' containing detailed information about the dishes of Himachal Pradesh must be prepared	4.601	0.653
6.	Restaurants offering special Himachali dishes on their Menus must be registered and inspected at regular interval by the officials of Department of Tourism and Civil aviation	3.988	1.283
7.	The close association between the Gastronomic aspects of Himachal Pradesh and Healthy eating practices must be highlighted	4.713	0.664
8.	Qualified persons must be properly trained about Himachali cuisine	4.552	1.148
9.	Department of tourism and civil aviation must initiate appropriate strategies for developing gastronomy tourism in the state	4.077	0.773
10.	Food culture of Himachal Pradesh should be promoted in the International arena	4.219	0.791
11.	Flavour mapping of Himachal Pradesh should be carried out and be supported with appropriate tours and culinary events	4.782	0.548

Table 6. Opinion about the requirement of specialized food guides

Sl. no.	Variables		Frequency (N)	Percentage (%)
1	Do you believe there is a necessity to specialize as a food/ Gastronomic guide?	Yes	94	78.3
		No	26	21.7
2	Would you like to participate in a specialized gastronomic guide course?	Yes	87	72.5
		No	33	27.5
3	The above course should be provided at which level?	High School	38	31.7
		Intermediate	46	38.3
		Graduation	25	20.8
		Post Graduation	11	9.2
4	You would prefer to undertake such training programme from______?	Private Institute	8	6.7
		MoT, GOI/ Dept. of tourism H.P.	112	93.3
5	What should be the type of specialization?	General	19	15.8
		Regional/local	101	84.2
		Total	120	100

OBJECTIVE-4

Tour Guides' Opinion About the Requirement of Specialized Food Guides

The analysis of tourist guides' opinion about the need for 'specialization in gastronomy tourism' revealed the following details *(Table-6)*. While 78.3% of the respondents believe that there is a necessity to specialize as a food guide, the rest 21.7% did not feel the importance of such specialized guides. It may be due to their current profession as a tourist guide who deal with several gastronomic products along with their regular tourism attractions.

Similarly 72.5% of the Guides showed their willingness to participate in specialized 'Food Guide Training Programs'. When they were asked about the educational level at which such courses should be provided, 38.3% them responded that it should be at Intermediate level, followed by 31.7% at High School level and 20.8% at Graduation level. While 93.3% of the guides were interested to undertake such courses from Ministry of Tourism, GOI or Department of Tourism & Civil aviation, Himachal Pradesh, only 6.7% showed their desire for private institutions. The higher demand for Government operated courses is basically due to their low fee structure as well as Government certification. In response to the final question related to the type of specialization, 84.2% feel that training should be given on regional / local cuisine of a particular destination instead of a general training on Himachali cuisine. It would not only allow them to have an in-depth knowledge of the local cuisine of a particular region but also the major eateries around the place which can be promoted as a part of gastronomic tourism.

SOLUTIONS AND RECOMMENDATIONS

With reference to the analysis carried out, it can be recommended that there is a need to attract more young people (21-30 years Age) as well as the 'women' to take up 'tour guiding as a profession' since they constitute only 21.7% and 7.5% of the total guides, respectively. Moreover there are only 19.2% of the guides who are able to communicate in English and hardly anyone who has a specialization in the area of gastronomic tour guiding. Hence both state department of tourism & the central MoT, must design "gastronomic tour guide training programs" and deliver it to the interested candidates through selected institutions. The course should not only comprise gastronomic aspects of the state but also language and behavioural training sessions. They should also be made aware about the use of internet for collecting information, suggestive selling of gastronomic products and importance of acquiring knowledge about 'regional delicacies' before guiding a new destination. Apart from this, there is a need to reorganise the course curriculum of the 'Tourism based U.G and P.G. courses' for incorporating a section on gastronomic tourism. As the survey depicts that offering of Himachali dishes is very limited in the hotel menus, the owners of such hotels & restaurants must be communicated about the importance of having regional cuisine on the menu, through capacity building programs. Further a list of such restaurants must be prepared and be promoted through brochure, leaflet & website of Department of tourism. The tour guides believe that food tourism has already been developed in the state up to some extent (3.741) and plays an effective role in the destination promotion (4.077) but it hasn't been given a specific Name/Brand till date. Though recently it has been identified under the theme of "Agro/Organic Tourism" in Himachal Tourism Policy-2019, there is a necessity for its proper organisation at the ground level. The belief that "Himachal Pradesh is a highly potential destination for gastronomic tourism (3.827) and may contribute significantly to the tourism segment", positively affected the opinion of tour guides about the branding of Himachali cuisine. As most of them agreed (mean value ^{3}4) to all the given statements, the strategies that would be helpful in this regard are celebration of Himachali Cuisine days, use of organic vegetables, practicing original recipes, printing of brochures/leaflets, listing of ethnic restaurants, highlighting the health aspects of the local dishes, preparation of gastronomic map and organisation of culinary tours and events. Finally the higher rate of interest among the guides to participate in 'Regional Food Guide Training Courses' and specialize as a 'Gastronomic Guide' supports the idea of providing such courses at High School or Intermediate level by the State Department of Tourism as well as MoT, GOI.

FUTURE RESEARCH DIRECTIONS

The growing popularity of gastronomy tourism across the country (India) including that of Himachal Pradesh certainly depicts a bright scope in the field of gastronomic tour guiding in the coming times. As far as formulation and implementation of 'food tourism strategies' and 'gastronomic tour guide training programs' is concerned, the major issues could be the bureaucratic system of operation in the state as well as on the national level which is very slow and subjected to political influence. Further lack of 'trained/ experienced food guides' in India will also be a major concern as it is always a tough task to create first line of tour guides in any field. Hence, keeping the above concerns in mind, future researches can be carried out for identifying the 'issues and challenges in the implementation of food guide training programs' as well as the real demand for such food guides can also be assessed from the tourists' perspective by directly interviewing them.

CONCLUSION

Gastronomy tourism has always been a part of the tourists' itinerary in Himachal Pradesh though it is being delivered to them as a component of the overall leisure tourism package. Hence, in the current tourism policy, the state government has recognized the importance of this segment of tourism and categorized them under a separate nomenclature of Agri/Organic tourism. This recognition as well as specific strategies formulated for its development and promotion is certainly going to increase the demand for gastronomic tour guides in the near future. So the present research tried to analyze the tour guides' knowledge about the ethnic delicacies of the region as well as their opinion about branding of Himachali cuisine and requirement of specialized food guides in the future. Analysis of the collected data suggested that there is a need to attract and engage 'younger people' as well as 'women' in the tour guiding job as their ratio is very minimal. Further a lower level of knowledge signified the importance of incorporating subjects related to 'gastronomic tourism' in the course curriculum. As it was observed that, the tour guides have a very positive opinion about the branding of Himachali cuisine, it will certainly be helpful in the development of gastronomic/ organic tourism in the state with the adoption of several unique strategies like celebration of Himachali cuisine days, listing of ethnic food outlets, practicing of indigenous recipes & food mapping of the region etc. Apart from these, the interested candidates could be provided with specifically designed 'food guide training programs' (as a part of capacity building initiative) in order to cope up with the future tourism demands.

REFERENCES

Abraham, S., & Ramraj, K. (2015). The Potential of Culinary Tourism in India. *ANVESHAK-International Journal of Management*, *4*(2), 133–148. doi:10.15410/aijm/2015/v4i2/67724

ANI. (2019, September 29). Two-day apple festival concludes in Shimla. *ANI News.* Retrieved on 14 October, 2019, fromhttps://www.aninews.in/news/national/ general-news/two-day-apple-festival-concludes-in-shimla20190929155529/

Au, L., & Law, R. (2002). Categorical classification of tourism dining. *Annals of Tourism Research*, *29*(3), 819–833. doi:10.1016/S0160-7383(01)00078-0

Banerjee, M. (2015). Food Tourism: An Effective Marketing Tool for Indian Tourism Industry. *International Journal of Scientific Research (Ahmedabad, India)*, *4*(6), 795–800.

Baruah, S. R. (2016). Promotion of Culinary Tourism as a Destination Attraction of North-East India. *IJIRSSC- Journal, 2*(1), 201-209.

Berry, K. (2014). Progress and Emerging Issues in Culinary Tourism: A Study with Special Reference to Punjab. *Journal of Hospitality Application & Research*, *9*(1), 28–46.

Bhalla, S., & Bhalla, T. C. (2007). Traditional foods and beverages of Himachal Pradesh. *Indian Journal of Traditional Knowledge*, *6*(1), 17–24.

Census of Himachal Pradesh. (2011). *Himachal Pradesh Religious Data.* Retrieved from Census of India website on 18 October, 2019, from https://www.census2011. co.in/census/ state/himachal+pradesh.html

Chuang, H. T. (2009). The rise of culinary tourism and its transformation of food cultures: The national cuisine of Taiwan. *The Copenhagen Journal of Asian Studies*, *27*(2), 84–108. doi:10.22439/cjas.v27i2.2542

Cook, I., & Crang, P. (1996). The world on a plate: Culinary culture, displacement and geographical knowledges. *Journal of Material Culture*, *1*(1), 131–154. doi:10.1177/135918359600100201

Corigliano, A. (2002). The route to quality: Italian gastronomy networks in operations. In A. M. Hjalager & G. Richards (Eds.), *Tourism and Gastronomy* (pp. 166–185). Routledge.

Csapó, J. (2012). The Role and Importance of Cultural Tourism in Modern Tourism Industry. In M. Kasimoglu & H. Aydin (Eds.), *Strategies for Tourism Industry - Micro and Macro Perspectives* (pp. 201–232). InTech Europe. doi:10.5772/38693

Cusack, I. (2000). African cuisines: Recipes for nation Building? *Journal of African Cultural Studies*, *13*(2), 207–225. doi:10.1080/713674313

Devi, K. S. & Amity University. (2017). *Cuisine Tourism in West Bengal.* Kolkata: Amity University.

Dube, M. P., & Padalia, M. (2002). *Democratic decentralization & Panchayati Raj in India*. Anamika Publishers.

Duttagupta, S. & University of Waterloo. (2013). *Foreign travellers' recommendation of culinary tourism in India based on cuisine image and satisfaction with experiences at culinary establishments: An exploratory study*. Waterloo: University of Waterloo.

Enright, M. J., & Newton, J. (2004). Tourism destination competitiveness: A quantitative approach. *Tourism Management, 25*(6), 777–788. doi:10.1016/j.tourman.2004.06.008

Frochot, I. (2003). An Analysis of Regional Positioning and Its Associated Food Images in French Tourism Regional Brochures. *Journal of Travel & Tourism Marketing, 14*(3-4), 77–96. doi:10.1300/J073v14n03_05

FSSAI. (2016). *Feasts Flavours and Fusion of India.* IGI Global India Pvt. Ltd.

Gonçalves, E. C., & Maduro, A. V. (2016). Complementarity and Interaction of Tourist Services in an Excellent Wine Tourism Destination: The Douro Valley (Portugal). In M. Peris-Ortiz, M. C. D. R. Rama, & C. Rueda-Armengot (Eds.), *Wine and Tourism -A Strategic Segment for Sustainable Economic Development* (pp. 123–132). Springer. doi:10.1007/978-3-319-18857-7_9

Hall, C. M., & Sharples, L. (2003). The consumption of experiences or the experience of consumption? An introduction to the tourism of taste. In C. M. Hall, L. Sharples, R. Mitchell, N. Macionis, & B. Cambourne (Eds.), *Food Tourism around the World. Development, management and markets* (pp. 1–24). Butterworth-Heinemann. doi:10.1016/B978-0-7506-5503-3.50004-X

H.P. Tourism Development & Reg. Act. (2002). *Himachal Pradesh Tourism Development & Registration Act-2002.* Retrieved from Himachal tourism website on 14 November, 2019 from https://himachaltourism.gov.in/wp-content/uploads/2018/09/58e76f7a08179_H.P.Tourism-Dev.-and-Registration-Act-2002.pdf

IITTM. (2019). *Regional Tour Guide training program.* Retrieved from IITTM website on 18 November, 2019, from https://www.iittm.ac.in

Iriguler, F. & Guler, M.E. (2017). Tourist Guides' Perceptions on Gastronomic Tourism in Turkey. *Journal of Tourism & Gastronomy Studies, 5*(1), 3-19.

ITWD. (2019, June 27). International Delhi Mango Festival begins (place or time): All you need to know. *India Today.* Retrieved on 11 October, 2019, from https://www. indiatoday.in/information/story/all-you-need-to-know-about-international-delhi-mango-festival-1557301-2019-06-27

Jain, S. (2018, February). Tripoto: Kasol- A mini Israel of India [Web log post]. Retrieved fromhttps://www.tripoto.com/trip/kasol-a-mini-israel-of-india-5993cd64c8f9e

Jalis, M. H., Che, D., & Markwell, K. (2014). Utilising local cuisine to market Malaysia as a tourist destination. *Procedia: Social and Behavioral Sciences, 114,* 102–110.

Jensen, O. (2010). Social meditation in remote developing world tourism locations–the significance of social ties between local guides and host communities in sustainable tourism development. *Journal of Sustainable Tourism, 18*(5), 615–633. doi:10.1080/09669581003615590

Jiménez-Beltrán, F. J., López-Guzmán, T., & Santa Cruz, F. G. (2016). Analysis of the Relationship between Tourism and Food Culture. *MDPI-Sustainability, 8*(5), 418. doi:10.3390u8050418

Jones, A., & Jenkins, I. (2002). A taste of wales–Blas Ar Gymru: Institutional malaise in promoting welsh food tourism products. In A. M. Hjalager & G. Richards (Eds.), *Tourism and Gastronomy* (pp. 115–131). Routledge.

Kaddi, A. K. (2015). A Study on Creation and Development of Wine Tourism Circuits in Maharashtra. *Atna Journal of Tourism Studies*, *10*(1), 1–12. doi:10.12727/ajts.13.1

Karimi, S. (2008, March). *What is culinary tourism?* Retrieved from http://traveltips .usatoday.com/culinary-tourism-1910.html

Kivela, J., & Crotts, J. C. (2005). Gastronomy tourism: A meaningful travel market segment. *Journal of Culinary Science & Technology*, *4*(2), 39–55. doi:10.1300/J385v04n02_03

Koroglu, A., Koroglu, O., & Sarioglan, M. (2007). An Empirical Study on the Role of Professional Tourist Guides in Promoting Turkish Cuisine. *1st National Gastronomy Symposium and Artistic Activities.*

Kumar, A., & Rana, V. S. (2016). Food as an Attraction for Tourism: A Study of Patiala City, Punjab. *IICMR Research Journal*, *11*(1), 7–11.

Laura, N. (2008). *Food tourism in India*. Retrieved December 10, 2013 from the website http: // traveltips.usatoday.com/food-tourism-india-18335.html

Lin, Y. C., Pearson, T. E., & Cai, L. A. (2011). Food as a form of destination identity: A tourism destination brand perspective. *Tourism and Hospitality Research*, *11*(1), 30–48. doi:10.1057/thr.2010.22

Ministry of Tourism -MoT. (2019). *Tour guide Training Programs*. Retrieved from MoT website on 18 November, 2019, from http://tourism.gov.in/search/node/guide

Mohanty, P. (2014). Culinary Tourism In India: An Emerging Concept. *Abhinav Journal*, *3*(5), 44–53.

Murgado, E. M. (2013). Turning food in to a gastronomic experience: Olive oil tourism. In N. Arcas, F. N. Arroyo López, J. Caballero, R. D'Andria, M. Fernández, E. R. Fernandez, A. Garrido, J. López-Miranda, M. Msallem, M. Parras, L. Rallo, & R. Zanoli (Eds.), *Present and future of the Mediterranean olive sector* (pp. 97–109). CIHEAM.

Murugadoss, K., Nagarathinam, S. R., Vinothkanna, S., & Soundarraj, H. (2016). A Study on Tourism Potential in Munnar Region, Kerala. *IJIRSET*, *5*(12), 262–271.

Nguyen, H. L. & University of Western Sydney. (2015). *The Impacts of Tour Guide Performance on Foreign Tourist Satisfaction and Destination Loyalty in Vietnam*. Sydney, Australia: University of Western Sydney.

Nicholson, R., & Pearce, D. (2000). Who Goes to Events: A Comparative Analysis of the Profile Characteristics of Visitors to Four South Island Events in New Zealand. *Journal of Vacation Marketing*, *6*(3), 236–253. doi:10.1177/135676670000600304

Onetikk. (2013, January 31). Archive.Today: Himachali Cuisine, A Pleasantly Surprising Fusion [Web log post]. Retrieved from https://archive.is/20130216144406/http:// www.onetikk.com/blog/1358/himachali-cuisine-a-plesantly-suprising-fusion/

Ontario Ministry of Tourism. (2009). *OCHTP: Ontario Cultural and Heritage Tourism Product Research Paper*. Retrieved from Ontario Ministry of Tourism website on 05 April, 2018, from http://www.mtc.gov.on.ca/en/publications/Ontario_ Cultural_and_Heritage_Tourism.pdf

Plummer, R., Telfer, D., Hashimoto, A., & Summers, R. (2005). Beer tourism in Canada along the Waterloo-Wellington Ale Trail. *Tourism Management*, *26*(3), 447–458. doi:10.1016/j.tourman.2003.12.002

Rand, G. D., Heath, E., & Alberts, N. (2003). The role of local and regional food in destination marketing: A South African situation analysis. *Journal of Travel & Tourism Marketing*, *14*(3-4), 97–112. doi:10.1300/J073v14n03_06

Raut, A., & Bhakay, J. (2012). Wine tourism in Maharashtra: Problems and solutions. *Asian Journal of Food and Agro-Industry*, *5*(2), 141–155.

Reynolds, P. (1993). Food and Tourism: Towards an understanding of sustainable culture. *Journal of Sustainable Tourism*, *1*(1), 48–54. doi:10.1080/09669589309450700

Sahoo, D. (2014). Social Architecture of Ethnic food tourism in India. *Indian Journal of Tourism & Hospitality Management*, *5*(1), 53–66.

Sahoo, D. & Kurukshetra University. (2018). *Religious Gastronomy: A case study of Jagannath Dhaam in Puri*. Kurukshetra: Kurukshetra University.

Sahoo, D. (2019). Religious Gastronomy as a Tourism Attraction: A case study of Hindu Temple Food Mahaprashada. *Research Review: International Journal of Multidisciplinary*, *4*(5), 919–927.

Sahu, P. (2018, July 23). Medium: Fabulous Trip to Mini-Israel(Kasol) [Web log post]. Retrieved from https://medium.com/@piyurisahu/fabulous-trip-to-mini-israel-kasol-ce17b530c399

Sapra, A. (2019). *In Delhi Food walk's Chandigarh to Kasauli Food Tour* [Video file]. Retrieved on 12 December, 2019, from https://www.youtube.com/watch?v=9J8 V0mH1Dic&t=41s

Sapra, A. (2019). *In Delhi Food walk's Shimla Food Walk - Part 1/3 I Amazing Mandyali Dham (Himachali Thali)* [Video file]. Retrieved on 12 December, 2019, from https://www.youtube.com/watch?v=kwPE_SW3CR8

Sehgal, M. (2017, July 5). Why Israeli tourists in Kasol, called mini-Israel. *India Today*. Retrieved on 11 October, 2019 from https://www.indiatoday.in/india/story/india-pm-modi-israel-benjamin-netanyahu-kasol-mini-israel-himachal-kullu-israeli-tourists-1022439-2017-07-05

Sharlene, M. (2012). *The Motivational Factors of the Role of Food in Tourism*. Paper presented at Conference: The many faces of the global tourism experience in the 21st century', 2012, TSVC. Retrieved from www.travelconference.co.uk /commentries.php?paper=182#.W3KzZc4zbX5

Sharma, P., Sharma, P., & Kaith, M. (2014). Traditional food, fairs and festivals of Himachal Pradesh. *International Journal of Asian Academic Research Associates*, *1*(21), 384–390.

Son, A., & Xu, H. (2013). Religious food as a tourism attraction: The roles of Buddhist temple food in Western tourist experience. *Journal of Heritage Tourism*, *8*(2), 248–258. doi:10.1080/1743873X.2013.767815

Tanwar, M., Tanwar, B., Tanwar, R. S., Kumar, V., & Goyal, A. (2018). Himachali Dham: Food, culture, and heritage. *Journal of Ethnic Foods*, *5*(2), 99–104. doi:10.1016/j.jef.2017.10.006

Tavakol, M., & Dennick, R. (2011). Making Sense of Cronbach's Alpha. *International Journal of Medical Education*, *2*(1), 53–55. doi:10.5116/ijme.4dfb.8dfd PMID:28029643

Tetik, N. (2006). Analysis of Expectations from Professional Tourist Guide and customers in Turkey: The Case of Kuşadası. In F. Irigüler & M. E. Güler (Eds.), *Tourist Guiding: "Cinderella" of the Tourism* (pp. 203-220). Retrieved December 16, 2019 from www.researchgate.net/publication/310957726_Tourist_Guiding_Cinderella_of_the_Tourism

Thomas, J. (2019, February). *Typical Food of India*. Retrieved from https://traveltips. usatoday.com/main-food-india-11361.html

Timothy, D. J., & Ron, A. S. (2013). Understanding heritage cuisines and tourism: Identity, image, authenticity, and change. *Journal of Heritage Tourism*, *8*(2-3), 99–104. doi:10.1080/1743873X.2013.767818

Tourism Policy, H. P. (2019). *The Himachal Pradesh Tourism Policy-2019*. Retrieved from Himachal tourism website on 14 November, 2019, from https://himachal tourism.gov.in/wp-content/uploads/2019/09/Himachal-Pradesh-Tourism-Policy-2019.pdf

Tourism Statistics, H. P. (2019). *Himachal Tourism Statistics-2019*. Retrieved from Himachal tourism website on 16 November, 2019, from https://himachaltourism. gov.in/wp-content/uploads/2019/03/edited_Tourist-statastics.pdf

United Nations World Tourism Organisation. (2011). *Religious Tourism in Asia & the Pacific*. Retrieved from UNWTO website on 27 May, 2018, from https://cf.cdn.unwto.org/sites/all/files/pdf/110325_religious_tourism_excerpt.pdf

United Nations World Tourism Organisation. (2012). *Global Report on Food Tourism. AM Report: Volume Four*. Retrieved from UNWTO website on 02 January, 2018, from http://cf.cdn.unwto.org/sites/all/files/docpdf/ amreports4-foodtourism.pdf

United Nations World Tourism Organisation. (2017). *Second Global Report on Gastronomy Tourism. Affiliate members Report: Volume Sixteen*. Retrieved from UNWTO website on 21 March, 2018, from http://cf.cdn.unwto.org/sites/all/files/pdf/ gastronomy_report_web.pdf

Upadhyay, Y., & Sharma, D. (2014). Culinary preferences of foreign tourists in India. *Journal of Vacation Marketing*, *20*(1), 29–39. doi:10.1177/1356766713486143

Varsha, K. (2018). *Visit Panchgani And Be A Part Of The Strawberry Festival*. Retrieved on 5 November, 2019, from https://www.yatrablog.com/strawberry-festival-in-panchgani

Vasudeva, S. (2014, June 19). *Indian Food Tours: The Growing Trend of 2014*. Retrieved from http://cooks.ndtv.com/article/show/indian-food-tours-the-growing-trend-of-2014-539675

Visit Scotland. (2016). *Scotland of Food & Drink Survey report: Whisky Tourism – Facts and Insights, March 2015*. Retrieved from Visit Scotland website on 15 January, 2018, from http://www.visitscotland.org/pdf/Whisky%20 Tourism%20%20Facts %20and%20Insights2.pdf

KEY TERMS AND DEFINITIONS

Agro/Organic Tourism: A term adopted in the Himachal Pradesh Tourism Policy-2019 for promoting gastronomic tourism in the state. It basically refers to any agriculture-based activity that brings tourists to a vegetable farm, fruit orchard or a ranch, operating in the state.

Culinary Tour: A gastronomic tour comprising of several activities like tasting of local delicacies, tour of vineyards, fruit orchards and wineries, attending cookery classes, participating in food /wine events and cooking competitions etc.

Ethnic Restaurants: These are specialty restaurants (not necessarily fine dining) which serve ethnic delicacies of a region, primarily in a natural setting inspired by the local culture.

Food Guide: A person who possesses ample knowledge about the gastronomic aspects of a region, state or nation and escorts the food loving tourists to such places of interest that are gastronomically rich or popular among the culinary tourists.

Gastronomic Map: An illustrative map of a region/state/nation showcasing their major gastronomic delights and brief information about them.

Gastronomic Tourism: It refers to the "travelling of tourists to new destinations specifically for enjoying the local delicacies and learning new culture through these food preparations". It is alternatively termed as food tourism or culinary tourism.

Himachali Cuisine: The food culture of north Indian state of Himachal Pradesh which basically refers to the local ingredients, cooking techniques, regional delicacies and the serving methods followed in the state.

Chapter 9
Best Practices for Tour Guides at Christian and Islamic Religious Sites:
A Case Study of Jordan

Sawsan Khreis
Yarmouk University, Irbid, Jordan

Jean Michel Chapuis
University of Paris 1 Panthéon-Sorbonne, France

Mohammed Shunnaq
Yarmouk University, Irbid, Jordan

EXECUTIVE SUMMARY

Tourism is a major economic pillar for many countries; it enables them to increase gross domestic product, improves the population lifestyle, and preserves natural, cultural, and religious tourism assets. Tourism appears as a magic tool that could change the future of a developing country. However, tourism requires qualified human resources who can deliver high-quality service. Tour guides assume a great responsibility as they communicate with tourists during their sojourn. Their mission is even more critical if they have to accompany tourists to sacred tourist destinations where tourists are engaged in religious activities. The best practices of guides embedded with tourist behaviors are expected to enhance a peaceful world and provide a good touristic experience.

INTRODUCTION

This chapter is a case study in which Authors with empirical research insight from various perspectives: geographical, economic perspectives (two members of IREST (Institut de Recherche ET d'Études Supérieures du Tourisme) along with the anthropological perspective (Yarmouk University.

DOI: 10.4018/978-1-7998-3725-1.ch009

The uniqueness of this chapter stems from the new amalgam of perspectives that has never been discussed before. Moreover, the strength of this chapter results from the authors' familiarity with the area of study, the complexity of the relationship between variables, the necessity of conducting pilot studies at three different religious sites, the elaboration of an objective survey and the necessity of making participant observations followed by pilot studies.

Tourism is a major economic pillar for many countries. It can contribute significantly to the gross domestic product, help in diversification of national industries, generate jobs, and increase the income of stakeholders in the tourism sector, aid in preserving the natural, cultural, and religious tourism assets. Therefore, tourism is one of the vital global economic activities that influence world economy.

For the tourism industry to thrive, qualified human resources who can deliver high quality services are required (Baum, 2006; Mok, Sparks & Kadampully, 2013; Nickson, 2013). Hence, tour guides assume a great responsibility as they communicate with tourists during their sojourns. Their mission becomes even more critical if they have to accompany tourists to sacred sites, where they may engage in religious activities. The qualification of domestic tour guides, their education, historical and cultural knowledge, and comprehension of the tourists' desires play a crucial role in achieving better tourist experiences (Shunnaq & Ramadan Shunnaq, 2012; Esen & Gülmez, 2017).

This chapter aims at answering the following questions:

What are the roles of tourist guides who accompany international visitors to different religious sites? What follows is a description of the current practices of tour guides at religious sites frequented by tourists. This study examines whether tour guides are professional in that they remain unbiased in terms of religious beliefs and whether they tend to expose tourists to dual narratives. This research closely examines the multiple roles tour guides play in the sustainability of religious tourism in Jordan. It contributes to a more in-depth understanding of tourism practices by studying what happens during a visit to a site related to different religions.

The first part of the study presents the religious tourism sites in Jordan, the second suggests hypotheses about the roles of tour guides, and the third shows empirical results from the field. Therefore, the authors through this study, could give planners a clear picture reflecting the current situation of the tour guides' performance at three main religious sites in Jordan. Besides, the comparative approach will enable decision makers to identify the differences between the tour guides' practices at the religious sites.

BACKGROUND

Many online guides and publications are available that outline best practices for the tourism sector (Harrill, 2005; Kişi, 2019). Previous studies have almost exclusively focused on best practices for tourism development, sustainable tourism, destination management, and other areas of tourism activities. However, available literature on the best practices of tourist guides for religious tourism in the Middle East, and particularly in Jordan, is insufficient. Earlier studies have called for a wider investigation in measuring the importance of tourist guides and the quality of their performance as co-creators of the tourism experience in different contexts because these studies are not generalizable and largely inconclusive.

According to academics Raj and Morpeth (2015), who explored the relationship between tourism, pilgrimage and religious tourism, state that religion has been studied by different academic disciplines. Gradually, the emerging academic disciplines integrated religion. Based on Raj and Morpeth's analysis, during the European Association for Tourism and Leisure Education's meeting in 2003, many issues were highlighted, including the necessity *to explore and refine the sacred-profane continuum to the role of guides, interpreters and interpretations at religious sites.*

Anthropology, geography and economy are the main disciplines, which add value in this field .The first book on this topic was published in 1996, titled, 'Tourism and religion' by Boris Vukonic. Based on Stauberg's research, previous studies related to the study of religious tourism, used two approaches. The first approach covers the structural relationship between religion and tourism while in the other, academics wrote about the dynamic interferences between cultural domains. Finally, religious tourism according to Stauberg began to expand among catholic intellectuals.

According to the World Federation Tourist Guide Associations (WFTGA, 2003), a tour guide is *a person who guides visitors in the language of their choice and interprets the cultural and natural heritage of an area in which the person normally possesses an area-specific qualification, usually issued and/or recognized by the appropriate authority.*

This study addresses the largely unexplored area of the best practices for tour guides who operate tours to religious sites in Jordan. This study will demonstrate the vital role of these guides in furthering world peace.

Jordan is a unique country in many ways, since it has an abundance of sites that attract tourists interested in cultural tourism (Abuamoud, Libbin, Green, & Al-Rousan, 2014; Bourdeau, Gravari-Barbas & Robinson, 2016). Not only is it rich in historical and archaeological sites, but also rich in tourist attractions sacred to followers of the monotheistic religions. This study focuses on religious tourism in Jordan. It is one of the few countries in the region that prides itself on a unique and peaceful coexistence between Muslims, who form the majority of the population, and Christians. Located between Mecca and Jerusalem, Jordan has played a central role in the history of the Ahl-Ul-Kitab (People of the Book i.e. Jews and Christians). Prophets of the Abrahamic religions have either lived or travelled through the land. It is believed that the land of modern Jordan forms part of the "blessed neighborhood" mentioned in the Holy Qur'an (Sura 17, verse 1).

Religious tourism has a particular place in the Middle East region, as it is the land of prophets. Jordan is blessed with a rich religious history and is, therefore, a holy land for Jews, Christians, and Muslims. The country prides itself on being able to offer its tourists, a number of important holy sites connected to the roots of Judaism, Christianity and Islam. It is generally believed that the history of religious tourism in Jordan dates back to the 4th century, when the land was Christian in its entirety. Many large churches were constructed, as is evident from the ruins of churches found in Jerash, Madaba, Amman and Petra. Monastic life thrived in the wilderness too, as is evident from the many monasteries in the Jordan Valley near the Jesus Baptism Site known as A*l-Maghtas* (immersion) (Sayegh, 2011). Christian pilgrims also sought out the church built on Mount Nebo (Fischer, Khan, Khemani, Mak & Najmi, 2009). Religion is considered one of the oldest motivations for travel. There is evidence that Christian pilgrims have traveled since medieval times (Webb, 2002; Mylod, 2013; Jawabreh, 2017).

More recently, Jordan realized the potential of this niche market and is proud to have one of the most sacred sites for the Christian faith, namely, the Baptism Site of Jesus Christ. Besides the Baptism Site, there are other Christian sites such as Mount Nebo, St. John the Baptist Roman Catholic Church in Madaba, Umm Qays that overlooks the Sea of Galilee, among others. There are more than 100 Christian sites in Jordan that are mentioned by name in the Bible. Five of these holy sites were officially recognized as pilgrimage sites by the Vatican in 2000 (World Religious Travel Association). Being able to welcome religious tourists to its land, Jordan becomes the embodiment of comprehensive brotherhood, friendly coexistence, and mutual respect.

Not only does Jordan have important Christian and Jewish religious sites, but it is also rich in Islamic sites such as shrines located in the suburbs of Amman and north of the Jordan Valley. According to a study of the holy sites of Jordan, there are more than 40 Islamic sites that include the shrines of Prophet Mohammad's companions (PBUH) who died or were martyred and buried in Jordan (Fakhry, Piccirillo & Saqqaf, 1996). Figure 1 shows a number of important religious destinations for cultural tourists.

Figure 1. Map of Jordan. Source: Ministry of Tourism and Antiquities

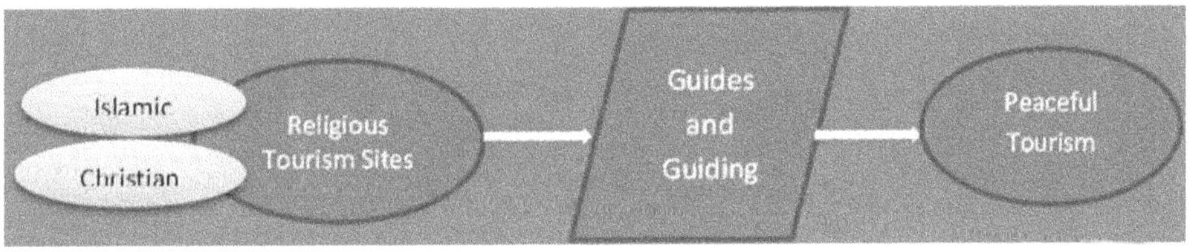

During the past decade, the tourism sector in Jordan suffered a sudden drop in tourist numbers. The drop was mainly due to the ongoing violence in neighboring countries. There were other reasons for the low tourist numbers too. Locally, the country was over-dependent on international tourists. The Jordan Tourism Board had been generally concentrating its marketing and development efforts on the satisfaction of international tourists. Additionally, tourism in Jordan is predominantly cultural tourism. This means that medical, religious and other types of tourism are niche markets that have not been prioritized until now. Religious tourism in Jordan is considered to be in its infant stage. Some research sheds light on this problem and suggests serious consideration of religious tourism as a good source of income (Briedenhann & Wickens, 2004; Kim & Chen, 2006; Licciardi & Amirtahmaebi, 2012). To develop the religious tourism sub-sector, the industry needs to focus on tour guides who are the drivers and an engine for supporting and enhancing visitors' experiences (Figure 2).

A BRIEF PRESENTATION OF RELIGIOUS SITES

Geographical Background, Tourism, and Religious Particularity

Before discussing religion and tourism, the study considers three sites namely, Petra, Mount Nebo and Baptism Site. Petra lies on the western slopes of the highlands overlooking Wadi Araba. The site is on the border of three bio-geographical zones, namely the Palearctic to the north, the Afrotropical in the south, and the Saharo-Arabian to the east and west. Mount Nebo is located to the west of Madaba and overlooks the mouth of the Jordan River and Jericho. The Baptism Site is located in the Jordan Valley, north of the Dead Sea and was designated a World Heritage site. For the majority of Christian denominations, this site is of immense religious significance

a. Petra

This ancient city is believed to be the one referred to in Luke 21:21-22 which says,

Figure 2. Vital role of tourist guides in religious tourism (Personal elaboration)

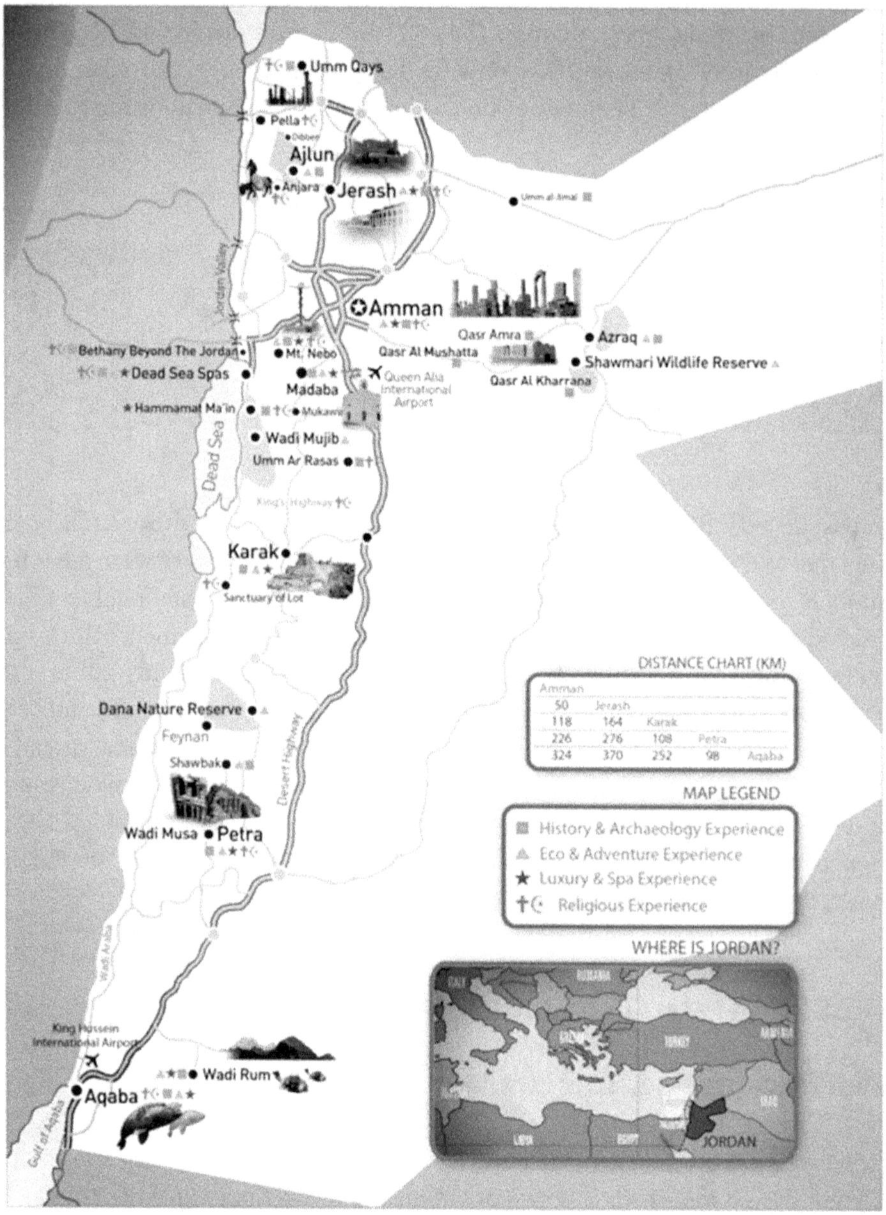

Then let those who are in Judea flee to the mountains, let those in the city get out, and let those in the country not enter the city. For this is the time of punishment in fulfillment of all that has been written.

Petra has long been an important city that not only welcomed travelers as a carrefour that relates trade routes extending from Far East of Asia and Europe, but also pilgrims. Epiphanius of Salamis, the bishop of Salamis, Cyprus (310-403) mentioned that in his day, Arab tribes from southern Palestine and

Transjordan performed a pilgrimage to a major sanctuary, the Equinox in Petra (Belmonte, Gonzalez-Garcia, Antón, & Betancor, 2019). Petra has retained its significance to date.

Presently, Petra is considered the barometer of the tourism sector's health in Jordan as the total number of visitors to Petra is the main indicator of tourism activity in the country. Petra has been listed on the world heritage list of UNESCO since 1985 and it was chosen as the second of the new Seven World Wonders in 2007. According to the Development and Tourism Region Authority (PDTRA), over one million tourists visited Petra and Aqaba in 2019. Figure 3 shows a map of Petra.

Figure 3. Plan of the Petra region showing the main sites discussed (Belmonte et. al, 2019), on an image courtesy of Google Earth's Baptism Site.

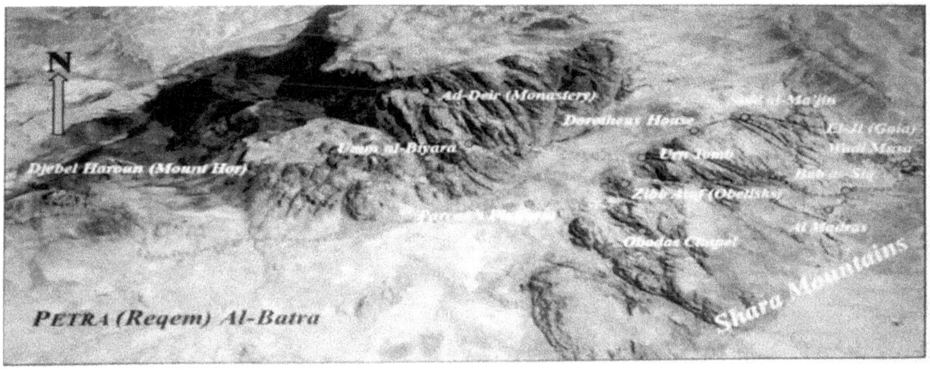

The Nabateans converted to Christianity in 423 A.D. after a Christian monk named Bar Sauma arrived in the pagan city of Petra with the intention of destroying Petra's pagan temples and Jewish synagogues. Petra had then suffered from a four-year drought and was relieved from its plight when it started to rain heavily upon the Christian monk's arrival. The pagan priests viewed this as a miracle, and some sort of supernatural intervention and converted to Christianity. With the conversion of the Nabateans, some tombs were turned into churches in and after 446 A.D (Konstantinos, 2001).

Petra has sites of old churches for Christian tourists and it also has significance in relation to the stories of Moses's brother, the Prophet Haroun, or Aaron, who according to the Bible (20: 22–29) died at the top of 'Mount Hor'. The prophets Moses and Aaron are recognized in all monotheistic religions. Mount Hor has been known as Jabal Haroun, one of the highest peaks of the mountains surrounding Petra, located 5 km south-west of the city of Petra. Jabal Haroun has most probably been a pilgrimage site for millennia. The biblical story of the Exodus has been attracting Christian pilgrims to the site since the late Roman period (Belmonte *et al*, 2019).

Pilgrimages continued in Islamic times as Muslim pilgrims also visited the shrine at the very top of the mountain. Muslim pilgrims were documented as early as the eighth century and have continued visiting the site to date. Until the twentieth century, some Bedouins in the vicinity of Petra continued to bury their dead in the Islamic fashion, although they broke from tradition by having the dead face Jabal Haroun instead of Mecca. Early twentieth-century ethnographic research carried out at El-Ji (Canaan, 1929) reported that twice a year, Bedouins of different tribes made a pilgrimage or *mawasim*, to the Prophet Aaron's sanctuary for prayers and offerings. One pilgrimage was undertaken in spring, however, the second more important one, called *Darb an-Nabi* Haroun (the path of Prophet Aaron), occurred during the grape season at the beginning of October (Al-Salameen & Falahat, 2009; Belmonte *et al*, 2019). This pilgrimage route from El-Ji (today Wadi Musa) to Jabal Haroun is plotted in red in Figure 3 in addition to some alternative routes to other sacred sites.

b. Mount Nebo

After 40 years of wandering in the Egyptian desert, Prophet Moses went up from the plains of Moab to Mount Nebo. There God showed him the Promised Land, and told him that he was only allowed to see it. Moses died in the land of Moab, however, his grave is unknown (Deuteronomy 34:6). The Christian monastery on Mount Nebo that is assumed to have been erected in AD 491–640 commemorates the place from where Prophet Moses viewed the Promised Land and the site of his burial.

Figure 4. Localization of Mount Nebo

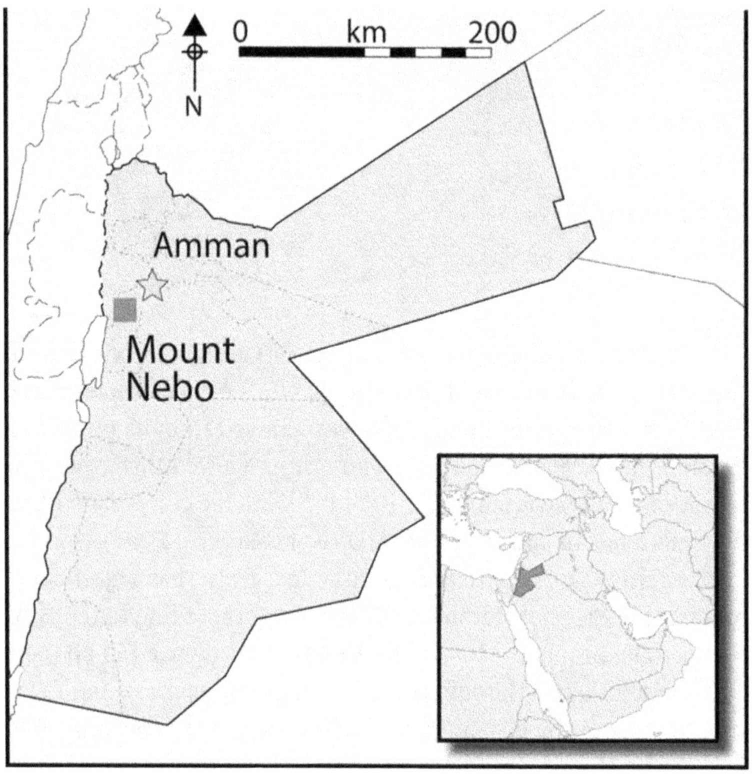

Until this day, Mount Nebo remains a vital destination for religious tourism associated with Prophet Moses. Mount Nebo has a strategic location as it is near Madaba, which is the closest town, and served as a crossroad for pilgrims and traders who passed between Egypt and Syria. The town is renowned for the outstanding value of its mosaic map of the Holy Land pilgrimage sites, which is the oldest mosaic map in the world. It is also one of the well-visited tourism destinations in Jordan.

c. The Baptism Site

As soon as Jesus was baptized, he went up out of the water. At that moment heaven was opened, and he saw the Spirit of God descending like a dove and lighting on him (Matthew 3:16-17)

The Baptism Site is located in the Jordan Valley, north of the Dead Sea and was designated a World Heritage site. For the majority of Christian denominations, this site is of immense religious significance. In A*l-Maghtas* in the Holy Land, God and His creatures meet in a journey of salvation. *Al-Maghtas* is the place where Jesus was baptized by John the Baptist in the Jordan River and/or in the springs around it. His baptism was symbolically an act of cleansing and one stage in his journey of salvation. The act also launched Jesus' public ministry. The Bible narrates that people used to go to John for baptism. They used to come from Jerusalem and Yahuda, and from the countries bordering Jordan. John recognized Jesus by inspiration from the Holy Spirit. Then he baptized Jesus of Nazareth in the river of Jordan.

Visiting the Baptism Site at the Jordan River and renewing the baptism promises has become an essential part of the worship rituals for many Christians. These religious rituals are habitually performed by biblical tourists especially at Pentecost and the remembrance of the descent of the Holy Ghost on Jesus and the Apostles (Sayegh, 2011). The Christian pilgrimage is always a time for prayer and reflection. On their way to the religious sites, the pilgrims are expected to view the Bible from geographical and historical perspectives.

Tour Guides as a Central Part in the Tourism Industry

Weiler and Davis's (1993) conceptual frameworks have drawn attention to tourist guiding roles. Their frameworks have been used as a basis to develop performance indicators. With the widespread adoption of sustainable tourism as an ideology, it is timely to acknowledge the importance of the roles tourist guides play in their work contexts. While specific guiding practices undertaken by a guide, such as a guide's use of heritage/nature interpretation or social/cultural mediation, are interconnected, they cannot be limited to one of the spheres outlined by Weiler and Davis (1993). The three spheres serve to collectively capture the key role dimensions of tour guiding. The tour guide has three roles: mediatory role (experience management), interpretive role (resource management) and instrumental role (tour management) (Weiler & Black, 2015). Professional tour guides play an essential role in forming the tourist's experiences of the sites they visit and should ensure the success of an authentic tourist experience (Harkin, 1995). It has also been pointed out that tour guides could modify tourist behaviors in that they persuade them, for instance, to be aware about environment issues (Pizam & Reichel, 1996; Chang, 2014).

This study selected managerial competence for investigation. The tour guide should know how to manage the tourist groups in terms of their distribution at the site, time management during the tourists' visit to the site, control of tourists' movements at the site, crowd management, and commitment to the tour program implementation. In addition to managerial roles, there are a variety of activities led by guides, and skills associated with these roles, including leadership and handling resources. Hirsh and Strebler (1994) identified three recurring features in the notion of competences. A competence is seen in the context of a particular job and the organization in which that job exists. Competences are positively associated with superior performance and can be described in terms of specific behaviors, which can be observed in the job. (Winterton & Winterton, 2002).

Tour guide qualification is a frequently-used term. All previous studies and stakeholders in the tourism sector are aware of the importance of providing qualified tour guides who are knowledgeable about the site in all its aspects; the local community, mastering of foreign languages, having verbal and non-verbal communication skills (Khreis, 2018). Moreover, many studies highlighted the outstanding role of the tour guide who is able to provide the visitor with an authentic experience, as he is the only person who could transform the tour into a tourist experience (Haynes, 2015). Based on this framework, two hypotheses are suggested at the end of this section and explored in this study.

Tour Guide Dimensions

The Ministry of Tourism and Antiquities (MoTA) and the Jordan Ammon Applied University College (AAUC) formerly known as Jordan Applied University College collaborate to offer tour guide programs and conduct examinations for participants to become licensed tour guides. The program duration is 6 months, divided into two semesters with a total of 360 hours. According to Article 4 of tourism law number 20 for the year 1988, tour guides should fulfill the following conditions. The applicant for a license to practice the profession of a tour guide must:

1. Have Jordanian nationality,aged between 22 and 45 years old.
2. Have good conduct.
3. Holds a first university degree or its equivalent from a recognized university or institute in the Hashemite Kingdom of Jordan.
4. Has knowledge of Arabic writing, reading and conversation.
5. Have passed the foreign language (to be used) exam, which enables him to practice the profession of tourist significance.
6. Have passed the qualification session organized by the Ministry.
7. Successfully complete the training requirements of the association, which are approved by the Ministry and which include field tours of not less than five rounds, accompanied by an experienced tour guide with at least five years of experience.
8. Have knowledge of tourism information and principles of tourism ethics as stipulated by the World Tourism Organization. Guides must also obtain a permit from the Ministry of Tourism and Archaeology to practice the profession, provided they meet the conditions stipulated in Tourism Law No. 20.

The authors consider a number of dimensions that, if applied, lead to a successful religious tourism experience. The tour becomes successful if the tour guide shows no bias to a particular religion and promotes mutual understanding between followers of different religions. The tour guide's attitude and performance during a tour to religious sites may contribute to an increase in the awareness of religious tolerance in the Holy Land. It may also boost the sense of spirituality of the place. It is of great importance that tour guides take into consideration all these dimensions on a religious tour. Any disregard of these dimensions may have an overall negative impact on the visitor's experience of the religious site. Finally, tour operators should be aware of the significance of the following:

1. The appearance of tour guides leaves an impact on the visitors' experience at the religious site. Appropriate clothes should be worn.
2. Signs at the sacred site help tourists comprehend historical, religious, and cultural information.
3. Religious symbols need to be interpreted with historical context.
4. Comprehensive knowledge about the religion, history, and possible archaeological data of the sacred site need to be communicated to the visitors.

5. Immersing visitors in the religious history of the site by creating an encounter that invokes past events, religious texts, or spiritual traditions.

Hypothesis One: Tour guides possess excellent verbal and non-verbal communication skills, in order to efficiently organize the tourists' activities at the site. According to Khreis (2018), integrating tourists in interesting activities according to their tastes and preferences contribute to creating a successful tourist experience. This leads us to hypothesis two that states:

Hypothesis Two: Tour guides not only take on the role of educators but are also neutral towards religion.

METHOD

The study relied on primary and secondary data collection. Data were collected through distribution of a questionnaire (Appendix) during the winter season that extended from 18 December**2019** to 20 February 2020. Each part of the questionnaire is designed to identify different competences of tour guides Section 1 contains eight questions, section 2 includes 12 questions, section 3 is composed of six questions, section 4 contains five questions and section 5 contains six questions. All questions are posed in a way that answers the study problem and clarifies ambiguity related to tour guides practices at religious sites. The authors distributed questionnaires directly among visitors to religious sites after the end of each tour, then questionnaires were collected to obtain the right impression and evaluation of the tour guides' performance. Participants were consent for being included in this study and approval from the organizational review board for conducting and publishing this study.

Questionnaires that aren't missing more than 20% of answers weren't excluded. This explains the difference of total number in each table. The difference of total number means that some visitors didn't answer some items.

Analysis

After collecting data from the returned responses, the authors used the Statistical Package for the Social Sciences (SPSS) to analyze the data with descriptive statistics. Secondary data are collected from literature related to religious tourism and tour guiding. The authors visited primary religious sites in Jordan to conduct open ended-interview questions with tour guides and religious travelers. They used a questionnaire to collect data from the religious travelers and used participant observation of the tour guides by registering for religious tour packages for both, Christian and Islamic religious sites.

In order to assess the reliability of the instruments, Cronbach's alpha value was calculated for each section. This value is considered acceptable if it is more than 0.60.

Statistical Treatment

After collecting data from the returned responses, the authors used SPSS (v23) to analyze the data. The authors used suitable statistical treatments for each question and hypothesis from the following tests:

1. Cronbach's alpha (α) to test reliability.
2. Percentage and frequency to describe the sample.

3. Arithmetic mean and standard deviation to answer the study questions.
4. Simple linear regression analysis with an F test
5. 5. Relative importance assigned using

$$\text{Class Interval} = \frac{\textit{maximum class - minimum class}}{\textit{number of classes}}$$

$$\text{Class Interval} = \frac{5-1}{3} = 1.33$$

According to calculations low degree <2.33, median degree from 2.34 – 3.66 and high degree ≥3.67.

FINDINGS

Socio-Demographic Variables of Visitors

In total, 228 questionnaires were distributed, of which 192 questionnaires were completed and returned. This represents the target sample, which encompasses international visitors of different nationalities who were selected randomly to evaluate the performance and skills of tour guides who accompanied them during the religious tour. Descriptive data for the socio-demographic variables of the tourists are shown in Table 1.

The reliability for each domain was tested by calculating Cronbach's alpha. Table 2 shows a Cronbach's alpha value of 0.77 – 0.96, and for the total questionnaire items, it was 0.98. This value is considered acceptable if it is more than 0.60.

Question One: Does the Tour Guide Have Management Competency?

To answer this question, we computed the mean and standard deviation for each item of managerial competence as shown in Table 3.

Table 3 shows the managerial competence level. The range of means for items in this domain is between 3.06 – 3.42, the highest mean is for item 'Tour guide assumes his ethical responsibility in that he respects group's preferences and beliefs' and the lowest mean is for item 'Overcrowding of visitors mitigates the value of touristic purpose.' The total mean was 3.27 by moderate agreement.

Difference Between Religious Sites

Table 4 shows the results of the one-way ANOVA for managerial competence level by religious sites. There is a statistically significant difference between religious sites for managerial competence (F=352.06, P<0.05).

Table 1. Socio-demographic variables of the tourists

Variable	Classification	Tourists	Percent %
Gender	Male	115	59.9
	Female	77	40.1
	Total	192	100.0
Education	Primary education	13	8.3
	Secondary education	18	11.5
	College degree	62	39.5
	Master's degree	51	32.5
	PhD degree	13	8.3
	Total	157	100.0
Monthly Income in US Dollars	<1000	17	10.8
	1000- 1499	14	8.9
	1500-1999	25	15.9
	2000-2499	25	15.9
	3000 3499	15	9.6
	3500- 3999	16	10.2
	More than 4000	45	28.7
	Total	157	100.0
Nationality	Australia	2	2%
	Brazil	29	26%
	Canada	2	2%
	France	6	5%
	Italy	1	1%
	Jordan	11	10%
	Mexico	4	4%
	Norway	2	2%
	Portugal	1	1%
	Spain	1	1%
	Switzerland	1	1%
	UK	3	3%
	USA	50	44%
	Total	113	100%
Religion	Atheist	2	2%
	Buddhist	1	1%
	Catholic	65	61%
	Christian	22	21%
	Muslim	10	9%
	Jewish	1	1%
	NONE	6	6%
	Total	107	100%

Table 2. Cronbach's alpha value for each domain

Variables	Cronbach's alpha
Managerial competence	0.96
Tour guide qualification (professionalism)	0.95
Tourists are satisfied	0.95
Further recommendations that will add a spiritual value to the religious tour	0.77
Total	0.98

Table 3. Means and standard deviations for each item of the 'managerial competence' level

Item no.	Statement	Mean	Standard Deviation	Order	Importance
3	Tour guide assumes his ethical responsibility in that he respects group's preferences and beliefs.	3.42	1.42	1	Moderate
1	Religious tour is well-organized in terms of time.	3.36	1.41	2	Moderate
7	Tour guide pay attention to security issues.	3.33	1.32	3	Moderate
2	Religious tour is well-organized in terms of program planning and implementation.	3.29	1.34	4	Moderate
8	Tour guide knows how to maximize visitor's satisfaction as he understands your preferences and needs.	3.25	1.36	5	Moderate
5	Time reserved to stay on religious site is sufficient to consume the place and enjoy your experience.	3.23	1.31	6	Moderate
4	Time reserved to stay on religious site is sufficient to do the religious rituals.	3.21	1.3	7	Moderate
6	Overcrowding of visitors mitigates the value of touristic purpose.	3.06	1.25	8	Moderate
Total		3.27	1.19	---	Moderate

Table 4. One -way ANOVA to test the mean difference for 'managerial competence' level by sites

Religious site	N	M	SD	F	P.
Petra	117	2.08	.587	352.06	.000
Mount Nebo	103	4.13	.686		
Baptism Site	59	4.11	.628		

N: number, M: mean, SD: standard deviation, F: F value, P: P-value

Table 5. Pairwise comparisons between religious sites

Religious sites	Petra	Mount Nebo	Baptism Site
Petra		2.04531*	2.03107*
Mount Nebo			.01423
Baptism Site			

Table 5 shows multiple comparisons between the religious sites using the Scheffe test. There is a statistically significant difference between Petra and Mount Nebo in favor of Mount Nebo, and between Petra and the Baptism Site in favor of the Baptism Site.

Question Two: Is the Tour Guide Qualified (Professional) in That he/she Fulfill all his/her Duties Towards Visitors?

To answer this question, we computed the mean and standard deviation for each item of the tour guide qualification (professionalism) (Table 6).

Table 6. Means and standard deviations for each item of 'the tour guide qualification (professionalism)' level

Item no.	Statement	Mean	Standard Deviation	Order	Importance
1	Tour guide is polite and respectful to visitor's religion and beliefs.	3.60	1.31	1	Moderate
2	Tour guide motivates visitors to become involved in peaceful tourism.	3.42	1.38	2	Moderate
8	Tour guide answers all questions posed by visitors.	3.41	1.36	3	Moderate
11	Tour guide has excellent communication skills.	3.41	1.27	4	Moderate
7	Tour guide is a good interpreter who masters well the history of this religious site.	3.36	1.39	5	Moderate
3	Tour guide enhances visitor's emotions to religion by presenting an impressive narrative.	3.34	1.31	6	Moderate
4	Tour guide's interpretations of the religious site align with visitor's beliefs.	3.25	1.27	7	Moderate
12	Tour guide encourage visitors to use all their senses to have a unique unforgettable experience recalling the history of saints and prophets.	3.24	1.24	8	Moderate
5	Tour guide exposes visitors to new information contradicting to your religious beliefs.	3.19	1.24	9	Moderate
6	Tour guide shifts opinions on behalf of peace in Jordan.	3.18	1.2	10	Moderate
10	Tour guide's behavior in this religious site differs from other tour guide's behavior in non-religious sites in Jordan.	2.78	1.13	11	Moderate
9	Tour guide talks all the time thus he hinders us from enjoying the spiritual moments and the splendor of the landscape.	2.64	1.25	12	Moderate
Total		3.23	1.03	---	Moderate

Table 6 shows the findings for 'the tour guide qualification (professionalism)' level. The range for this domain is 2.64 – 3.60. The highest mean is for the item 'Tour guide is polite and respectful to visitor's religion and beliefs' while the lowest mean is for the item 'Tour guide shifts opinions on behalf of peace in Jordan.' Total mean was 3.23 by moderate agreement.

Difference Between Religious Sites

Table 7. One-way ANOVA to test the mean difference for the 'tour guide qualification (professionalism')' level by religious sites

Religious site	N	M	SD	F	P.
Petra	117	2.23	.597	300.19	.000
Mount Nebo	103	3.97	.595		
Baptism Site	59	3.93	.518		

N: number, M: mean, SD: standard deviation, F: F value, P: P-value

Table 7 shows the results of the one-way ANOVA to test mean differences for the tour guide qualification (professionalism) level by religious sites. It shows a statistically significant difference between religious sites for tour guide qualification (professionalism) (F=300.19, P<0.05).

Table 8. Pairwise comparisons between religious sites

Religious sites	Petra	Mount Nebo	Baptism Site
Petra		1.74172*	1.70002*
Mount Nebo			.04170
Baptism Site			

Table 8 shows multiple comparisons between religious sites using the Scheffe test. The result shows a statistically significant difference between Petra and Mount Nebo in favor of Mount Nebo, and between Petra and the Baptism Site in favor of the Baptism Site.

Table 9. Means and standard deviations for each item of 'the tour finds success' level

Item no.	Statement	Mean	Standard Deviation	Order	Importance
3	Tour enhances mutual understanding between religions.	3.41	1.27	1	Moderate
1	Tour increases positive perception about the religious tolerance in Jordan.	3.34	1.22	2	Moderate
2	Tour changes perceptions towards Jordanian in that they respect visitor's beliefs.	3.31	1.23	3	Moderate
4	Tourism is a magic tool towards a peaceful tourism.	3.28	1.25	4	Moderate
6	Tour boosts a sense of the spirituality of the place.	3.25	1.25	5	Moderate
5	I don't feel that there is any bias to a particular religion even though the Christians form a minority in Jordan.	3.13	1.21	6	Moderate
Total		3.29	1.10	---	Moderate

Table 10. One-way ANOVA to test mean differences between 'the tour finds success' levels by religious sites

Religious sites	N	M	SD	F	P.
Petra	117	2.26	.634	239.42	.000
Mount Nebo	103	4.07	.717		
Baptism Site	59	3.96	.648		

N: number, M: mean, SD: standard deviation, F: F value, P: P-value

Question Three: Are Tourists Satisfied?

To answer this question, we computed mean and standard deviation for each item of 'the tour finds successes shown in Table 9.

In 'the tour finds success' level, the range of means for items in this domain is 3.13 – 3.41. The highest mean is for item 'Tour enhances mutual understanding between religions' and the lowest mean is for item 'I don't feel that there is any bias to a particular religion even though the Christians form a minority in Jordan.' The total mean was 3.29 by moderate agreement.

Difference Between Religious Sites

Table 10 shows the results of the one-way ANOVA for 'the tour finds success' level by sites. There is a statistically significant difference between religious sites for tour finds success ($F=239.42$, $P<0.05$).

Table 11. Pairwise comparisons between religious sites

Religious site	Petra	Mount Nebo	Baptism Site
Petra		1.81155*	1.70687*
Mount Nebo			.10468
Baptism Site			

Table 11 shows multiple comparisons between the religious sites using the Scheffe test. There is a statistically significant difference between Petra and Mount Nebo in favor of Mount Nebo, and between Petra and the Baptism Site in favor of the Baptism Site.

Question Four: Do These Recommendations add a Spiritual Value to the Religious Tour?

To answer this question, we computed the mean and standard deviation for each item of 'further recommendations that will add a spiritual value to the religious tour' as shown in Table 12.

Table 12. Means and standard deviations for each item of 'further recommendations that will add a spiritual value to the religious tour' level

No.	Statement	Mean	Standard Deviation	Order	Importance
3	Duration of tour should be extended	3.03	0.4	1	Moderate
2	Duration of stay should be extended to enjoy more the visit!	3.02	1	2	Moderate
1	Time dedicated to interpret the site along with its outstanding value is sufficient	2.98	0.67	3	Moderate
4	Tour guide shouldn't expose us to a dual narrative related to two religion beliefs (Islam and Christianity).	2.8	1.04	4	Moderate
5	Tour guide should take the Italian experience applied, in the Christian pilgrimage tours in leading religious visitors accompanied by a religious man and wearing like religious men.	2.73	0.96	5	Moderate
Total		2.91	.52	---	Moderate

Table 12 shows the 'further recommendations that will add a spiritual value to the religious tour' level. The range of means for items in this domain is 2.73 – 3.03. The highest means are for item 'How much time do you think should be dedicated to this visit?' The lowest means are for item 'Tour guide should take the Italian experience and apply it in the Christian pilgrimage tours in leading religious visitors accompanied by a religious man, respect the religious dress code exactly like religious men' The total mean was 2.91 by moderate agreement.

Table 13. One-way ANOVA to test mean differences for 'further recommendations that will add a spiritual value to the religious tour' level by sites

Religious site	N	M	SD	F	P.
Petra	117	2.59	.397	52.32	.000
Mount Nebo	103	3.13	.469		
Baptism Site	59	3.15	.469		

N: number, M: mean, SD: standard deviation, F: F value, P: P-value

Table 14. Multiple comparisons between religious sites

Religious site	Petra	Mount Nebo	Baptism Site
Petra		.54082*	.55599*
Mount Nebo			.01517
Baptism Site			

Table 15. Frequency and percentage for the socio-demographic variables

Variable	Classification	Frequency	Percent %
Gender	Male	115	59.9
	Female	77	40.1
	Total	192	100.0
Education	Primary education	13	8.3
	Secondary education	18	11.5
	College degree	62	39.5
	Master's degree	51	32.5
	PhD degree	13	8.3
	Total	157	100.0
Monthly Income in USD	<1000	17	10.8
	1000- 1499	14	8.9
	1500-1999	25	15.9
	2000-2499	25	15.9
	3000 3499	15	9.6
	3500- 3999	16	10.2
	More than 4000	45	28.7
	Total	157	100.0
Nationality	Australian	2	2%
	Brazilian	29	26%
	Canadian	2	2%
	French	6	5%
	Italian	1	1%
	Jordanian	11	10%
	Mexican	4	4%
	Norwegian	2	2%
	Portuguese	1	1%
	Spanish	1	1%
	Swiss	1	1%
	British	3	3%
	American	50	44%
	Total	113	100%
Religion	Atheist	2	2%
	Buddhism	1	1%
	Catholic	65	61%
	Christian	22	21%
	Islam	10	9%
	Judaism	1	1%
	NONE	6	6%
	Total	107	100%
Religious site	Petra	117	41.9
	Mount Nebo	103	36.9
	Baptism Site	59	21.1
	Total	279	100.0

Difference Between Religious Sites

Table 13 shows the results of the one-way ANOVA to test differences for 'further recommendations that will add a spiritual value to the religious tour' level by sites. There is a statistically significant difference between sites 'for further recommendations that will add a spiritual value to the religious tour' ($F=52.32$, $P<0.05$).

Table 14 shows multiple comparisons between religious sites using the Scheffe test. There is a statistically significant difference between Petra and Mount Nebo in favor of Mount Nebo, and between Petra and the Baptism Site in favor of the Baptism Site.

Question Five: What is Socio-Demographic Profile of Visitors?

To answer this question, we computed the frequency and percent for socio-demographic variables (Table 15).

Table 15 shows the socio-demographic variables of this study sample. There were 115 (59.9%) male tourists and 77 (40.1%) female tourists. The highest category for education was a college degree by frequency (62, 39.5%), and the lowest category by frequency was primary education and a PhD degree by frequency (13, 8.3% and 13, 8.3%, respectively). For monthly income (USD), the highest categories were USD 1500-1999 and USD 2000-2499 by frequency (25, 15.9% each) and the lowest category was USD 1000- 1499 by frequency (14, 8.9% and 14, 8.9%, respectively). By nationality, the highest number (50, 44%) was American (USA) and the lowest numbers were Italian, Portugal, Spain, Swiss, UK (1, 1% for each nationality). Catholic tourists were the highest in number (65, 61%) while the lowest frequency of tourists were Buddhist and Jewish (1, 1% for each religion). Considering the religious site of the tours, the highest category was Petra by frequency (117, 41.9%), and the lowest category was the Baptism Site (59, 21.1%).

Question Six: Give Comments Related to This Tour 'Exploring Reasons of Satisfaction or Disappointments of Visitors?'

1- Very great but some religious sites should look after cleaning especially Petra, because of the horses.
2- I'm very impressed with the guide, bus drivers' skills, tour sites, hotel accommodations meals.
3- Fantastic guide.
4- Jordan is a beautiful country.
5- The Jordanian people are more than wonderful; the police are always cooperative with tourists.
6- Jordan is a country of religions peace between Muslims and Christians.
7- My only issue was not getting enough information on what to do when getting food or other instructions when at the rest stops.
8- I was totally satisfied in Jordan.
9- Wonderful experience.
10- Very long drive to Petra.
11- Not a guide from this place.

The study reveals that the majority of visitors are highly educated middle-class American Catholics. According to the results, Mount Nebo occupies first rank, where tour guides are qualified as they fulfill their duties so they exemplify the best managerial competences. Therefore, tourists were satisfied. The Baptism Site occupies the second rank and the last-ranked site is Petra. This is expected as the majority of visitors ignore its religious value. Moreover, tour guides do not interpret its religious value as they focus only on the Nabatean civilization, which is considered a cultural tour, and presentation of its monuments. Visitors think religious tours need more time and they do not care about the attire of the tour guide. Visitors to the Baptism site are more interested in recommendations that add a spiritual value to the site. As it is the place where Jesus Christ was baptized, it has an outstanding value for them. After careful consideration of all these results, the authors had the following recommendations.

DISCUSSION

The data gathered in this study, shows that tour guiding impacts the value and emotions of religious tourism. Religious tourists need enough time to recall the past, to contemplate their feelings, to consider their spiritual experience, and to live the moment. The authors revised the dominant practices of tour guides in developed countries. Religious sacred sites are usually given high priority and visitors are given the chance to stay for days at the site to pray and practice their religious rituals. Thus, religious tours should be planned to permit visitors to choose to become pilgrims so that they can spend more time at the site. Moreover, visitors should not be exposed to dual narratives related to two religions in the chosen sites (Christianity and Islam).

The results for managerial competence level by religious sites show a statistically significant difference between the religious sites (F=352.06, P<0.05). The multiple comparisons between religious sites using the Scheffe test, show a statistically significant difference between Petra and Mount Nebo in favor of Mount Nebo, and between Petra and the Baptism Site in favor of the Baptism Site.

Is the Tour Guide Qualified (Professional) in That he/she Fulfills all his/her Duties Towards Visitors?

To answer this question, we computed the mean and standard deviation for each item of the tour guide qualification (professionalism) using a one-way ANOVA by sites. The result shows that there is a statistically significant difference between religious sites for the tour guide qualification (professionalism) (F=300.19, P<0.05). Table 7 shows the multiple comparisons between religious sites using the Scheffe test. The result shows a statistically significant difference between Petra and Mount Nebo in favor of Mount Nebo, and between Petra and the Baptism Site in favor of the Baptism Site.

Has the Tour Find Success?

The one-way ANOVA to test mean differences for 'tour finds success' level by sites shows there is a statistically significant difference between religious sites. For 'tour finds success' the multiple comparisons between sites using Scheffe test, the result shows there is a statistically significant difference between Petra and Mount Nebo in favor of Mount Nebo, and between Petra and Baptism Site in favor of the Baptism Site.

Do These Recommendations add a Spiritual Value to the Religious Tour?

There is a statistically significant difference between religious sites for further recommendations that will add a spiritual value to the religious tour. The result shows there is a statistically significant difference between Petra and Mount Nebo in favor of Mount Nebo, and between Petra and Baptism Site in favor of the Baptism Site.

Give Comments Related to this Tour 'Exploring Reasons of Satisfaction or Disappointments of Visitors?

12- **Majority of comments shows an admiration towards the religious sites, the country** Jordan is beautiful country, **the competence of tour guides "**I'm very impressed with the guide, bus drivers' skills, tour sites, hotel accommodations meals.", "Fantastic guide» and **the religious tolerance** "Jordan is a country of religions peace between Muslims and Christians" . **However, some visitors criticize the cleanliness of religious sites "**some religious sites should look after cleaning especially Petra, because of the horses", the accessibility "Very long drive to Petra" and sustainability issues "Not a guide from this place" and the tour guides "My only issue was not getting enough information on what to do when getting food or other instructions when stopping at the rest"

RECOMMENDATIONS FOR DECISION MAKERS IN TOURISM SECTOR

There were many obstacles while conducting this study, especially at the Baptism Site, which is managed under the patronage of a member of the Royal family. he authors have valuable recommendations, which should be considered by representatives of the Ministry of Tourism and Antiquities, and the Jordanian Tour Guides Association, in order to deliver a high quality service and a memorable religious tour.

1) Tour guides who accompany religious visitors need more training, especially for the Baptism Site. According to the results of this study, tour guides at Mount Nebo are more qualified.
2) Tour guides who are responsible for religious tours should be given additional training and taught etiquette as seen in Italian and French tour guides working with religious visitors.
3) Tour operators who organize religious tours to Jordan should extend the duration of these tours as it is recommended that they should be at least three days long. The authors reject the idea of visiting a sacred place like the Baptism Site for one hour in a hurry.
4) Religious tours must have priority over other types of tourism such as adventure tourism, as religious tourism is a promising market for Jordan.
5) Jordan should apply rigid requirements in the selection of tour guides for religious tours.

FUTURE RESEARCH DIRECTIONS

Further studies should be conducted in the future, using the same methodology to obtain a follow-up of the tour guides performance.

Comparative studies should be conducted to position Jordan as a niche market for religious tourism in comparison with similar competitor destinations such as Israel and Lebanon.

CONCLUSION

This research contributes to the theory of tourism by structuring the literature on religious tourism with the perspective of tour guiding. It also expands knowledge related to the world of tour guiding by understanding the role of tour guides in specific spiritual locations. The results are useful for stakeholders, planners and developers of this niche market. After conducting a pilot study, conducting interviews and participant observations, the results strengthen the role of tour guides in the experience of religious tourism. The framework and the results support arguments that religious tourism, if planned and developed, will help Jordan from the dilemma of future fluctuations in tourist numbers. Religious tourism is less affected by political situations or other external influences. Besides, the Tour Guide Association in Jordan will detect points of weaknesses related to the tour guides' performance in religious sites so that tour guides will avoid them in the future. This study will give stakeholders a profound understanding of the current performance of tour guides at religious sites and whether they fulfill their mission.

Tourism means peace, respect and mutual understanding. Tourism also encourages peace between nations and acceptance of the other, on the one hand, and peace encourages the flow of tourists to tourist destinations around the world, on the other hand. Jordan is considered an oasis of peace and security amid an atmosphere of conflicts and wars in the surrounding countries (Syria, Iraq and Israel). Political stability in Jordan has a direct impact on the prosperity of the tourism sector.

ACKNOWLEDGMENT

We would like to express our gratitude to Mr.Ali Dweikat, Mr.Awni Kawar, Mr,Wael Janineh, Mr.Ra'ed Haq and Mr.Ali Massadeh who facilitated our mission.

REFERENCES

Abuamoud, I. N., Libbin, J., Green, J., & ALRousan, R. (2014). Factors affecting the willingness of tourists to visit cultural heritage sites in Jordan. *Journal of Heritage Tourism*, *9*(2), 148–165. doi:10.1080/1743873X.2013.874429

Al-Salameen, Z., & Falahat, H. (2009). Religious practices and beliefs in Wadi Mousa between the late 19th and early 20th centuries. *Jordan Journal of History and Archaeology*, *3*(3), 175–204.

Baum, T. (2006). *Human resource management for tourism, hospitality and leisure: An international perspective*. Cengage Learning EMEA.

Belmonte, J. A., González-García, A. C., Rodríguez-Antón, A., & Betancor, M. A. P. (2019). Equinox in Petra: Land- and Skyscape in the Nabataean Capital. *Nexus Network Journal*, 1–23.

Bourdeau, L., Gravari-Barbas, M. M., & Robinson, M. (Eds.). (2016). *World heritage, tourism and identity: inscription and co-production*. Routledge. doi:10.4324/9781315546315

Briedenhann, J., & Wickens, E. (2004). Tourism routes as a tool for the economic development of rural areas—Vibrant hope or impossible dream? *Tourism Management, 25*(1), 71–79. doi:10.1016/S0261-5177(03)00063-3

Canaan, T. (1929). Studies in the Topography and Folklore of Petra. *Journal of the Palestine Oriental Society, 9*, 136–218.

Chang, K. C. (2014). Examining the effect of tour guide performance, tourist trust, tourist satisfaction, and flow experience on tourists' shopping behavior. *Asia Pacific Journal of Tourism Research, 19*(2), 219–247. doi:10.1080/10941665.2012.739189

Esen, F. Ö., & Gülmez, M. (2017). View of Academicians and Professional Tourist Guides on Domestic Practice Tours. *International Journal of Commerce and Management, 5*(7), 222–250.

Fakhry, Piccirillo, & Saqqaf. (1996). *The Holy Sites Of Jordan*. Turab.

Fischer, J., Khan, I., Khemani, T., Mak, D., & Najmi, R. (2009). Jordan tourism cluster. *Report For Microeconomics of Competitiveness, 8*.

Harkin, M. (1995). Modernist anthropology and tourism of the authentic. *Annals of Tourism Research, 22*(3), 650–670. doi:10.1016/0160-7383(95)00008-T

Harrill, R. (2005). *Guide to Best Practices in Tourism and Destination Management* (Vol. 2). The Education Institute of the American Hotel & Lodging Association.

Haynes, C. (2015). *The role of indigenous tour guides in providing authentic experience on guided tours in the Caribbean* (PhD dissertation). Pennsylvania State University.

Hirsh, W., & Strebler, M. (1994). Defining managerial skills and competences. Gower Handbook of Management Development, 4.

Jawabreh, O. A. (2017). Distinction of Jordan as a destination for religious tourism. *Journal of Environmental Management and Tourism, 8*(22), 1171–1182.

Khreis, S. (2018). Evaluation of social communication skills of the tour guide in Jordan as perceived by international tourists: an exploratory study in the archaeological site of Petra. *Dirasat, Human And Social Sciences, 45*(4).

Kim, H. J., Chen, M. H., & Jang, S. C. S. (2006). Tourism expansion and economic development: The case of Taiwan. *Tourism Management, 27*(5), 925–933. doi:10.1016/j.tourman.2005.05.011 PMID:32287716

Kişi, N. (2019). A Strategic Approach to Sustainable Tourism Development Using the A'WOT Hybrid Method: A Case Study of Zonguldak, Turkey. *Sustainability, 11*(4), 964. doi:10.3390u11040964

Konstantinos, D. (2001). *The World of the Nabataeans. Volume 2 of the International Conference: The World of the Herods and the Nabataeans held at the British Museum*. Academic Press.

Licciardi, G., & Amirtahmasebi, R. (Eds.). (2012). *The economics of uniqueness: investing in historic city cores and cultural heritage assets for sustainable development*. The World Bank. doi:10.1596/978-0-8213-9650-6

Mok, C., Sparks, B., & Kadampully, J. (2013). *Service quality management in hospitality, tourism, and leisure*. Routledge. doi:10.4324/9780203047965

Mylod, E. J. (2013). *Latin Christian Pilgrimage in the Holy Land, 1187-1291* (Unpublished Dissertation). University of Leeds.

Nickson, D. (2013). *Human resource management for hospitality, tourism and events*. Routledge. doi:10.4324/9780080966496

Pizam, A., & Reichel, A. (1996). The effect of nationality on tourist behavior: Israeli tour-guides' perceptions. *Journal of Hospitality & Leisure Marketing*, *4*(1), 23–49. doi:10.1300/J150v04n01_03

Raj, R., & Morpeth, N. D. (Eds.). (2015). *Religious Tourism and Pilgrimage Management: An International Perspective*. CABI Publishing. doi:10.1079/9781780645230.0000

Sayegh, B., Yagi, S., & Connell, K. (2011). *Christian Antiquities in Jordan*. Jordan Book Center.

Shunnaq, M., & Shunnaq Ramadan, S. (2012). Camels in Asia and North Africa: Interdisciplinary Perspectives on their Past and Present Significance. In Camel Trekking in Jordan: Between Oriental Perception and Authentic Tourist Attraction. Verlag der Österreichischen Akademie der Wissenschaften (Austrian Academy of Sciences Press).

Vukonic, B. (1996). *Tourism and Religion*. Pergamon, Elsevier Science Ltd, Tourism Social Science Services.

Webb, D. (2002). *Medieval European Pilgrimage, c.700–c.1500*. Palgrave. doi:10.1007/978-1-4039-1380-7

Weiler, B., & Black, R. (2015). *Tour guiding research: Insights, issues and implications* (Vol. 62). Channel View Publications.

Weiler, B., & Davis, D. (1993). An exploratory investigation into the roles of the nature-based tour leader. *Tourism Management*, *14*(2), 91–98. doi:10.1016/0261-5177(93)90041-I

Winterton, J., & Winterton, R. (2002). *Developing managerial competence*. Routledge. doi:10.4324/9780203020395

ADDITIONAL READING

Abumanssur, E. S. (2003). Turismo Religioso, (Paulo, S. Ed.) Ensaios antropológicos sobre Religião e Turismo, Papirus.

Cohen, E. (1985). The tourist guide: The origins, structure and dynamics of a role. *Annals of Tourism Research*, *12*(1), 5–29. doi:10.1016/0160-7383(85)90037-4

De Lima, I. B. (2016). Pivotal Role of Tour Guides for Visitors' Connection with Nature: Conceptual and Practical Issues. *International Journal of Humanities and Applied Sciences*, *5*(1), 18–22.

Graburn, N. (1977). Tourism: The Sacred Journey. In V. Smith (Ed.), *Hosts and Guests: The Anthropology of Tourism* (pp. 21–36). University of Pennsylvania Press.

Griffin, K., & Raj, R. (2017). The Importance of Religious Tourism and Pilgrimage: Reflecting on definitions, motives and data. *International Journal of Religious Tourism and Pilgrimage*, *5*(3), 2.

Pew Research Center. (2015). *The future of world religions: Population growth projections, 2010–2050.* Retrieved from http://www.pewforum.org/2015/04/02/ religious-projections-2010-2050/

Tsaur, S. H., & Teng, H. Y. (2017). Exploring tour guiding styles: The perspective of tour leader roles. *Tourism Management*, *59*, 438–448. doi:10.1016/j.tourman.2016.09.005

Wong, J. Y., & Lee, W. H. (2012). Leadership through service: An exploratory study of the leadership styles of tour leaders. *Tourism Management*, *33*(5), 1112–1121. doi:10.1016/j.tourman.2011.11.022

APPENDIX

Dear Guest,
You are part of a touristic trip in Jordan that includes the visit of religious sites. This survey is conducted for scientific purposes in order to develop a doctoral research. Your evaluation of the tour in religious sites is really valuable for this research.
Please enjoy your visit to your second home "Jordan".

V) What is socio-demographic profile of visitors?
What is your age? ______________
Your Gender?
a) Female
b) Male
Education:
a) primary education
b) secondary education
c) college degree
d) Master degree
c) Phd degree
Monthly Income in USD ?
a) less than 1000
b) 1000- 1499
c) 1500-1999
d) 2000-2499
e)3000 3499
f) 3500- 3999
g) More than 4000
Your Nationality? _______________
Your Religion? _______________

VI) Please give your personal comments related to this tour explaining reasons of satisfaction or disappointment.

Table 16.

No	Item	Totally Agree	Agree	Neutral	Disagree	Totally Disagree
I)	**Managerial competence**					
1	Rural tour is well organized in terms of time.					
2	Rural tour is well organized in terms of program planning and implementation.					
3	Tour guide assumes his ethical responsibility in that he respects group's preferences and beliefs.					
4	Time reserved to stay on rural destination, is sufficient to do the religious rituals.					
5	Time reserved to stay on rural site is sufficient to consume the place and enjoy your experience.					
6	Overcrowding of visitors mitigates the value of touristic purpose.					
7	Tour guide pay attention to security issues.					
8	Tour guide knows how to maximize visitor's satisfaction as he understands your preferences and needs.					
II)	**Tour guide qualification(professionalism)**					
1	Tour guide is polite and respectful to visitor's culture and prefernces.					
2	Tour guide motivates visitors to become involved in preparing local food.					
3	Tour guide enhances visitors to taste food while insisting on its healthy ingredients .					
4	Tour guide's interpretations of the history of traditional food is exciting in that it persuade tourists to enjoy eating it.					
5	Tour guide exposes visitors to new experience contradicting to your expectations.					
6	Tour guide shifts opinions on behalf of local community's hospitality .					
7	Tour guide is a good interpreter who masters well the history of this rural destination.					
8	Tour guide answers all questions posed by visitors.					
9	Tour guide talks all the time thus he hinders us from enjoying the spiritual moments and the splendor of the landscape.					
10	Tour guide's behavior in this religious site differs from other tour guide's behavior in non-religious sites in Jordan.					
11	Tour guide as excellent communication skills.					
12	Tour guide encourage visitors to use all their senses to have a unique unforgettable experience recalling the history of Saints and Prophets.					
III)	**Tourists are satisfied in that:**					
1	Tour increases positive perception about the religious tolerance in Jordan.					
2	Tour changes perceptions towards Jordanian in that they respect visitor's beliefs.					
3	Tour enhances mutual understanding between religions.					
4	Tourism is a magic tool towards a peaceful tourism.					
5	I don't feel that there is any bias to a particular religion even though the Christians form a minority in Jordan.					
6	Tour boosts a sense of the spirituality of the place.					
IV)	**Further recommendations that will add a spiritual value to the religious tour:**					
1	Duration of tour should be extended					
2	Duration of stay should be extended to enjoy more the visit!					
3	Time dedicated to interpret the site along with its outstanding value is sufficient					
4	Tour guide shouldn't expose us to a dual narrative related to two religion beliefs (Islam and Christianity).					
5	Tour guide should take the Italian experience applied, in the Christian pilgrimage tours, in leading religious visitors accompanied by a religious man and wearing like religious men.					

Chapter 10
The Social Significance of Religious Guides:
The Development of the Flow of Religious Capital in the Islamic Religious Tourism Industry

Shin Yasuda
Takasaki City University of Economics, Japan

EXECUTIVE SUMMARY

As religious tourism developed, some stakeholders have focused on the religious guide as a spiritual leader of the pilgrimage to supply religious knowledge and deepen the spiritual experience of the participants. In fact, the customers of religious tour operators seem to recognise religious guides as the important element in selecting a religious tour, and religious tour operators actively promote these religious figures as a significant element in their tour services. This chapter, therefore, considers the social context of religious guides in the Islamic religious tourism industry by mapping them in the structure of the marketplace from the perspective of the flow of 'religious capital'. Particular focus is placed on clarifying the flow of the religious guide's religious capital in the religious tourism industry, and the social networking emerging from the religious tourism industry through considering the novel use of financial and physical resources in the marketplace for religious tourism.

INTRODUCTION

The worldwide tourism industry has actively begun promoting related religious pilgrimage activities, which are usually based on charity and reciprocity among religious followers and community members, to enhance the commitment of various stakeholders in the field. The rapid expansion of religious tour operators and other tourism stakeholders, who organise religious tours to religious sites, has encouraged an expansion in the number of religious tours to these places and in the number of participants, regardless of religion. As a result, this new form of pilgrimage activity, which can be called 'religious

DOI: 10.4018/978-1-7998-3725-1.ch010

tourism', has become dominant in contemporary pilgrimage practices, visiting religious sites of any religion (Rinshede, 1992; Timothy & Olsen eds., 2006; Raj & Griffin eds., 2015; Olsen, 2018). Although some religious figures have condemned such developments, believing that they diminish the religious significance of pilgrimages, a majority of pilgrims have begun to participate in these tourism activities, which have become a part of their social life.

As religious tourism developed, some stakeholders have focused on the religious guide as a spiritual leader of the pilgrimage (they can also be called 'spiritual guides') to supply religious knowledge and deepen the spiritual experience of the participants (Shinde, 2010; Green, 2011, 2014; Reader, 2013; Kawashima, 2016; Yasuda, 2018). Some of the religious guides are firmly situated in a certain religious institution or religious hierarchy, while others are indigenous religious figures who sometimes remain ambiguous in any religious hierarchy. Although some of their activities are based on the traditional way of religious practices, other activities are recognised as innovative ones that go beyond the religious traditions. In this situation, people begin to choose their own religious guide according to his or her status in social life and religious preferences. In fact, the customers of religious tour operators seem to recognise religious guides as the important element in selecting a religious tour, and religious tour operators actively promote these religious figures as a significant element in their tour services to attract more customers. Thus, religious guides in religious tourism has become a significant topic to examine in this field of research.

In the case of Islam, the development of the religious guides has become a significant element in the practices of the pilgrimage to Makkah and other religious places. Many religious tour operators work competitively in the market to improve both the quality and the quantity of their services in the religious tourism industry (Yasuda, 2013, 2018). In this situation, some tour operators and religious guides have actively promoted religious tours to other fields like cultural tourism, leisure activities, and other entertainment, which are usually described as the core elements of secular cultural tourism in the international tourism industry, and not generally a destination for pilgrimage and religious tourism. As the tours of religious tour operators have expanded towards cultural heritage sites and leisure destinations, religious guides have begun to take on a significant role and have become popular in this type of tourism. This odd combination of religious guides and marketplace among religious tour operators has become a part of their core marketing strategy, and it seems to enhance the religious tourism industry as well as the self-enlightenment of their customers. In this sense, religious guides in the emerging religious tourism industry reflect the development of the market principle in the industry, and their marketing practices have a social significance by constructing new way of social meanings and contexts in Muslim societies. Describe the general perspective of the chapter. End by specifically stating the objectives of the chapter.

LITERARY REVIEW: MARKETPLACE AND RELIGIOUS CAPITAL IN THE RELIGIOUS TOURISM INDUSTRY

The development of religious tourism has influenced various aspects of pilgrimage activities according to the spread of mass tourism and the international travel market, and pilgrimage activities are commercialised or commodified by the expansion of the involvement of the tourism industry. Some studies on religious tourism focus on the transformation of traditional religious sites and activities in conjunction with the contemporary tourism industry, and they limit the impact of religious tourism to traditional religious spheres (Eade and Sallnow eds., 1992; Shackley, 2001; Kitiarsa 2010; Reader, 2013; Olsen,

2013). Thus, previous studies emphasise that religious tourism has spread into secular cultural tourism through the expansion of the activities of religious guides as spiritual leaders of tours.

Some studies focus on marketing approach and consumer behaviour analysis to clarify individual motivation and consumer demands for the religious tourism and religious guides, through the analysis of religious branding, advertising, and marketing of these commodities (Haq, 2010; Haq and Wong, 2010; Temporal 2011; Sandıkcı and Rice, 2011; Alserhan, 2017; Melewar and Alwi, 2017; Hall and Prayag eds. 2019; Jamal, Raj and Griffin eds, 2019). These researches focus on the consumption of religious commodities like products and services and the marketing strategy of suppliers that reflect consumers' personal motivation and religious preference.

On the other hand, some other researchers deepen the discussion of commodification of religion itself through the analysis of its social context, and show the transformation of pilgrimage and the development of religious tourism from the perspective of the concept of 'marketplace' (Einstein, 2007; Kitiarsa, 2010; Sinha, 2011; Reader, 2013). Reader (2013) indicates that *now pilgrimages are embedded in a context of markets, consumer activity, publicity and promotion, and how they operate not just in the marketplace but through it* (Reader, 2013, p. 8). He also describes how the commercialisation of pilgrimage promotes the involvement of new stakeholders like the tourism industry, governmental institutions and other related figures and organisations as well as religious ones, which leads to an 'intrusion of market forces' (Reader, 2013, p, 11). As a result, pilgrimage activities became heavily dependent on the tourism industry and its market principle, and pilgrimage activities are radically transformed into market activities, whereas they were previously described as activities based on charity and reciprocity among religious followers and community members. He indicates that *pilgrimage is located in and operates through the marketplace, and it is through the mechanisms of the marketplace that constructs such as the sacred are materialised and successful pilgrimage sites emerge* (Reader, 2013, p. 195). Kitiarsa (2010) also indicates that *the highly dynamic market economy and religion convergence is created through market mechanism, technological advancement, and global flows of people, capitals, and information, including religious symbols and institutions* (Kitiarsa, 2010, p. 564).

In the process of the emerging marketplace for pilgrimage activities, *taking religion as consumer goods has enabled individuals to personalize their piety and religious sensibility* (Kitiarsa, 2010, p. 570). In this situation, some entrepreneurs are able to *benefit from the often competitive nature of the pilgrimage service industry and its agents, whose very existence and success are highly dependent on their ability to attract such custom* (Reader, 2013, p. 115). In this situation, various entrepreneurs begin to offer innovative religious services in the pilgrimage marketplace, as Ian Reader shows, and this has transformed the traditional uses of financial and physical resources in religious capital. In this sense, some researchers focus on the importance of religious guides as entrepreneurs in religious tourism (Shinde, 2010; Yasuda, 2013, 2018; Kawashima, 2016). Kiran Shinde indicates that they are *serving as indigenous 'religious entrepreneurs' who engage in religious tourism to shape the "inner experience" of their customers; they direct financial and physical resources in novel ways and create new economic combinations by introducing products and services, thus opening new markets and reorganising the industry* (Shinde, 2010; Yasuda, 2018, p. 21), as well as those acting as 'spiritual leader' or 'spiritual guides' to deepen their religious capital, which can be defined as *[the] degree of mastery of and attachment to a particular religious culture* (Stark & Finke, 2000, p. 120).

As the theory of religious economy and the discussion of religious capital conceptualise, the concept of 'religious capital' is described as an individual's investment to embody *in terms of spiritual knowledge, abilities, skills, tastes and other dispositions in the individual level* (Schilderman, 2011a, p. 43).

When introducing the market model approach, *the investment in religious capital is a function of the perceived benefits as return on investment* (Schilderman, 2011b, p. 183). However, religious capital is located in the contexts shared among the surrounding stakeholders. Hans Schilderman mentioned that *it is also evident in religious art, texts, and objects of devotion, or institutionalized in churches, convents, seminaries and the like* (Schilderman, 2011a, p. 43).

Religious capital is created through the consumption of new 'products' and 'markets' on the supply side, which respond to the demands of specific religious values (Finke & Iannaccone, 1993). Green (2014) argues that *the heart of religious economy is the transactional process of exchange between producers and consumers of religious goods, services and values* (Green, 2014, p. 7). This kind of exchange emerges through the negotiation and communication between two or more parties within a given market terrain' (Green, 2014, p. 7). In sum, religious capital brings to light the interactive dynamics of religious life through negotiation among the stakeholders (Green, 2014, p. 8).

In this situation, religious capital *stipulates its character as a commodity that has a certain value in exchange, and therefore may contribute to some shared or common good* (Schilderman 2011a, p. 46). *The presence of religious goods and some enormous efforts to 'sell' religious faith via market and media channels have demonstrated the growing mundaneness of religious teaching and belief at the outset* (Kitiarsa, 2010, p. 569). By negotiating with these religious guides, therefore, followers have been encouraged to change their financial and physical resources in the provided context. Thus, religious guides as spiritual 'entrepreneurs' or 'firms' seem to encourage the specific flow of religious capital through recontextualising financial and physical resources into novel ways in the marketplace through providing religious knowledge and spiritual experiences, and to relocate religious knowledge and spiritual experience in the market principle (Green, 2011, 2014).

Green (2014) conceptualises that religious entrepreneurs and firms promote 'adaption' and 'innovation' in the process of social interaction between producers and consumers, and create a new public sphere, which embodies the social interactions in the marketplace, through the development of organisations, theologies, books, revenues, schools, shrines, rituals, networks, and partnerships. This marketplace as a public sphere promotes competition, emulation, and adaptation among stakeholders, and *religious fusions and hybridity are particularly characteristic of intensified market interactions* (Green, 2014, p. 11). As a result, religious entrepreneurship creates new forms of consumer flows of time, money, and social networks in the market through the redistribution of customers' resources and networks, as entrepreneurs redistribute the actors, resources, and networks into new forms.

In this situation, religious guides can be described as a social existence embedded in Muslim societies. As Hans Schilderman indicates, *religion has characteristics of religious dispositions of persons and their associations in groups and communities at the micro-level, but it also manifests itself as mediated by organisations and institutions at the macro-level* (Schilderman, 2011a, p. 40). Moreover, religious guides are strongly embedded in the market principle in the religious tourism industry. As the activities of religious guides are provided in the religious tourism industry, their practice adheres to the social context in the market, such as norms, knowledge, and experience. As the discussion of religious capital shows the importance of *habitus* as French Sociologist Pierre Bourdieu mentioned (Verter, 2003; Schilderman, 2011a, 2011b), social contexts like shared and common good in the market strongly regulate the nature of individual behaviours, interactions, and communication among the stakeholders in the field.

Moreover, the value of products and services are recognised in the context shared among stakeholders like customers and producers of the services, which can be described as the concept of 'value-in-context'. As Vargo and Lusch (2008) and other related studies indicate the importance of value-in-context in service-dominant logic[1] (Vargo and Lusch, 2008; Chandler and Vargo, 2011; Eid and el-Gohary, 2015), the consumption of products and services in the marketplace reflects and forms shared contexts among the stakeholders through the continuous negotiation among them. In this situation, every commodity gains different meanings in the process of negotiation, and the consumption practice is regarded as the making process of social contexts. In this sense, religious guides embody shared religious capital as 'value-in-context' based on the market principle and social negotiations among the stakeholders, which seem to make religious guides socially significant in the religious tourism industry.

Although religious guides reflect the social meanings and contexts embedded in the marketplace, previous studies on religious guides emphasise the importance of the micro-level relationship between religious guides and their followers in the perspective of the concept of religious capital. As studies have mentioned, religious guides are strongly embedded in the social contexts in the marketplace, which emerged from continuous interaction with the stakeholders in the marketplace. In this situation, considering religious capital as value-in-context in the religious tourism industry is the significant perspective in the discussion of religious tourism.

MAIN FOCUS OF THE CHAPTER

This chapter considers the social context of religious guides in the Islamic religious tourism industry by mapping them in the structure of the marketplace from the perspective of the flow of 'religious capital'. Particular focus is placed on clarifying the flow of the religious guide's religious capital in the religious tourism industry, and the social networking emerging from the religious tourism industry through considering the novel use of financial and physical resources in the marketplace for religious tourism.

The methodology is a conceptual study using a qualitative content analysis approach. It is based on the flow of religious capital from the perspective of qualitative social network analysis (Ahrens, 2018), which reflects the relationship among actors surrounding religious guides in the marketplace, through collecting their representations in written documents like related travelogues, journals, magazine articles, leaflets, and posts on SNS (social networking services) inside and outside Islamic world such as Malaysia, Indonesia, UAE, UK and more. Moreover, participant observation of religious tours has also been adopted to sustain the research topic.

The chapter first reviews the literature on religious guides in religious tourism in order to conceptualise their social significance in the marketplace. Religious capital in the marketplace is a special focus. Then, the development of the religious tourism industry in Islam and the activities of religious guides in the field are described. From the description, the social context of religious guides in the religious tourism industry is determined by focusing on the consumption in the marketplace as value-in-context practice and on the emerging public sphere which can be called the 'Islamic cultural sphere' based on social context as value-in-context in the market.

The Development of the Religious Guide in the Islamic Religious Tourism Industry

In the case of Islam, the development of religious tourism can be seen from the middle of the twentieth century. Thanks to the development of the international aviation network and motorisation in the 1970s, the pilgrimage to Makkah (*hajj* and *umrah*) and religious visits to religious places (*ziyara*) became popular among Muslims, and some religious places became famous beyond its community members. For instance, Sunni people also organised such religious journeys to Madina, Jerusalem, Konya, and other related places to obtain their religious capital (Timothy & Iverson, 2006; Kessler, 2015, Kessler & Raj, 2018; Olsen 2018). Especially, some tariqa (Islamic school of Sufism) have a long tradition of religious journeys to visit their related places for *tasawwuf* (Sufism); the members have actively organised religious tours to the various Islamic places. Moreover, Shi'ite people has long tradition for making *ziyara* toward religious places in Iraq, Iran, Syria, and Lebanon, and their activities has influenced local politics and economics (Cole, 2002; Yasuda, 2013). This religious mobility based on religious motivations has had significant meaning for the establishment of 'Islamic tour operators' that facilitate people visiting these religious places by providing religious tours as 'commodity' in order to satisfy the religious needs of their customers. Some entrepreneurial figures inside and outside the Muslim world have actively participated in the management of religious tourism and the establishment of Islamic tour operators, and their practices have gradually influenced pilgrimages and religious visits as a whole.

The change in international religious mobility is also driven by the strong demand for the expansion of such tour operators to involve the pilgrimage to Makkah and other religious visits. Due to the rapid expansion of the number of pilgrims to Makkah from the 1970s, the Saudi government introduced a hajj quota system in 1988 that *permits every country to send one thousand pilgrims for each one million people enumerated in its national census* through the issuance of pilgrimage visas (Bianchi, 2013; Sardar, 2014, p. 336). In this environment, governments became involved in the pilgrimage to Makkah through the Organisation for Islamic Cooperation (OIC) countries, and various governmental institutions such as ministries of religious affairs and religious organisations assisted the operation by providing large budgets to reduce the cost to pilgrims (Bianchi, 2004, 2013; Sardar, 2014). However, the expansion of the number of pilgrims placed an enormous financial burden on OIC countries. In this situation, some countries allowed a number of hajj-visas to be used by private companies in order to reduce the cost of the operation. As a result, the pilgrimage to Makkah also became a 'commodity' in the market, promoted by private tour operators in the 1990s, and their activities had a strong influence on pilgrimage activities (Bianchi, 2004, 2013).

As there is little governmental regulation for the content management of these operators, the number of Islamic tour operators and their package tours have become dominant in the pilgrimage activities in religious places. As a result, there was strong competition in the market, which created a variety of services, and the customers could choose their favourable services that fit their religious preferences. In this highly competitive environment, Islamic tour operators have struggled to gain high customer satisfaction and a good reputation through their management strategy, by focusing on improving the quality and quantity of their religious and touristic services. Especially, Islamic tour operators emphasise the importance of religious services, which is the core element of their religious tours, and innovate their religious services through communication with their customers. As a result, religious services provided by Islamic tour operators reached a certain consensus and norm in the market through the interaction among stakeholders (Yasuda, 2013).

Figure 1. The Office of Islamic Tour Operator
Source: Photo by author, Sharjah in UAE, February 2020

On the other hand, Islamic tour operators have encouraged cooperation among stakeholders to reduce transaction costs and to ensure the efficient use of their management resources. Some Islamic tour operators reach cooperation agreements and revitalise the exchange of their management resources like finance, human resources, and management knowledge. Through the development of cooperation as well as competition among Islamic tour operators, these agencies became the hub of the network, and formed a firm industry, which can be called the 'religious tourism industry' in the Islamic world (Yasuda, 2013).

In this situation, most of the Islamic tour operators adopt religious guides as the symbol of their religious tours and the core of their management resources (Mzakki, 2008; Rudnyckyj, 2009a, 2009b, 2010; Hoesterey, 2015). These religious guides are called *ustaz*, *shaykh*, *murshid*, or other names related to 'leader' or 'spiritual leader' in the Islamic context, and they accompany clients on the religious tours to provide religious knowledge and experience which lead to individual religious capital (Hoesterey, 2015). Although they play a central role in the religious tours, most of them are not employees of the company; rather these figures are independent, and are providing religious services like sermons and they preach in public spheres like publications, TV and other mass media, the internet, religious events, and seminars. In this sense, the work of the religious guide is to some extent described as extra revenue in compensation for providing their religious capital (Hoesterey, 2015).

The backgrounds of these religious guides vary; some are religious figures who read higher Islamic education in university and madrasa (religious school), others are figures who accumulated experience of religious journeys, or have some other religious ability to attract customers without reading Islamic higher education. For instance, AA Gym (Abdullah Gymnastiar[2]), a popular TV preacher in Indonesia, is one of the figures who gained a high reputation among Indonesian Muslims for his high religious capital such as spiritual knowledge and experience, even though he has no Islamic higher education (Hoesterey, 2015). He is constantly accompanying religious tours to Makkah and other religious places organised by MQ Travel[3] and interacts with the customers during the travels (MQ Travel, 2020). Some other famous religious guides in the religious tourism industry gain a high reputation among their customers due to the deep religious capital they have without having an Islamic higher education.

Figure 2. Digital Leaflet of Religious Tour with Religious guides
Source (Ziyarah Tours, 2020)

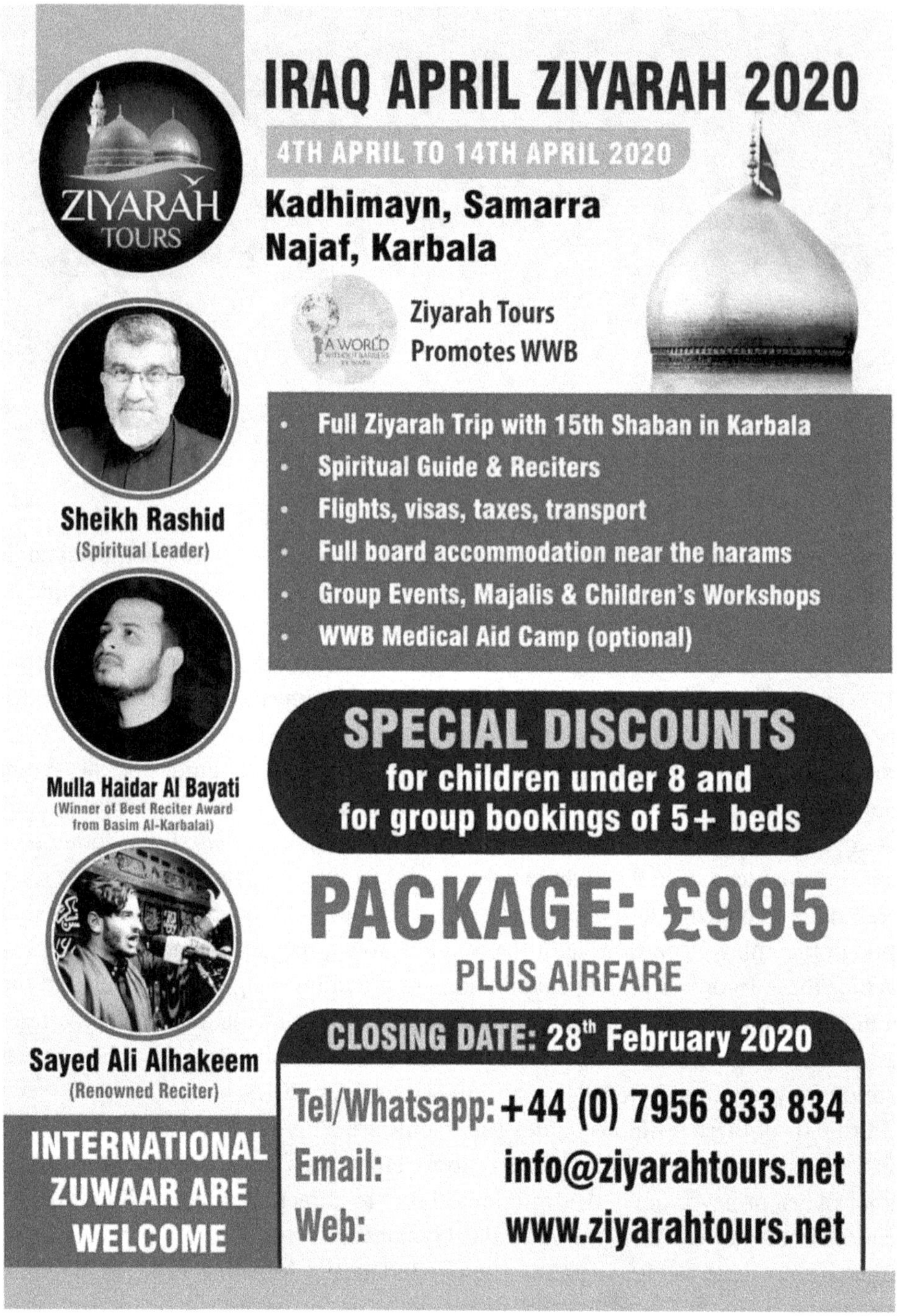

As for their activities during religious tours, most of the religious guides do not conduct travel services like guiding the participants and negotiating with other stakeholders. Rather, they conduct religious services during the tours; holding religious sermons and seminars, conducting other religious programmes, and communicating with their customers. In their activities, religious guides use their religious capital like knowledge and experience to develop their customers' religious capital. These figures, therefore, act as religious guides for the Islamic tour operators to share their religious capital with their customers.

Figure 3. Religious Tour Leaflet with Religious guide
Source (MQ Travels, 2020)

Management Strategy, Market Norm, and the Flow of Religious Capital

Although the activities of religious guides in the Islamic religious tourism industry are heavily dependent on their own religious capital, these interactions between religious guides and their customers are strongly regulated by the market norms among the stakeholders. Especially, religious guides are strongly dependent on the management strategy of Islamic tour operators; their activities are provided as a 'commodity' like religious and other types of services for the consumption in compensation for their owning religious capital, as studies of the theory of religious economy show (Stark & Finke, 2000; Schilderman, 2011). As a result, religious guides and their tours are priced through the demands of their customers. In this situation, the rank of religious guides become visible through the price of their religious tours. Moreover, the customers' need of religious tour is strongly dependent on the religious capital of religious guides that customers want to share with.

As religious guides reflect the quality of management resources as well as the religious capital that an Islamic tour operator holds, some Islamic tour operators use religious guides and their religious capital beyond religious tours to firm up their portfolio. In this situation, religious guides widen their activities beyond religious tourism to use their religious capital on their customers. As Islamic tour operators have strongly promoted religious tours for their essential tour products, they have also encouraged to conduct cultural tours to Islamic and other cultural heritage sites, as well as leisure travel tours to resorts and other entertainment facilities from the beginning of their establishment. For instance, Tabung Haji Travel and Services[4], a Malaysian government Islamic tour operator, actively organises cultural tours and leisure travel as well as *hajj, umrah,* and *ziyara* tours from 2012, as the rise of customers' needs to enjoy another type of tours in their life (*New Streits Times*, 19 Mar 2014). They organise package tours to Bali Island in Indonesia and other resorts for leisure and entertainment activities, tours to watch football matches outside the country, as well as cultural tours to Turkey, Spain, Bosnia and Herzegovina, Jordan, and Egypt. Some of the cultural tours include prominent *ustaz* (religious guides) to use their religious capital to ensure the fulfilment of customers' needs.

Figure 4. Spiritual Tours to Ache in Indonesia with Ustaz
Source (TH Travel, 2020)

In the case of MQ Travel in Indonesia, the tour operator is actively organising cultural tours to Spain, Turkey, and Uzbekistan with AA Gym and other religious guides with whom the company has a relationship. On these tours, customers enjoy the communication and interaction with a religious guide as well as the religious experiences, as is evidenced by what the customers of MQ Travel as well as the tour operator share on SNS like Facebook, twitter, WhatsApp and other services. Their customers also use these devices, and also communicate in the digital spaces.

These examples show that the rank of religious guides, which is based on the religious capital the religious guides seem to own, strongly influences the customer trend and the price of their services and tours. This customer trend indicates that the tendency shows the content of religious capital that customers require to gain through the consumption of religious tours. In this sense, religious guides are strongly embedded in the market principles of the religious tourism industry.

Religious Investment and Social Context in the Value-in-Context Concept

As the religious guides are embedded in the market principles of the religious tourism industry through the demand for religious capital according to customers' preferences and needs, the rank of religious guides strongly reflects the need for religious capital in the market. In this situation, the consumption of religious products and services assures more than consumption, rather it is described as an 'investment' to obtain religious capital and sustain the rank of religious guides.

In this situation, Islamic tour operators have actively invested their collected recourses from their customers in religious and social practices in the communities they belong to from the beginning of their establishment. For instance, Tabung Haji Travel and Services invests revenue from its customers in religious activities and rural community development in Malaysia as well as in the development of pilgrimage services in the country from the perspective of Islamic financial scheme (TH Travel, 2020). In this scheme in cooperation with Islamic finance institutions, Malaysia has developed various industries like agriculture, hospitality, and other sectors as well as religious services through investing collected resources into the field. Some other local Islamic tour operators are active to invest their profits in community development activities like poverty reduction, improvement of public health, and the development of religious education and activities (Yasuda, 2018). Through the investment in community activities, Islamic tour operators have encouraged community members to get involved in the community's religious activities for creating shared or common good. These investment activities, therefore, can be described as creating social context by assuming invested resources in the Islamic context.

This type of investment directs a new flow of individual financial and physical resources in novel ways and creates a new economic field. As a result, their consumption practices and expended resources like time, money, and the physical expense of customers obtain a social context and meaning through connecting a social network in the market, which is described as 'investment' rather than 'consumption'. In their narratives, the stakeholders like producers and customers of religious services emphasise that the consumption of these services are for the sake of physical and spiritual self-enlightenment for the future by deepening their religious capital, and it is also the reflection of the result of physical and spiritual self-enlightenment in the past, which physical and spiritual success is the result of the past investments. In the narrative of religious guides, they frequently mention the rewards of spiritual investment for the better life in this world as well as the afterlife (Rudnyckyj, 2009a, 2009b, 2010).

In this sense, consumers in the marketplace approve the piety and morality reflected in the market products through the consumption of the products. Moreover, the social interaction between religious guides and consumers in the marketplace links the consumption of individual time, money, and social networks to leisure activities, which results in shared experiences among consumers in the Islamic context. This promotes strong communication and social ties among the experienced consumers and leads to new models of lifestyle and communities based on the consumption experiences (Haenni, 2009). Leisure culture is therefore *a feeling related to shared sensibilities about appropriate morals and social behaviour, perceptions of class congruity, a sense of security linked to safety from physical and verbal harassment, and territorial belongings* (Deeb & Harb, 2013, p. 9). The Islamic way of culture in the religious tourism industry can be described as an important social practice that 'sanctifies' individual leisure experiences as being Islamic, as evidenced in the religious practices of *halal* (lawful) and *haram* (unlawful) in the traditional way of Islamic contextualisation of practices (Izutsu, 2002, p. 235). The investment of the stakeholders, therefore, reflect religious context embedded into a social network such as shared or common good in the market. In this situation, this new capital flow forms a certain consensus among the stakeholders which creates social contexts promoted by religious guides and Islamic tour operators. For creating social context in the market, some Islamic tour operators use religious guides to show this social context to direct capital flow based on the investment for religious capital.

Making Social Contexts of Religious Capital in the Islamic Cultural Sphere

In their narratives and activities, the religious guides stress the novel way of leisure and entertainment in the context of Islamic piety. Although some studies show that leisure and entertainment is *a tension between hedonism and reinvigorated religiosity, and from that tension we see the emergence of multiple lifestyles within Muslim societies, expressed and produced through consumption* (Haenni, 2009; Deeb & Harb, 2013, p. 217), Islamic principle is *an essentially positive view of leisure and recreation* (Ibrahim, 1982; Martin & Mason, 2004, p. 5; Bayat, 2007; Van Nieuwkerk, 2008, 2013). In this situation, the micro-level practices of religious guides create an Islamic morality that fits contemporary society and promotes new forms of lifestyles for them through the creation of social context as value-in-context. Accordingly, religious guides *must continue to negotiate with consumers to form morally acceptable practices and to promote 'adaption' and 'innovation' that fits with their own cultural practices* (Yasuda 2018, p. 27).

As the religious tourism industry has widened their activities into various field through the production of social context as value-in-context in the market, religious guides continually recontextualise the travel experience as tools for self-enlightenment within the religious narratives, with their social context embedded in Islamic piety. These activities can be described as the spread of the 'Islamic cultural sphere' through the growing influence of Islamic sensibilities in the public space through market activities, as Van Nieuwkerk indicates (Van Nieuwkerk, 2008, 2013). Engaging in these cultural spheres encourages people to transform their dispositions and tastes, or 'habitus', in an Islamic way. Piety is, therefore, *about the construction of definite and distinctive lifestyles of new religious tastes and preferences* through experiences in the cultural public sphere (Turner, 2008, p. 2; Van Nieuwkerk, 2008, p. 172). Consequently, the development of these newly formed public spheres has encouraged people to consider *a possibility for new publics and new voices to express new ideas* (Van Niewskerk, 2008, p. 175), and to transform contemporary Muslim social life as a whole. The physical and spiritual comfort provided by religious tours is achieved through the social recognition of others engaging with Islamic axioms and principles in the cultural public sphere, which offer various benefits both in everyday life and in the Islamic notion of the afterlife (*akhira*).

Although people's active engagement in religious tourism as a cultural sphere is recognised as an Islamisation or pietisation of individual Muslims in the contemporary world and as self-enlightenment, the development of religious guides can alternatively be described as the transformation of a set of axioms and Islamic principles derived from the economic order. In this sense, the development of an Islamic cultural sphere in the name of religious tourism industry has the potential for enhancing a certain capital flow based on the moral economy in Muslims' lifestyles through the social incentives provided by the cultural sphere. Mehmet Asutay and other researchers in Islamic economics indicate that a new paradigm has emerged to transform *the organisation of decision-making arrangements, mechanisms for the provision of information and co-ordination, property rights, and incentive mechanism into economic sphere* (Asutay, 2007, p. 4). They state that the Islamic economic system has encouraged people to engage in a public sphere they call the 'Islamic moral economy', which has Islamic moral incentives promoting the religious norms and rules of economic practices. The development of religious guides and religious tourism into other tourism fields is, therefore, described as the enhancement of religious capital flow that promotes financial and physical resources as well as religious knowledge and spiritual experiences according to market economy principles.

In this sense, the development of the religious tourism industry and the practices of religious guides have contributed to form a certain realm of social meanings and contexts based on Islamic piety, which has created social networking among the stakeholders in the marketplace.

CONCLUSION

This chapter considered the social context of religious guides in the Islamic religious tourism industry by mapping them into the structure of the marketplace from the perspective of the flow of 'religious capital'.

The flow of religious guides' religious capital has emerged in the marketplace of the religious tourism industry. In the case of Islam, the development of Islamic tour operators has promoted the cooperation with religious guides as the symbol of their managed religious capital in the field. In this environment, each customer can easily find suitable religious services that reflects their religious preferences. On the other hand, Islamic tour operators and religious guides have actively encouraged people to participate in the field by using their religious services in order to deepen their religious capital for physical and mental self-enlightenment.

In this market environment, the flow of religious capital has gradually obtained social context among the stakeholders through the expansion and strengthening of social networking that emerged from the religious tourism industry through considering the novel use of financial and physical resources in the market. This leads to the emergence of an 'Islamic cultural sphere' in which stakeholders reach consensus about social context based on market principles and Islamic piety in their practices as adaptations and negotiations.

In conclusion, therefore, the social context of religious guides in the religious tourism industry is seen as creating a public sphere called 'Islamic cultural sphere' through creating social context and a network based on their flow of religious capital in the marketplace. This cultural sphere in the market embeds individual consumption into Islamic piety as social context through enhancing the flow of religious capital among the stakeholders, with religious guides located at the hub of this social network. As the flow of religious capital is sustained through consumption in the marketplace, the consumption of religious commodities constructs firm religious piety embedded in this social network as well as individual religious capital. In this sense, religious guides are described as the significant hub for the flow of religious capital for creating an Islamic cultural sphere in society.

FUTURE RESEARCH DIRECTIONS

Although the study indicates the social significance of religious guides from the perspective of the flow of religious capital, there are few case studies for this topic based on qualitative field research to show the specific social practices, narratives and contexts related to religious guides. There are high demands to research on this topic.

ACKNOWLEDGMENT

This research is supported by JSPS (Japan Society for Promotion of Science) Grant-in-Aid for Young Scientists 18K18283 'A Study of Tourism Experiences and 'Context Practices' in Islamic Tourism' (Research Leader: Shin Yasuda, Takasaki City University of Economics, Japan), and Grant-in-Aid for Scientific Research (A) 19H00564 'Anthropological Studies on Veneration of Saints and Holy Relics in Islam and Christianity' (Research Leader: Masayuki Akahori, Sophia University, Japan).

REFERENCES

Ahrens, P. (2018). Qualitative network analysis: A useful tool for investigating policy networks in transnational settings? *Methodological Innovations*, *11*(1). Advance online publication. doi:10.1177/2059799118769816

Alserhan, B. A. (2017). *The principles of Islamic marketing*. Routledge. doi:10.4324/9781351145688

Asutay, M. (2007). A political economy approach to Islamic economics: Systemic understanding for an alternative economic system. *Kyoto Bulletin of Islamic Area Studies*, *1*(2), 3–18.

Bayat, A. (2007). *Making Islam democratic: Social movements and the post-Islamist turn.* Stanford University Press. doi:10.5117/9789053569832

Bianchi, R. (2004). *Guests of God: Pilgrimage and politics in the Islamic world.* Oxford University Press. doi:10.1093/0195171071.001.0001

Bianchi, R. (2013). *Islamic Globalization: Pilgrimage, Capitalism, Democracy, and Diplomacy*. World scientific. doi:10.1142/8810

Chandler, J. D., & Vargo, S. L. (2011). Contextualization and value-in-context: How context frames exchange. *Marketing Theory*, *11*(1), 35–49. doi:10.1177/1470593110393713

Cole, J. (2002). *Sacred place and holy war: The politics, culture and history of Shi'ite Islam*. I. B. Taubis.

Deeb, L., & Harb, M. (2013). *Leisurely Islam: negotiating geography and morality in Shi 'ite South Beirut*. Princeton University Press.

Eade, J., & Sallnow, M. J. (Eds.). (2000). *Contesting the sacred: the anthropology of pilgrimage*. University of Illinois Press.

Eid, R., & El-Gohary, H. (2015). The role of Islamic religiosity on the relationship between perceived value and tourist satisfaction. *Tourism Management*, *46*, 477–488. doi:10.1016/j.tourman.2014.08.003

Einstein, M. (2007). *Brands of faith: Marketing religion in a commercial age*. Routledge. doi:10.4324/9780203938874

Fealy, G. (2008). Consuming Islam: Commodified religion and aspirational pietism in contemporary Indonesia. In G. Fealy & S. White (Eds.), *Expressing Islam: Religious life and politics in Indonesia* (pp. 15–39). Institute of Southeast Asian Studies. doi:10.1355/9789812308528-006

Gibbons, R. (1992). *A Primer in Game Theory*. Coleshill: Harvester Wheatsheaf.

Green, N. (2011). *Bombay Islam: The religious economy of the west Indian ocean, 1840–1915*. Cambridge University Press. doi:10.1017/CBO9780511975165

Green, N. (2015). *Terrains of exchange: Religious economies of global Islam*. Oxford University Press. doi:10.1093/acprof:oso/9780190222536.001.0001

Haenni, P. (2005). *L'islam de marché: l'autre révolution conservatrice*. Seuil.

Hall, M., & Prayag, G. (Eds.), *The routledge handbook of halal hospitality and Islamic tourism*. Routledge. doi:10.4324/9781315150604

Haq, F., & Wong, H. Y. (2010). Is spiritual tourism a new strategy for marketing Islam? *Journal of Islamic Marketing*, *1*(2), 136–148. doi:10.1108/17590831011055879

Hoesterey, J. B. (2015). *Rebranding Islam: Piety, prosperity, and a self-help guru*. Stanford University Press.

Ibrahim, H. (1982). Leisure and Islam. *Leisure Studies*, *1*(2), 197–210. doi:10.1080/02614368200390161

Izutsu, T. (2002). *Ethico-religious concepts in the Qur'an* (Vol. 1). McGill: McGill-Queen's Press-MQUP.

Jamal, A., Raj, R., & Griffin, K. (2019). *Islamic tourism: Management of travel destinations*. Wallington: CABI.

Kawashima, T. D. (2016). Travel agencies and priests as spiritual leaders: The merits of collaboration. *Tourist Studies*, *16*(1), 40–56. doi:10.1177/1468797615588430

Kessler, K. (2015). Conceptualizing mosque tourism: A central feature of Islamic and religious tourism. *International Journal of Religious Tourism and Pilgrimage*, *3*(2), 11–32.

Kessler, K., & Raj, R. (2018). The Rise of Heritage and Religious Tourism to Sacred Sites in Oman. In S. Yasuda, R. Raj & K. Griffin (Eds.), Religious tourism in Asia: Tradition and change through case studies and narratives (pp. 10-20). Wallington: CABI. doi:10.1079/9781786392343.0010

Kitiarsa, P. (2010). Toward a Sociology of Religious Commodification. In B. S. Turner (Ed.), *The new Blackwell companion to the sociology of religion* (pp. 563–579). Wiley-Blackwell. doi:10.1002/9781444320787.ch25

Martin, W. H., & Mason, S. (2004). Leisure in an Islamic context. *World Leisure Journal*, *46*(1), 4–13. doi:10.1080/04419057.2004.9674343

Melewar, T. C., & Alwi, S. F. S. (2017). *Islamic marketing and branding: Theory and practice*. Routledge. doi:10.4324/9781315590035

Mzakki, A. (2008). Islam as a symbolic commodity: Transmitting and consuming Islam through public sermons in Indonesia. In P. Kitiarsa (Ed.), *Religious commodification in Asia: Marketing gods* (pp. 205–219). Routledge.

New Straits Times. (2014, Mar. 19). Attractive tour packages with Tabung Haji Travel. *New Straits Times*.

OIC (Organization of Islamic Corporation). (n.d.). Retrieved from https://www.oic-oci.org/

Olsen, D. H. (2013). A scalar comparison of motivations and expectations of experience within the religious tourism market. *International Journal of Religious Tourism and Pilgrimage*, *1*(1), 1–5.

Olsen, D. H. (2018). Religion, pilgrimage and tourism in the Middle East. In D. Timothy (Eds.), Routledge Handbook on Tourism in the Middle East and North Africa (pp. 109-124). London: Routledge.

Pink, J. (Ed.). (2009). Muslim Societies in the Age of Mass Consumption: Politics, Culture and Identity between the Local and the Global. Tyne Newcastle: Cambridge Scholars.

Raj, R., & Griffin, K. A. (Eds.). (2015). *Religious tourism and pilgrimage management: An international perspective*. Cabi. doi:10.1079/9781780645230.0000

Reader, I. (2013). *Pilgrimage in the marketplace*. Routledge. doi:10.4324/9781315885704

Rinschede, G. (1992). Forms of religious tourism. *Annals of Tourism Research*, *19*(1), 51–67. doi:10.1016/0160-7383(92)90106-Y

Rudnyckyj, D. (2009a). Market Islam in Indonesia. *Journal of the Royal Anthropological Institute*, *15*(1), 182–200.

Rudnyckyj, D. (2009b). 'Spiritual Economies: Islam and Neoliberalism in Contemporary Indonesia'. *Cultural Anthropology*, *24*(1), 104–141. doi:10.1111/j.1548-1360.2009.00028.x

Rudnyckyj, D. (2010). *Spiritual economies: Globalization and the afterlife of development*. Cornell University Press.

Sandıkcı, Ö., & Rice, G. (Eds.). (2011). *Handbook of Islamic Marketing*. Edward Elgar Publishing. doi:10.4337/9780857936028

Sardar, Z. (2014). *Mecca: The sacred city*. Bloomsbury Publishing USA.

Schilderman, H. (2011a). Religious capital and public accountability: challenges to empirical theology. In L. Francis, & H. G. Ziebertz (Eds.), The Public Significance of Religion (pp. 41-63). Reiden: Brill. doi:10.1163/ej.9789004207066.i-495.24

Schilderman, H. (2011b). Varieties of Religious Solidarity. In L. Francis, & H. G. Ziebertz (Eds.), The Public Significance of Religion (pp. 169-190). Reiden: Brill. doi:10.1163/ej.9789004207066.i-495.66

Shackley, M. (2001). *Managing sacred sites: Service provision and visitor experience*. Cengage Learning EMEA.

Shinde, K. A. (2010). Entrepreneurship and indigenous enterpreneurs in religious tourism in India. *International Journal of Tourism Research*, *12*(5), 523–535. doi:10.1002/jtr.771

Sinha, V. (2011). *Religion and commodification:'Merchandizing'diasporic Hinduism*. Routledge. doi:10.4324/9780203842799

Stark, R., & Finke, R. (2000). *Acts of faith: Explaining the human side of religion*. University of California Press.

Tabung Haji Travel and Services. (n.d.). Retrieved from https://www.thts.com.my/

Temporal, P. (2011). *Islamic branding and marketing: Creating a global Islamic business*. John Wiley & Sons.

Timothy, D. J., & Iverson, T. (2006). Tourism and Islam: considerations of culture and duty. In *Tourism, religion and spiritual journeys* (pp. 202–221). Routledge. doi:10.4324/9780203001073

Timothy, D. J., & Olsen, D. H. (Eds.). (2006). *Tourism, religion and spiritual journeys* (Vol. 4). Routledge. doi:10.4324/9780203001073

TravelM. Q. (2020). Retrieved from https://mqtravel.co.id/

Turner, B. S. (2008). Introduction: The price of piety. *Contemporary Islam*, *2*(1), 1–6. doi:10.100711562-007-0034-7

Van Nieuwkerk, K. (2008). Creating an Islamic cultural sphere: Contested notions of art, leisure and entertainment. An Introduction. *Contemporary Islam*, *2*(3), 169–176. doi:10.100711562-008-0059-6

Van Nieuwkerk, K. (2013). *Performing piety: singers and actors in Egypt's Islamic revival*. University of Texas Press.

Vargo, S. L., & Lusch, R. F. (2008). From goods to service (s): Divergences and convergences of logics. *Industrial Marketing Management*, *37*(3), 254–259. doi:10.1016/j.indmarman.2007.07.004

Verter, B. (2003). Spiritual capital: Theorizing religion with Bourdieu against Bourdieu. *Sociological Theory*, *21*(2), 150–174. doi:10.1111/1467-9558.00182

Yasuda, S. (2013). Commitment for Strategy: Religious Entrepreneur Networks in Syrian Shi'ite Religious Tourism. *The Journal of Sophia Asian Studies*, *31*, 35–49.

Yasuda, S. (2018). Entrepreneurship for Religious Tourism in Mumbai, India. In S. Yasuda, R. Raj & K. Griffin (Eds.), Religious tourism in Asia: Tradition and change through case studies and narratives (pp. 21-29). Wallington: CABI.

Ziyarah Tours. (2020). Retrieved from https://www.ziyarahtours.net/

ADDITIONAL READING

Bianchi, R. (2013). *Islamic Globalization: Pilgrimage, Capitalism, Democracy, and Diplomacy*. World scientific. doi:10.1142/8810

Green, N. (2015). *Terrains of exchange: Religious economies of global Islam*. Oxford University Press. doi:10.1093/acprof:oso/9780190222536.001.0001

Hoesterey, J. B. (2015). *Rebranding Islam: Piety, prosperity, and a self-help guru*. Stanford University Press.

Reader, I. (2013). *Pilgrimage in the marketplace*. Routledge. doi:10.4324/9781315885704

Rudnyckyj, D. (2010). *Spiritual economies: Globalization and the afterlife of development*. Cornell University Press.

Shackley, M. (2001). *Managing sacred sites: Service provision and visitor experience*. Cengage Learning EMEA.

Van Nieuwkerk, K. (2013). *Performing piety: singers and actors in Egypt's Islamic revival*. University of Texas Press.

Yasuda, S. (2018). Entrepreneurship for Religious Tourism in Mumbai, India. In S. Yasuda, R. Raj & K. Griffin (Eds.), Religious tourism in Asia: Tradition and change through case studies and narratives (pp. 21-29). Wallington: CABI.

KEY TERMS AND DEFINITIONS

Cultural Sphere: A kind of public sphere driven by cultural activities in the process of communication and negotiation among the actors in the field. The concept is coming from Jurgen Harbermas's concept of 'public sphere'.

Islamic Tour Operator: It is a kind of tour operators which organise Islamic religious tours (*hajj*, *umrah* and *ziyara*, and halal journeys) for earning profit from their customers by providing travel services.

Religious Capital: It is the concept that shows the degree of mastery of and attachment to a particular religious culture.

Religious Guide: It is a kind of guide who provides religious or spiritual experiences toward their customers.

Religious Tourism: It is a type of tourism based on religious values. There is variety of definitions and contents of the concept in comparison with pilgrimage and other type of tourism like cultural tourism and heritage tourism.

Ustaz: It means 'leader' or 'teacher' in Islamic context, who supplies spiritual experiences in the process of religious tours. The word originally means 'teacher' in Arabic, which do not directly mean religious meanings.

Value-in-Context: It is the concept that context is defined as a set of unique actors with unique reciprocal links among them.

ENDNOTES

1 Value-in-context is the concept that context is defined as *a set of unique actors with unique reciprocal links among them* (Chandler & Vargo, 2011, p. 40).

2 Abdullah Gymnastiar, or AA Gym (born in 1962), is Indonesian Islamic preacher, author, and businessman. He became popular by introducing unique way of preaching in TV programmes, seminars and publications based on self-enlightenment of individual spirituality. Although he did not read Islamic higher education, his way of speech and lifestyle gain high reputation among Indonesian Muslims, and have various business in spiritual market (Hoesterey, 2015).

3 MQ travel is Indonesian tour operator based in urban areas in the country. The tour operator strongly tie with AA Gym and other ustaz related to him. The company frequently organise cultural and leisure tours as well as hajj and umrah tours (MQ travel, 2020). AA Gym sometimes accompanies with the tours to provide special spiritual experiences for their customers (MQ Travel, 2020).

4 Tabung Haji Travel & Services Sdn Bhd is Malaysian tour operator to organise Muslim travel packaged tours as well as *hajj* and *umrah* package tours. It was established in 1972 subsidiary of Lembaga Tabung Haji. As the rise of customers' needs to organise cultural and leisure packaged tours, TH Travel also provided these kinds of tours in their managements, which gain high reputations among Malaysian Muslims (*New Streits Times*, 19 Mar 2014; TH Travel, 2020).

Chapter 11
Tourist Guiding in Halal Tourism:
Case of Southeast Asia

Nor Aida Abdul Rahman
Universiti Kuala Lumpur, Malaysia

EXECUTIVE SUMMARY

The growing interest in Halal tourism activity represents a prolific lens for scholars and practitioners in the tourism arena to study its concept, its role, and its potential to affect the sustainability of Halal tourism industry. The success of tourism industry is highly dependent on traveler experience with respect to customer service, experience, facilities, comfort, and delight. The notion of Halal tourism is connected and geared to the adoption of Islamic principles and practice in all aspect of tourism activities such as in Halal hotel, Halal restaurant, Halal spa, Halal package tour, Halal transportation, Halal warehouse, and Halal retail. Halal tourism market reached USD 181 billion in year 2018 and is expected to grow every year. The number of Muslim travelers to Southeast Asian countries has rapidly increased as Southeast Asia (SEA) is known as a dynamic environment with a total Muslim population around 240 million.

INTRODUCTION

Globally, international tourism activity shown substantial growth of Muslim traveler across the globe. This include both Muslim tourist arrivals and Muslim tourist receipt in many countries. As published by Global Muslim Travel Index (GTMI) 2017, the number of Muslim tourist has increased from 121 million to 156 million in year 2016. In fact, it is projected that Muslim tourist to spend US$220 billion by year 2020. Many countries (both Muslim and non Muslim countries) are now start to attract Muslim travelers. Recently, Whitehead (2019) stress that Muslim tourist spend will be increase to $380 billion by year 2023. Earlier report by Thomson Reuters and Dinar Standard (2013), Halal tourism is valued at USD 137 billion in 2013 and expected to reach USD 181 billion by USD 181. With the steady growth on Halal tourism, offering Halal tourist guide, Halal tour or Halal operator is substantial.

DOI: 10.4018/978-1-7998-3725-1.ch011

According to Mohsin et al (2016) and Rahman (2019), Halal tourism is the tourism activity that adhere to value and Islamic principle in Islam. Halal tourism can be defined as a demand of Muslim traveler that need convenient to observe their religion during their holiday. At present, Halal tourism and hospitality is widely recognized and enhanced especially with regards to Halal inflight catering, Halal entertainment, Halal hotel, Halal food services, halal medical services, Halal Spa and Saloon and many more (Rahman 2018a). As such, developing Halal market segments could attract more Muslim tourist to come and develop further the Halal tourism industry. This is vital since halal tourism is dependent on tourist experience during the holiday and looking forward for their experience towards Halal customer service, Halal guidance, as well as Halal facilities.

With the recent development on Halal tourism and significant growth of Muslim travel market, it is essential for tour operator and tour guides to have comprehensive understanding of the special needs of Muslim travelers. This is necessary to provide Muslim travelers with a substantial service for their comfort during their holiday. In fact, it is important for tour guide and operator to fully understand the preference of Muslim travelers as it will also help the travel guide and tour to improved their service performance and sustain in the business. The term of tourist guide was preferred for use in this study in compliance with the World Federation of Tourist Guide Association (WFTGA) adopted terminology.

The guiding activity may include activity such as a place of attraction, religious places, food attraction, shopping places and many more. A recent study by Tatar et al (2018) and Rabotic (2010) highlights the importance of multifaceted role of tourist guiding in contemporary tourism. Tourist guiding term is preferred to use in this study as it is the terminology used by World Federation of Tourist Guide Association. The 'guide' word represent the manage and arrange activity for the travelers in enhancing traveler or tourist experience and enhance the destination of the tourist. Align with the Halal development worldwide with the increased number of Muslim travelers globally, therefore this paper aim to elaborate in detail on tourist guiding in Halal tourism in Southeast Asia (SEA) as a main reference case. This paper will start with discussing the definition on Halal tourism and provide generic information about Muslim travelers in Southeast Asia, followed by Muslim tourist guiding core elements. In the next section, the researcher then focuses on the Emerging Tourist Guide Association in all eleven countries in SEA region as a main case in this research. In conclusion, the paper also presents some recommendation for future research with highlighting the key implication to the social, scholars and practitioners.

LITERATURE REVIEW

Halal Tourism: Concept and Provision

Halal tourism can be define as tourism activity that consist of product or services that meets the Islamic requirement and the need of Muslim travelers. This is not limited to Halal food or dietary that conform to Islamic teachings and Islamic values, but also other Halal activities during the holiday such as Halal hotel, halal spa, Halal swimming, Halal medical and others. From the academic point of view, research on Halal tourism starts gaining popularity about a decade ago. Research on halal tourism is also increased which cover many areas and many issue. With regards to Halal Tourism definition from scholar's point of view, below Table 1 present a number of Halal tourism definition from past studies.

Table 1. Definition on Halal Tourism

Author (Year)	Definition	Name of Journal/ Book
Bon and Hussain (2010)	Refer Halal tourism as a subtype of religious tourism and deal with Muslim oriented tour	Bridging tourism theory and practice (2010)
Battour et al (2011)	Any action that is permissible to be used or engaged in tourism activity, and fitting Islamic teachings	International Journal of Tourism Research
Duman (2012)	Activity in tourism that follow Islamic principles	Islam and Civilization Renewal
El Gohary (2016)	Is an activity in tourism that deeply rooted within the Islamic Shariah. Also known as religions tourism	Tourism Management Perspective
Mohsin et al (2016)	Activity in tourism that adhere to Islamic value	Tourism Management Perspective
Battour and Ismail (2016)	An activity that covers every activity in tourism including hotel, food and restaurants, and Halal facilities	Tourism Management Perspective
Ryan	Refer Halal tourism is defer from Islamic tourism. Pilgrimage tourism – made to Mecca by adherents to the Islamic faith. Halal tourism is mostly undertaken for recreational, leisure and social purpose but according to Islamic law	Tourism Management Perspective
Rahman et al (2018b)	Halal tourism in this research refer to halal in aviation namely halal in flight catering	International Journal of supply Chain Management
Rasul 2019	Halal tourism include perceivable and non-perceivable aspects of Shariah (Islamic teaching)	Tourism Recreation Research
Rahman 2019a	Refer Halal tourism as any activity that abide Islamic rules such as Halal Hotel, Halal restaurant, halal kitchen, Halal logistics, Halal clothing, Halal spa and Halal in flight menu	Book chapter in Tourism Events in Asia

Source: The author

Having review the literature on the notion of Halal Tourism, Halal tourism actually covers many aspect. Among the provision of Halal tourism are Halal food and restaurants, Halal logistics (transportation and warehouse), Halal entertainment, Halal shopping and spa, friendly Muslim facilities such as clean toilet and mosque for prayer, and also Muslim friendly tourist guide or service staff (see Figure 1). Timothy and Iverson (2006) recommends hotels should educate their staff on cross-cultural communication to allow them to treat Muslim tourists with respect and consider recruiting religious staff. As highlighted by very recent article by Jia and Chaozi (2020), the provision of Halal tourism in Islamic definition with non-Islamic definition are similar. All Muslim travelers are looking for the same Halal tourist guidance and the same Halal facilities. Both also mention the six area of Halal tourism namely Halal food, halal transport, Halal entertainment, Halal shopping place, Halal toilet and knowledgeable service staff.

Figure 1. Halal Tourism Provision
Source: Drawn by the author

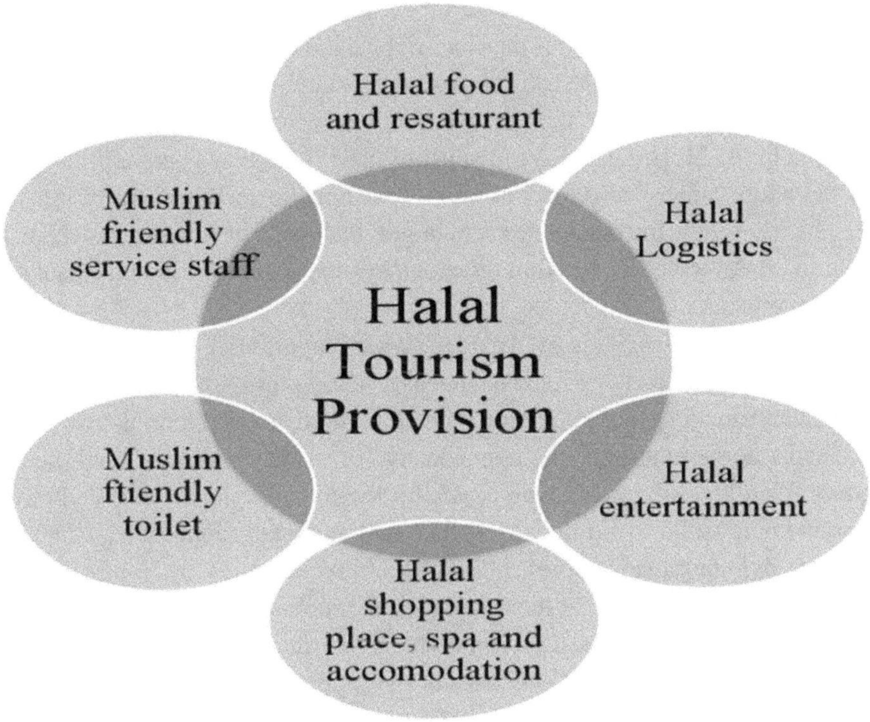

Muslim Travelers in Southeast Asia, Halal Tourist Guide and Elements of Tourist Guidance

Globally, Muslim now is acknowledged as the fastest-growing segments of the travel industry. In fact, as report by New York Times (2019), the rise of Halal tourism activity leads to the increased number of Muslim travelers across the globe. In parallel, tourism operators are continuously and increasingly trying to meet Halal travelers demands to have Halal travel activities during their vacation. As a result, Halal tour guiding has become a good profession in many countries including Southeast Asia region. As mentioned by Alfonso and Maria (2018), Muslim travel market will be one of the fastest growing segments of the global travel industry. This is align with the increase number of Muslim population across the world. As highlighted in Pew Research Centre (2015), by the year 2070, Muslim population will be the largest population in the world. The increasing number of Muslim population reflects to the high demand of Muslim product and services. It is expected that with the growing number of Muslim population, there will be an increased to Halal tourism activity as well.

Halal tourism or Islamic tourism has been used interchangeably by the tourism scholars ((Battour and Ismail, 2016; Hamza et al., 2012; Khan and Callanan, 2017). It can be refer to the activity, action, attitude or people habits during travel with an Islamic style (Elaziz and Kurt, 2017; Rahman, 2019). In addition, as highlighted by Rahman et al (2018a), it also includes Halal meat request with the airlines during onboard, or it may relates on the need of Halal transportation and Halal warehouse for their Halal goods during their travel (Rahman et al 2018b). Similar arguments has been emphasized by Shafaei and

Mohamed (2015), the availability of these Halal attributes activities during travel (Halal food restaurants, Halal hotel, Halal spa, halal transport etc) will create a good brand image for Muslim travel destination and attract Muslim travelers to come and increased their satisfaction and perception of quality.

In Halal tourism, Halal tourist guiding is the central issue as Halal tourism is based on the best global tourism standards and covers four main sectors namely destination, marketing, industry and institution. Essentially, all Halal tourism provider should have detail information on the needs of Muslim travelers. This is vital to help the Muslim travelers to abide to their shariah or Islamic principles during their vacation. Halal tourist guiding or also refer as Halal tour operator or Halal travel agent is a licensed tour operator that has thorough and deep understanding of the needs of Muslim travelers. They not only provide the place for Halal travel for Muslim travelers, but they also provide tour package that covers the needs of Muslim travelers.

Earlier study on Halal traveler needs by Amadeus which reported in British "Context Consulting", their findings highlight three key criteria for Halal need travelers, namely maximizing trip value by hiring Halal tour guide; Halal friendly accommodation with Halal food and no alcohol, as well as Halal facilities (for instance separate swimming pools, gym and spa for women and men); and the third criteria is about Halal family friendly destination which is safe for them to go out and ease them for prayers such as halal food restaurant, musolla or mosque. Musolla refer to a prayer room for Muslim provided by any organization such as shopping mall, airport, offices etc. In recent study by Rahman et al (2018b) and Rahman et al (2018a), they mention that halal logistics is becoming an issue for many Halal travelers when they travel as they want their Halal products status that they carry during travelling is maintain halal from point of origin to point is Halal logistics of destination. As such, the author would like to enhance the fourth criteria for basic Halal traveler needs and supply chain. This include halal transportation and halal warehousing activity to store the Halal traveler's goods or carriage. Recently, Halal medical also has been mention as one of the criteria needed by Muslim travelers during their holiday. Issue on the Halal medical tourism is on the rise and becoming a key focus for many scholars and practitioners. Among many countries in the world, Malaysia has become the main leader in Halal medical tourism. Halal medical could be define as a medical services that has option to Muslim patients to have a medical alternative. It is refer to gelatin free and porcine free medical products.

Another interesting part for Halal tourist guiding is Halal family destination. Halal family destination may refer to the tourism activity to the historic Islamic countries as well as to Mecca and Madinah to perform Hajj and pilgrimage. Figure 2 below illustrated the six core elements on Halal travelers need for Halal tourist guiding activity in Halal tourism. While table 2 below highlights Muslim traveler's facilities and activities.

Figure 2. Core elements of tourist guiding in Halal tourism
Source: Drawn by the author

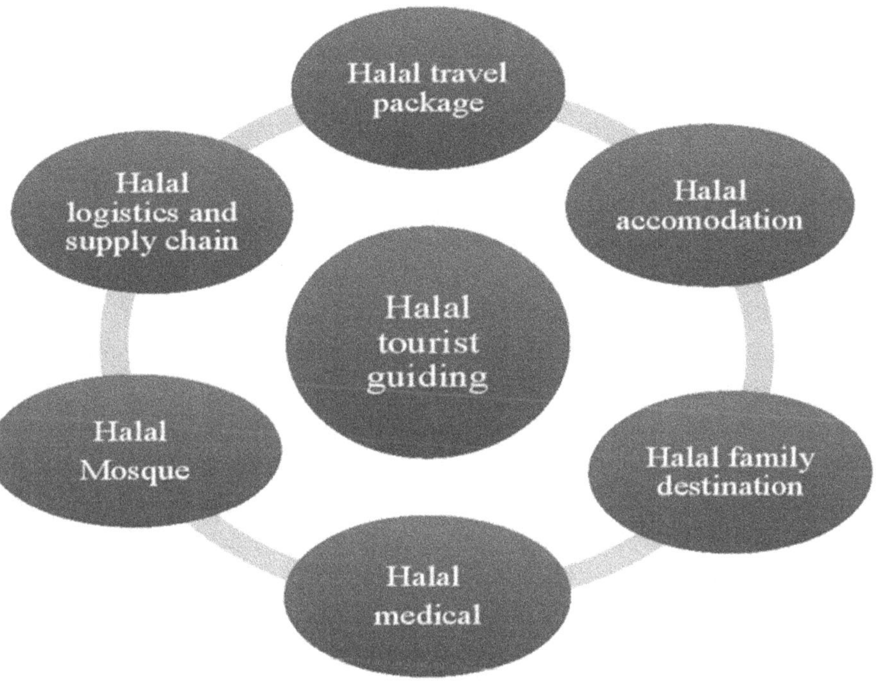

Table 2. Muslim traveler's facilities and Halal Tour Operator Guiding

Muslim travelers facilities	Explanation	References
Halal travel package	Availability on the Halal food and restaurant, availability on hall in flight food	Battour et al (2011); Rahman et al. (2008a)
Halal Accommodation – Hotel	Refer to hotel that provide Holy Quran in a room, and show Qiblah direction, separate swimming pool for male and female	Mansfield et al (2000); Henderson (2003); Alserhan et al. 2018; Razalli et al., 2013
Halal family destination	Availability of the Halal destination undertaken by Muslim during holiday such as for recreational, leisure, social and specific purpose. Specific purpose here refer to performing Hajj or pilgrimage	Rodrigo and Turnbull (2019) Syan (2016)
Halal logistics and supply chain	Refer to Halal supply chain activity such as Halal transportation and Halal warehousing. The purpose of Halal logistics and supply chain is to ensure the integrity of the Halal status is prudent	Rahman et al. (2018b) Majid et al. (2019)
Halal medical services	Refer to Halal medical treatment from Islamic friendly Hospital. For example, female patient being treated by female doctor. According to this article, satisfaction on Muslim travelers towards the hospital during their holiday is dependent on doctor and the hospital roles. Attitudes factor mediated the effect of hospital halal practice (Halal medical service) and patients satisfaction	Zailani et al (2016)
Halal mosque	Concern of Muslim travelers on the availability of the mosque. Location on nearby mosque, or location of mosque in the area of tourism	Weidenfield (2006); Mohsin (2005)

Source: The author

Given the Muslim travelers and Halal tourism is a niche market and good research area to enhance, Halal tourist guide should pay more attention to this Halal tourism elements and core activities as it has a huge potential for Halal tour operator to sustain in the industry. In fact, one of the key findings from Battour et all (2011, p. 537) states that *"easy access to worship places (Masjid/prayer room), easy to find Halal food, availability of the Quran and the Qiblah direction pointer. Islamic toilets and Islamic entertainment were reported as highly prioritized tangible attributes. It is also concluded that the intangible aspects are limited only to Muslim countries because these aspects are not realistic within the context of Westernized non-Muslim societies. Moreover, the application of the proposed intangible aspects may represent challenges facing the tourism industry in Muslim countries willing to satisfy non-Muslim tourists…"*

It is a necessity for Halal tour operator to have a good understanding on Halal information and its availability, location and vicinity of the mosque and other Halal facilities. Another vital factor Halal tour operator need to explore is Muslim friendly halal activities such as Halal shopping, Halal entertainment, sightseeing visiting to a historic places or nature reserve. Halal Tour guides should therefore definitely include a shopping stop in the Halal tourist activity itinerary. Places such as crowded beaches are not ideal for Muslim travelers, however one that is family-friendly, private and less crowded; so that the women will be comfortable, will definitely be appreciated. The needs of Muslim tourists are quite specific but they are not hard to satisfy. Tour guides should therefore know the special needs of this segment of travelers in order to effectively tailor tour packages that will please the Muslim tourist. The next section will explore Halal tourist guiding activities in Southeast Asia (SEA) countries. The discussion will start with the emergence of Halal Tourist Guide Association in SEA, followed by within case analysis of Halal tourist guiding in every country in SEA.

Emerging Tourist Guide Association: Halal Tourist Guide Perspective

As mentioned earlier, it is acknowledged that tourism industry worldwide has recognized the interest of Halal tourism from the view of industry and research. The success of developing Halal tourism industry is not only from the initiative of government, but also from the Halal industry players, as well as non-profit organization like NGOs. According to World federation of Tourist Guide Association(WFTGA) which based in Vienna, tourist guide can be define as a person who involved with giving guidance to the travelers in their language, with the choice of cultural and heritage of the area. They are also person who satisfy the travelers and giving the beautiful memory and experience to the travelers (WFTGA, 2005). Tourist Guide Association was established with the aim to establish contact with Tourist Guide Associations throughout the world. In general, there are 61 countries who registered with WFTGA. In southeast Asia, there are also a number of tourist guide association that officially registered with WFTGA. Among all 11 countries in SEA region, only six countries have registered with WFTGA as shown in Table 3 below. There are Malaysia, Indonesia, Philippines, Singapore, Thailand and Myanmar.

Table 3. Examples of SEA countries that register with WFTGA

Countries	Tourist Guide Association (TGA) register with WFTGA	Tourist Guide Association Name
Malaysia	Yes	1. Malaysian Women Tourist Guides Association 2. Sabah Tourist Guide Association
Indonesia	Yes	1. Indonesian Tourist Guides Association (ITGA)
Philippines	Yes	1. Multi Lingual Accredited Tour Guides Association of the Philippines (MULATGAP) 2. Philippine Federation of Professional Tour Guide Association
Myanmar	Yes	1. Mandalay Tourist Guide Society
Singapore	Yes	1. Society of Tourist Guide
Thailand	Yes	1. Professional Tourist Guide Association of Thailand
Timor-leste	No	
Vietnam	No	
Brunei	No	
Cambodia	No	
Laos	No	

Source: The author

Within Case Analysis

Malaysia Country (Halal Tourist Guiding)

Malaysia is known as a leading country in Halal industry including Halal tourism. Malaysia consist of 13 states and three federal territories which separated by the South China Sea. Halal tourism in Malaysia starts as early as 1990s when Malaysia starts to introduce 'Tahun Melawat Malaysia' (Rahman, 2019a). The Halal industry in Malaysia starts established in year 1974 when the Research Centre for the Islamic Affair Division in the Prime Minister's Office started to issue Halal certification letters for Halal products. The rapid development of Halal industry has led to the growing Halal ecosystem in Malaysia country and its neighboring countries. The Halal ecosystem has been expanding to many areas from Halal food and restaurants, to Halal banking, Halal hotel and tourism, Halal cosmetics, Halal medical, Halal logistics, Halal entertainment, Halal cleaning and many more.

Malaysia is a popular Muslim countries and received high numbers of both Muslim and non-Muslim travelers. In another word, Muslim are now becoming more mobile which constantly need for travel for both leisure, work or specific Islamic purpose. As published in Islamic tourism (2019), Malaysia has been chosen as the worlds's top destination for Muslim travelers for three (3) consecutive years in recent survey conducted by Singapore's Crescent Rating and United States' Dinar Standard. Following Table 4 shows the Halal tourist guide resources in Malaysia with its web link, as well as popular yearly Malaysia Halal Tourism activities.

Table 4. Halal Tourist Guiding in Malaysia

Halal tourism	Islamic Institution in Malaysia	Web link
Halal Tourist Guide Resources	Islamic Tourism Centre	http://itc.gov.my
	Islamic Arts Garden Complex	https://www.malaysia-traveller.com › selangor-islamic-arts-garden
	Islamic Arts Museum Malaysia	http://www.iamm.org.my
Halal Tourism Activity	Malaysia International Halal Showcase (MIHAS)	http://www.mihas.com.my
	World Halal Forum (WHF)	http://www.worldhalalforum.org

Source: Drawn by the author

Singapore Country (Halal Tourist Guiding)

Singapore is a neighboring country to Malaysia. It was reported that Singapore is recognized as the 6th most popular travel destination for Muslim millennials (The Straits Times, 2017). Apart from Malaysia, Singapore has been listed as the top 10 countries for Muslim travelers including Indonesia, Japan, Thailand, Australia, United Arab Emiriah (UAE), Britain, India and United States. As reported by The Straits Times, many Muslim travelers stop and visit Singapore country on their way to Thailand or Indonesia. The below Table 5 shows the Tourist Guiding resources available in Singapore for Muslim travelers information.

Table 5. Singapore Tourist Guiding

Tourist Guide Association	Web link	Example of guide activity in Singapore	Remark and Potential
Society of Tourist Guides Singapore	https://societyoftouristguides.org.sg/	Educational tours, maritime tours, heritage tours etc	Could explore and develop further on Halal tourism in Singapore especially on Halal food and restaurants guidance, Halal hotels, Halal entertainment, Halal facilities etc.
Official Singapore Tourism Site	https://www.visitsingapore.com/	Heritage tours etc	Could add more info for Muslim travelers information and Halal tourist guidance

Source: Author

Table 6. Halal travel guide in Brunei

Tourist Guide Association	Web link	Example of guide activity in Brunei	Remark
Halal Trip	https://www.halaltrip.com/other/blog/10-things-a-muslim-traveler-should-keep-in-mind-when-visiting-brunei/	Halal meal, mosque, nearby airports etc.	Provide details information for Halal travel places in Brunei including Halal food, hotels, mosque, and halal sightseeing
Brunei Private Tours	Toursbylocals.com/Brunei	Hotels, Restaurants etc	A friendly Muslim local guide for Halal tour private design
Brunei Tourism - Ministry of Primary Resources and Tourism	www.mprt.gov.bn	Gastronomy etc	Brunei government agency that promote Halal tourism in Brunei

Source: The author

Brunei Country (Halal Tourist Guiding)

Brunei or known as Brunei Darussalam is an Islamic country located on a tiny island of Borneo, surrounded by Malaysia and the South China Sea. Brunei is known for its beaches and biodiverse rainforest. Brunei has taken a serious look into Islamic tourism or Halal tourism. As reported by Puvaneswary (2017), Halal tourism in Brunei more towards Islamic tourism especially on the local way of life or sightseeing at some of the mosque that represent the culture and heritage of Brunei. In the recent report by the Asean post (2018), Malaysia and Indonesia have consistently been among the Brunei's largest tourism market. In year 2016, it was reported the number of tourist that flew in by air to Brunei are

Table 7. Halal Travel Guide in Indonesia

Tourist Guide/ Halal Travel	Web link	Example of guide activity in Indonesia	Remark
Indonesia Halal Travel	https://www.indonesia.travel/gb/en/home	Halal stall, mosque etc	This is the official website of Indonesia Tourism
Indonesia Private Tours	https://www.toursbylocals.com/Indonesia-Tours?s	Halal food guide, accommodation, budget travel etc	Personal guide for Halal travelers to explore Indonesia Halal tourism activities

Source: The author

218,809 tourists with 24.8% are coming from Muslim travelers from Malaysia, and 9.5% came from Indonesia. In order to develop further Halal tourism industry in Brunei, there is a strong need to have a solid resource for the traveler to get some information to plan on their holiday. However, since Brunei is a Muslim country, it is not a critical issue as most of the food or restaurants, as well as facilities in Brunei are Halal and Muslim friendly. Below table 5 shows the Halal travel guide available in Brunei.

Table 8. Halal Travel Guide in Thailand

Tourist Guide/ Halal Travel	Web link	Example of guide activity in Thailand	Remark
Halal Zilla	https://www.halalzilla.com/halal-food-bangkok/84962	Halal food restaurants	Halal famous food in Thailand. Provides 15 Halal places for Muslim travelers in Bangkok
Official Tourism Thailand	https://www.tourismthailand.org/landing	Travel map, travel information, accommodation etc	Provide Halal tourism information in Thailand.

Source: The author

Indonesia (Halal Tourist Guiding)

Indonesia is known as the largest Muslim population with holding about 12.7% of the world Muslim population. Indonesia or officially known as Republic of Indonesia is located between the Indian and Pacific oceans. Ministry of Tourism Indonesia are very confident that Indonesia can emerge as the among the leading Halal tourist destination in future. As reported by Suradin (2018), the official website homepage that promotes Halal tourism are mostly focus on several regions in Indonesia such as Aceh, West Sumatra, Riau, Riau Archipelago, South Sulawesi, East Java, Central Java, West Java, Jakarta and West Nusa Tenggara. Furthermore, among the Islamic attributes of destinations in Indonesia are mainly in Islamic Muslim festival celebration, Islamic Halal food, Islamic muslim prayers, Muslim dress and clothing, Mosque and other Islamic architecture. Below Table 6 shows Halal travel guide available in Indonesia.

Thailand (Halal Tourist Guiding)

Thailand is one of the popular Halal tourism country in Southeast Asia. Thailand is popular with tropical beaches, opulent royal palaces, seafood, iconic temples, and certain famous dishes. As reported in Bangkok Post (2019) recently, the Global Muslim Travel Index (GMTI) shows that Thailand ranked second highest on the preferred destination for Muslim travelers. This report indicate that Halal tourist guiding is significant in Thailand as it is expected that Thailand will receive more Muslim tourist in the near future. In fact, it is also vital for Halal tour operator or Halal tourist association to have full or deep understanding on Islamic tourism activities and prepare for the needs of Muslim tourist such as the washroom facilities, prayer or mosque facilities, gender based facilities at the hotel (for instance swimming pool), halal restaurant and other halal entertainment. Table 7 below elaborates Halal tourist guiding for Muslim travelers in Thailand.

Table 9. Halal Travel Guide in Philippines

Tourist Guide / Halal Travel	Web link	Example of Guide Activity in Phillipines	Remark
Philippines Halal will Travel	https://www.havehalalwilltravel.com/blog/tag/philippines/	Friendly destination, best beaches, popular islands etc	Provide some key halal activities for Muslim travelers
Mr and Mrs Halal Travel	https://www.mrandmrshalal.com/travel-guide-to-manila-for-muslims/	Halal food guide, honeymoon travel, budget travel etc	Provide information for Halal travel in Manila
Philippines Muslim Tours and Holidays	https://www.seduniatravel.com/muslim/asia/philippines	Halal food, mosque, etc	Provide Halal tourist guidance for Muslim travelers

Source: The author

Table 10. Halal travel Guide in Vietnam

Tourist Guide / Halal Travel	Web link	Example of guide activity in Vietnam	Remark
The Official Tourism Website of Vietnam	https://vietnam.travel/home	Heritage guide, local food, siland	Is an official government tourism website in Vietnam
Muslim Travel Vietnam	http://www.havehalalwilltravel.com	Popular hills, popular island, Halal food, etc	Is a Halal tour operator in Vietnam that focus on Halal tourist in Vietnam
Vietnam Muslim Tour	www.islamvietnam.com	Hotels, resort, transport etc.	Are local Halal tour operator that provide private Halal trips for Muslim travelers

Source: The author

Table 11. Halal Travel in Myanmar

Tourist Guide / Halal Travel	Web link	Example of guide activity in Myanmar	Remark
Myanmar Tours and Holidays	https://www.seduniatravel.com › muslim › asia › myanmar-exposure	Food, accommodation, transport etc	Provide Halal tourism activities in Myanmar for Islamic or Muslim travelers
Myanmar Tripfes	https://www.tripfez.com/packages/asia/myanmar	Festivals in Myanmar etc	Provide halal tour guides for Muslim travelers

Source: Author

Philippines (Halal Tourist Guiding)

Philippines or known as Republic of Philippines is recognized as archipelagic country in Southeast Asia. According to UN data, there are about 108, 116, 615 populations in Philippines with 10% Muslim population. The main religion of Philippines is Christian. Even though it is a non-Muslim country, but still they received Muslim tourist from all over the world. The government of Philippines has a planned to accelerate Halal tourism industry. As reported by Crescent (2016), The Philippines Department of Tourism (PDOT) has initiative to develop long term infrastructure across the country to support Halal travel activity. Philippines is now partnered with Crescent, the world leading authority on Halal travel to improve its facilities and develop further its capabilities and resources to achieve its objective to boost Halal tourism activity in Philippines. In fact, it is recognized that tourism sector is a key indicator for Phillipi9nes economic development which projected to reach $200 billion by year 2020. Table 8 display Halal tourist guide information available in Philippines.

Table 12. Halal Travel in Cambodia

Tourist Guide / Halal Travel	Web link	Example of Guide in Cambodia	Remark
Cambodia Halal Tourism	http://www.cambodianmuslimtour.com/	Halal food, backpack etc	Is a leading halal tour provider for Muslim traveler
Cambodia Muslim Tripfrez	https://www.tripfez.com/packages/asia/cambodia	Festivals etc	Provide Halal tour activities to Muslim operator

Source: The author

Table 13. Halal Travel in Timor Leste

Tourist Guide / Halal Travel	Web link	Example of guide in Timor Leste	Remark
Sedunia Muslim Tour	Seduniamuslim.com	Accommodation, destinations etc	Provide Halal tour activities in Timor Leste
Timor Leste Travel Guide	https://www.vanguardvoyager.com/	Destination, food, hotel etc	Halal travel guide in Timor leste

Source: The author

Vietnam (Halal Tourist Guiding)

Vietnam is the easternmost country in Southeast Asia. Vietnam is located near to China border to the north and Laos and Cambodia to the West. According to Vietnam Religious Department, there are minority Muslim in Vietnam with a population around 72,000 Muslim people. Vietnam has become attractive to the Islamic visitor or Muslim travelers to travel as Vietnam is popular with delicious foods, markets and shopping. Below Table 9 display the web link that provide Halal tourist guiding in Vietnam.

Myanmar (Halal Tourist Guiding)

Myanmar or formerly known as Burma is a unique country in Southeast Asia as it has more than 100 ethnics group. Myanmar is located at the border of India, Bangladesh, China, Laos, Thailand and Yangon. The population of Myanmar is around 54 million as estimated by United Nation data. There is always a question post by Muslim travelers whether is it safe to travel to Myanmar especially for Muslim. As reported, even though Myanmar is popular with Rohingya case, however important to note that Myanmar has significant number of local Muslims that live peacefully with Buddhist. In fact it is also safe for Muslim travelers to holiday in Myanmar. For Muslim travelers, there are some information provided by Halal tour operator that provide detail information on Halal trip available in Myanmar. Table 10 below shows the Halal tourist guide for detail information for Muslim travelers.

Cambodia (Halal Tourist Guiding)

Cambodia or also recognized as Kampuchea is a country that located in the southeastern part of the IndoChina Peninsula in Southeast Asia. Cambodia has a population around 15 million and its capital city is known as Phnom penh. Muslim tavelers could easily find good Halal food in Cambodia in Phnom Penh and Siem Reap. These two location is a hot place that received Muslim tourist every year.

Timor Leste (Halal Tourist Guiding)

Timor Leste or also called as East Timor is located near to eastern side of the island of Timor. It is known as a small country and it is not far from Australia. The official language for Timor Leste is Portuguese and Tetum. In term of tourism perspective, Timor Leste is still at developing stage of its tourism industry. It would take time to develop Halal tourism activity in Timor Leste. As reported in Timor Leste Tourism

Development Report in June 2014, the common activity by tourist to Timor Leste mostly those who visit for business reasons, to see family and friends and to work for international or local government organizations in Dili. As highlighted in the report, there are very few that came to Timor Leste for the purpose of holiday, same goes to Halal trip or Halal tourism. Even though it is under development stage, however it is still have tourist guiding information for Halal Travelers that travel to the country. Below table 12 shows the Halal tourist guiding available in Timor Leste.

CONCLUSION

With the growth of the Muslim population in Southeast Asia, Halal tourist guiding is significant to be further developed to support the need of Muslim travelers. Each of government in Southeast Asia countries also aware the importance of Halal tour operator and Halal travel guidance to have a deep understanding and knowledge with regards to Halal Tourism. This is important to support the industry to sustain and calling for more tourist to visit the countries. In fact, the Halal activities provided by halal tour is actually the key Halal tourism sustainability. In general, this study has contributed in providing description on the Halal tourism, Halal trip activities as well as halal tourist guiding in a great detail related to halal information for Muslim travelers that travels to Southeast Asia countries.

REFERENCES

Alserhan, B.A., Wood, B.P., Rutter, R., Halkias, D., Terzi, H., & Serhan, O.A. (2018). *The transparency of Islamic Hotels: "Nice Islam" and the "self-Orientalizing"*. Academic Press.

Asean Post. (2018). Could Islamic Tourism be Brunei's answer? *The Asean Post*. https://theaseanpost.com/article/could-islamic-tourism-be-bruneis-answer

Bangkok Post. (2019). *Thailand ready to tap busy Muslim Travel Market*. https://www.bangkokpost.com/business/1696452/thailand-ready-to-tap-busy-muslim-travel-market

Battour, M., & Ismail, M. N. (2016). Halal tourism: Concepts, practices, challenges and future. *Tourism Management Perspectives*, *19*, 150–154. doi:10.1016/j.tmp.2015.12.008

Bon, M., & Hussain, M. (2010), Tourism in the Muslim world N. Scott (Ed.), Bridging tourism theory and practice (2010), vol. 2, pp. 47-59

Charmaz, K., & Belgrave, L. (2002). *Qualitative interviewing and grounded theory analysis T* (2nd ed.). Sage.

Crescent. (2014). *Why Should Tour Guides Be Aware of the Special Needs of Muslims?* https://www.crescentrating.com/magazine/opinion/3656/why-should-tour-guides-be-aware-of-the-special-needs-of-muslims.html

Crescent. (2016). *Philippines accelerates plans to develop Halal Tourism*. https://www.crescentrating.com/magazine/press-releases/3832/philippines-accelerates-plans-to-develop-halal-tourism.html

Duman, T. (2010). The value of Islamic tourism: Perspectives from the Turkish experience Islam and Civilizational Renewal (ICR). Academic Press.

El-Gohary, H. (2016). *Halal Tourism, is it really Halal? Tourism Management Perspective* (Vol. 19). Academic Press.

Elaziz, M. F., & Kurt, A. (2017). Religiosity, consumerism and halal tourism: A study of seaside tourism organizations in Turkey. *Tourism (Zagreb)*, *65*(1), 115–128.

Hamza, I. M., Chouhoud, R., & Tantawi, P. (2012). Islamic tourism: Exploring perceptions and possibilities in Egypt. *African Journal of Business and Economic Research*, *7*(1), 85–98.

Henderson, J. C. (2003). Managing tourism and Islam in Peninsular Malaysia. *Tourism Management, 24*(4), 447– 456.

Islamic Tourism. (2019). *Experience worry-free vacations with facilities adhering to Islamic Principles.* https://www.tourism.gov.my/niche/islamic-tourism

Jia, X., & Chaozhi, Z. (2020). "Halal tourism": Is it the same trend in non-Islamic destinations with Islamic destinations? *Asia Pacific Journal of Tourism Research*, *25*(2), 189–204. doi:10.1080/10941665.2019.1687535

Khan, F., & Callanan, M. (2017). The halalification of tourism. *Journal of Islamic Marketing*, *8*(4), 558–577. doi:10.1108/JIMA-01-2016-0001

Mansfeld, Y., Ron, A., & Gev, D. (2000). *Muslim Tourism to Israel—Characterization, Trends and Potential.* University of Haifa: Center for Tourism, Pilgrimage, and Recreation Research. (in Hebrew)

Mohsin A. (2005). Tourist attitudes and destination marketing— the case of Australia's Northern Territory and Malaysia. *Tourism Management, 26*(5), 723– 732.

Mohsin, A., Ramli, N., & Alkhulayfi, B. A. (2016). *Halal Tourism: Emerging Opportunities, Tourism Management Perspectives* (Vol. 19). Part B.

Pew Research Centre. (2015). *The Future of World Religions: Population Growth Projections, Year 2010-2050.* https://www.pewforum.org/2015/04/02/religious-projections-2010-2050/

Puvaneswary, S. (2017). *Brunei takes serious look at Islamic Tourism.* https://www.ttgasia.com/2017/11/16/brunei-takes-serious-look-at-islamic-tourism/

Rabotic, B. (2010). Tourist guides in contemporary tourism. *International Conference on Tourism and Environment*, 1-10.

Rahman, N.A., Abdul, M.F., Rahim, S.A., & Noh, H.M. (2018b). Implementing Air Cargo Halal Warehouse: Insight from Malaysia. *Journal of Islamic Marketing*, *9*(3).

Rahman, N. A. A. (2019a). Prominent and Innovative Tourism Events in Peninsular Malaysia in Routledge Advances in Event Research. In *Tourism Events in Asia, Marketing and Development*. Routledge.

Rahman, N. A. A. (2019b). Halal Business and Halal Technology in Korea: History, Concepts and Research Opportunities. *Test Engineering and Management*, *81*(Nov- Dec), 5782–5790.

Rahman, N. A. A., Mohamad, M. F., Muda, J., Majid, Z. A., & Noh, H. M. (2018a). Linking halal requirement and branding: An examination of halal flight kitchen provider in Malaysia. *International Journal of Supply Chain Management, 7*(3), 208–215.

Razalli, M. R., Yusoff, R. Z., & Roslan, M. W. M. (2013). A framework of halal certification practices for hotel industry. *Asian Social Science, 9*(11), 316.

Razzaq, S., Hall, C. M., & Prayaq, G. (2016). *The Capacity of New Zealand to accommodate the Halal Tourism market- or not*. Academic Press.

Rodrigo, P., & Turnbull, S. (2019). Halal Holidays: How is value perceived by Muslim tourist? *International Journal of Tourism Research, 21*(5), 675–692. doi:10.1002/jtr.2290

Ryan, C. (2016). *Halal Tourism, Tourism Management Perspective* (Vol. 19). Part B.

Shafaei, F., & Mohamed, B. (2015). Involvement and brand equity: A conceptual model for Muslim tourists. *International Journal of Culture, Tourism and Hospitality Research, 9*(1), 54–67. doi:10.1108/IJCTHR-06-2014-0050

Suradin, M. (2018). Halal Tourism Promotion in Indonesia: An Analysis on Official Destination Websites. *Journal of Indonesian Tourism and Development Studies, 6*(3), 143–158. doi:10.21776/ub.jitode.2018.006.03.01

Tatar, C. F., Herman, G. V., & Gozner, M. (2018). Tourist Guides'. *Contribution to Sustainability in Romania, Geo Journal of Tourism and GeoSites, 21*(1), 282–287.

The Straits Times. (2017). *Singapore is the 6th most popular travel destination for Muslim Millenials.* https://www.straitstimes.com/singapore/singapore-is-the-6th-most-popular-travel-destination-for-muslim-millennials-study

Timothy, D., & Iverson, T. (2006). Tourism and Islam: consideration of culture and duty. In D. Timothy & D. Olsen (Eds.), *Tourism, Religion and Spiritual Journeys* (pp. 186–205). Routledge Taylor & Francis Group. doi:10.4324/9780203001073

UNWTO Would Tourism Organization. (2019). Weidenfeld A, Ron A. 2008. Religious needs in the tourism industry. Anatolia. *International Journal of Tourism and Hospitality Research, 19*(2), 18–22.

Whitehead, R. (2019). *Sun, Sea and Saudi: How countries around the world are courting Muslim tourist.* https://cnaluxury.channelnewsasia.com/experiences/how-countries-are-courting-muslim-tourists-11571148

Zailani, S., Ali, S. M., Iranmanesh, M., Moghavvemi, S., & Musa, G. (2016). Predicting Muslim medical tourists' satisfaction with Malaysian Islamic friendly hospitals. *Tourism Management, 57*, 159–167. doi:10.1016/j.tourman.2016.05.009

Chapter 12
As an Alternative:
Cultural Routes Education for Tourism Guides – A Model Suggestion From Turkey

Canan Tanrisever
Kastamonu University, Turkey

EXECUTIVE SUMMARY

Tourist guidance training/education is different in many countries of the world. Tourism types or tourism choices affect the specialization areas of tourist guides. Tourist guides that should be one of the components of an ecologically based integrity are not yet included in all routes. This area, which is not included in the tourist guidance curriculum, has been proposed as a training module for specialization in this study. In this context, the aim of the study is to contribute to the field and literature by creating the infrastructure of the cultural routes education program as a module. The study is a heuristic research. The interview form used in the research was formed by utilizing expert opinions on qualitative data collection methods based on literature review. According to the results of this research, this module is divided into seven main topics.

1. INTRODUCTION

Tourist guidance training/education is different in many countries of the world. The tourist guide profile that countries need varies according to the geography, culture and the share of the country in the tourism pie. Tourism types or tourism choices of tourists affect the specialization areas of tourist guides. The changing world, the same as a result of globalization, has pushed individuals to search for originality and differentiated them in holiday preferences as in many choices. In recent years, the increase in the number of people who adopt the motto of healthy living, the philosophy of turning to nature and turning to one's self has increased the demand for nature walks. Trekking, which is difficult for rural people to make sense of, is preferred as an escape therapy from stress by citizens. Even the meaning of the word trekking has been differentiated by the perception that has changed over the years. Merriam-Webster Dictionary, trekking is used in Africa to migrate with a cane or by foot, whereas in the process it has

DOI: 10.4018/978-1-7998-3725-1.ch012

evolved into a long walk in nature that has changed in use. In nature, brisk walks to go from one point to another within a specified time period have also diversified over time and the concept of route has emerged. It has gained an economic dimension in alternative tourism activities. When the ancient cities, ancient roads and the roads passed by the important people on the routes are considered, the names of the route were called "cultural routes". When the demand for cultural routes, the main starting point of which is trekking, is not intense, those who walk on individual routes have received support from locals, climbers and other hikers from time to time. Today, this type of tourism, which includes both solo-hikers and groups, has become a professionalization area and the need for tourist guides has increased. If cultural routes are not seen as a specialty, the gap of this need will quickly be filled with other alternatives. The most important objective in the preparation of this section is to create a substructure related to this area of expertise, which is not yet known even among the tourist guides, and to prepare an auxiliary resource. Tourist guides that should be one of the components of an ecologically based integrity are not yet included in all routes. This area, which is not included in the tourist guidance curriculum, has been proposed as a training module for specialization in this study. In this context, the aim of the study is to contribute to the field and literature by creating the infrastructure of the cultural routes education program as a module.

If the travel contains a purpose and date such as hiking, mule path, faith, trade, or following the old paths, it can be defined as a cultural route (if it is suitable for walking by nature). According to different organizations, this concept was developed by the Council of Europe. According to the European Council, the purpose of the Cultural Routes Program is to demonstrate that Europe's different countries and cultural heritage is a common cultural heritage, and to show the way to travel in time and space. The basic principles of the European Council are human rights, cultural democracy, cultural diversity and identity, dialogue, mutual exchange and enrichment that transcend time and space (Tanrısever, 2016).

In this context, the components of cultural routes are;

- Historical cities,
- Rural Settlements,
- Traditional life,
- Ancient settlements,
- Singular historical buildings,
- Natural environment and wildlife.

Common features of routes are;

- Having a theme (Culture, History, Nature, Ecology, Geology, Belief, etc.),
- To be marked (Road signs, road start signs, maps, etc.)
- Documentation (web page, guide book, GPS coordinates, web page) (Robert & Eryurt, 2013).

Cultural routes can be completed on foot or by bicycle or horseback. The markings of cultural routes are made with the marking of the international standard Grande Randonne, meaning safe road. In this marking, the least damage to the nature takes precedence. Marking is made on dirt roads, trees, stones, rocky grounds. The walking paths are marked with red-white marking and the bicycle paths are marked with red-yellow marking. Start signs are placed at the start of the road. When creating routes, tangible and intangible cultural heritage should be identified; geographical locations should be defined and marked within the scope of a specific theme. Within a theme, different data sets are combined into the route to create continuity (Sugio, 2005).

Cultural routes are classified according to their identified characteristics as follows (Suárez-Inclán Ducassi, Rosa, 2005):

- According to regional scope: Local, national, regional, continental or transcontinental.
- According to their cultural content: different geographical areas within a given cultural region or sharing the process of mutual influence in the formation or evolution of cultural values.
- According to goals or functions: Social, economic, political or cultural. These features can also share a multidimensional context.
- According to temporal periods of use: those that are no longer used due to the impact of socio-economic, political and cultural changes.
- According to their structural configurations: Linear, circular, cruciform, radial or mesh.
- According to their geographical location: Land, water, mixed or other physical environments. Routes expressed linearly are defined according to linear, belt, network, cyclic and cluster spatial form.

Routes are classified with A and B theme. Theme A refers to the routes used for a specific purpose in a historical period and Theme B refers to the routes prepared for the purpose of developing tourism and preserving the concrete and intangible heritage (Table 1).

- *It was compiled by Görmüş's (2017) in New Approach to Preserve Cultural Landscape Values: Cultural Routes*

According to the classification listed in Table 1, cultural routes in Europe and in Turkey, are also included in Tables 2 and 3.

The concept of cultural route in the world was born in Europe. In 1954, in Paris, the European Council started to discuss cultural routes by the Europe working group called "Europe continues" and in 1985 this concept was developed in Grenada. In 1987, the Council of Europe established the European Cultural Routes Program and aimed at demonstrating the heritages of different countries and that countries in Europe have a common culture. In realizing this aim, the basic conditions of the European Council; human rights, acquisition of cultural identity, democracy, cultural diversity and dialogue were determined. The aim of mutual exchange and enrichment, which transcends time and space, were highlighted (COE, 2019; Bastemur, 2013; Ban vd. 2016). Other main objectives of this program are;

- to demonstrate that the common heritage of different countries and countries in Europe have a common culture,
- to travel in time and space,
- to raise awareness of European citizenship,
- to promote intercultural and inter-religious dialogue through a better understanding of European history,
- to develop and preserve cultural and natural heritage,
- to help sustainable development.

Table 1. Classification of Culture Routes

	Cultural Routes				
	Theme A		Theme B		
	• Routes Used for a Special Purpose in a Period of History		• The Defined Routes for the Use of Cultural Heritage and Landscape and Promotion of Tourism Activities, Ensuring Sustainable Development, Preservation of Natural and Cultural Heritage		
	A.1	A.2	B.1	B.2	B.3
Types	• Trade Trails • Pilgrim Routes • Migration Routes • Military Trails	• Routes of people who play an important role in history	• Historical Routes	• Historical and Nature Routes	• Naturel Routes
General Features	• Used regularly during a period of history; - Cultural exchanges and travels - Population movements - Continuity in time and space • It has dynamic and historical functionality with a well-defined purpose for route use • Existence of both tangible and intangible elements	• Only physical routes of travel for a special purpose, such as discoveries and occupations • Population and cultural changes without mobility during a period of history	• Thematic material related to actual / ideal connection defined / defined by historical and cultural heritage • Unused route in a period of history	• Present-day thematically defined by both historical and natural elements of geography associated with real or ideal bond •	• Observation of both historical and natural elements, as well as outdoor sports such as hiking and mountaineering.

The historical process that underpins the importance of the European Cultural Route Program is as follows.

1954- in Paris, the European Cultural Convention, the continuation of European culture (Turkey ratified in 1957)
1987- Determination of the Santiago de Compostella Route as the first route in Spain
1994- The first conference of UNESCO and ICOMOS in Madrid, the recognition of routes as cultural heritage
1997- Establishment of the European Cultural Routes Institute by the Council of Europe
1998- Establishment of ICOMOS International Committee on Cultural Routes
2000- Florence, European Landscape Convention, protection and planning of landscapes in Europe
2005 - UNSECO recognizes cultural routes as one of the four heritage categories
2005 - Faro, establishing the link between the Council of Europe Framework Convention on the Value of Cultural Heritage for Society, cultural heritage and human rights
2012- Colmar Declaration within the scope of 25th year of European Cultural Routes

The first pilgrimage route in 1987, the Santiago de Compostella Pilgrimage Route, was an example of subsequent cultural routes (Nillson & Mekonnen, 2016). Today, it is still one of the most popular routes. This route met the following conditions;

- Follow a route with a date or a newly designed route,
- Increasing multilateral cooperation projects such as scientific research, preservation, development and enrichment of historical heritage, cultural and educational exchange among young Europeans, modern cultural and artistic practices, culture and sustainable tourism,
- It is expected to be governed by one and more independent, organized associations and federated networks (COE, 2019; Bastemur, 2009).

The role of the European Institute of Cultural Routes in the program, which was established after the routes were regarded as assessing cultural heritage; evaluating project applications, monitoring the activities and activities of the collaborating organizations, disseminating information and archiving documents. Today there are 47 + 17 member states (council members and non-members) and 38 routes.

- *Prepared by the author.*

The concept of cultural route differs in the Americas and Asia. In America, hiking trails are generally located within the borders of national parks and are natural heritage rather than cultural features. According to researches, 92% of the American people spend their free time in national parks with walking and jogging (Americancamper Report, 2017). In the Far East countries, instead of the concept of cultural route, the term walking paths is more commonly used, especially in Japan with the name Shrin Yoku. Turkey is not a member of European Union but it is a member of Cultural Routes Council. The cultural routes in Turkey are given in the second chapter.

2. CULTURAL ROUTES IN TURKEY

The beginning of the cultural routes in Turkey dates back to 1999 when Garanti Bank launched a contest for this purpose. Lycian Way project created by British Kate Clow living in Turkey is selected. Turkey's first trail is the Lycian Way, between routes is still one of the most preferred route of the marchers. The second route led by Kate Clow is St. Paul's Route (Clow, 2000; Clow, 2017; Bastemur, 2011). Because Turkey is not a European Union member, routes are not included in the European Union Cultural Routes list. Taking place on the list, passing routes from Turkey are Jewish Heritage, Olive Tree and Historical Thermal Routes. The objectives in the planned cultural route as follows:

- To preserve, interpret and present natural and cultural heritage in a holistic way
- To provide the perception of regional identity by going beyond single heritage areas
- Revive and promote the original values of rural areas
- To develop projects that will mobilize the economic potential of lost values such as traditional production and local architecture
- To develop new tourism areas alternative to mass tourism and all-inclusive tourism types
- To alleviate the pressure at places where tourism activities are concentrated, such as the coastline, sensitive archaeological sites and historic city centers
- To support regional development by creating small scale investment areas in the service sector
- To establish horizontal cooperation between local stakeholders
- To raise awareness in the local community and to create awareness
- To provide a qualified and developing travel experience for those who will experience the area / region (Bastemur, 2013; Campolo vd. 2016; Clow, 2011).

Table 2. The cultural routes of the different categories which are certified by the Council of Europe are as follows.

	Name	Certificate Year	Content	Number of Countries	Type
1	Santiago de Compostela Piligrim Route	1987	Pilgrim	9	A1
2	The Hansa	1991	Trade	16	A1
3	Viking Route	1993	Sea, trade	14	A1
4	Via Francigena	1994	Religion	4	A1
5	El Legado Andalusi Route	1997	Architectural, tangible and intangible cultural heritage	8	B1
6	Phoenicians Route	2003	Sea, tangible and intangible cultural heritage	9	B1
7	Pyrenean Iron Route	2003	Architecture and art	3	B1
8	Mozart Way	2004	Architecture and art	10	B1
9	European Route of Jewish Heritage	2004	Jewish heritage	19	B1
10	St. Martin of Tours Route	2005	Religion	10	A1
11	Cluniac Sites in Europe	2005	Tangible and intangible heritage	8	B1
12	Olive Trees Route	2005	Gastronomy	14	B3
13	Via Regia	2005	Tangible and intangible heritage	7	
14	Transromanica	2007	Tangible heritage	9	B1
15	Iter Vitis Route	2009	Vinitourism	19	B3
16	Cistercian Abbeys Route	2010	Tangible heritage	11	B1
17	European Cemeteries Route	2010	Art	20	B1
18	Prehistoric Rock Trails	2010	Prehistoric art	6	B2
19	European Route of Historic Thermal Towns	2010	Health	15	B3
20	Route of St. Olav Ways	2010	Religion	3	A1
21	European Route of Ceramics	2012	Ceramic Heritage	5	B1
22	European Route of Megalithic Culture	2013	Megalithic Heritage	7	B1
23	Huguenot and Waldensian Trail	2013	Culture and history	4	B2
24	ATRIUM	2014	Architecture	5	B2
25	Réseau Art Nouveau Network	2014	Art	16	B2
26	Via Habsburg	2014	History and art	4	B1
27	Roman Emperors and Danube Wine Route	2015	History	4	B1 B2
28	Emperor V. Charles Route	2015	Tangible and intangible heritage	10	A2
29	Destination Napoleon	2015	History	12	A2
30	In the Footsteps of Robert Louis Stevenson	2015	Culture and History	4	A2
31	Fortified Towns of the Grande Region	2016	History and Architecture	3	B1
32	Impressionism Route	2018	Art	7	B2
33	Via Charlemagne	2018	Art	7	B2
34	Iron Curtain Trail	2019	History	23	B1
35	Liberation Route Europe	2019	History	11	B1
36	European Industrial Heritage Route	2019	Industrial Heritage	24	B1
37	Le Corbusier Destinations: Architectural Promenades	2019	Architecture	6	B1
38	Reformation Route	2019	History	8	B1

While planning, establishment of infrastructure, determination of financial resources, expense items, stakeholders and collaborators, establishment of working units, preparation of process work schedule, obtaining full control of the work area, receiving expert support and meeting with the concerned persons, drawing up of land studies, values and regions, itinerary and alternative routes, determination of special project areas, evaluation of tourism activities on the routes, identification of promotion strategy and route marking, mapping, GPS, route book, information and guidance systems should be established within this scope (Karataş, 2015; Kervankıran & Çuhadar, 2014; Valeru & Malaescu, 2016). By passing through these stages, the accepted cultural routes in Turkey are as follows.

2.1. Abraham's Path

Aim: To bring holy and historical places together in a single route and to commemorate Abraham's legacy of hospitality and friendship.

Walking Time: Possible to from February to May and from October to November.

General Information: Abraham, meaning the father in Hebrew, is the historical personality that made the Kaaba known as the ancestor of the prophets with the events that took place in the Torah and the Qur'an. Sanliurfa, known as the place of birth and the place of childhood, has been accepted as a sacred land for more than a thousand years. Part of the route is 170 km long in Turkey. As the region is very hot, morning hours are generally preferred for walks (Cultural Routes Society, 2019).

2.2 Between Two Sees

Aim: To create an opportunity to observe Istanbul's past and present.

Walking Time: Possible to walk in all seasons depending on weather conditions.

General Information: The four routes, each consisting of 4 tracks, each with an average of 15 kilometers, are 60 kilometers in total. On the route from rural areas to the city center, it is possible to see areas such as today's lignite quarries, the 3rd bridge, the airport, as well as the historical Yarımburgaz Cave. It is a walking route of four days between the Black Sea and the Marmara Sea (Cultural Routes Society, 2019).

2.3 Carian Trail

Aim: To create an alternative tourism activity parallel to today's holiday centers by creating a hiking route that follows the traces of the Carian civilization that lived in ancient times.

Walking Time: Possible to walk from March to April and from October to November.

General Information: Its 820 km length is Turkey's long-distance hiking trail. Caria Civilization gives the name to the route. It has been known as Caria since 2000 and the civilaztion lived its most independent period BC. 11.-6.. Halicarnassus (Bodrum) is one of the most important cities. The tomb of King Mousolos, one of the seven wonders of the ancient world, is located in Bodrum today. The route is not only for its natural beauty, but also for its historical value, which all professional or non-hikers want to experience. It is marked in accordance with international standards (Karyayolu.com).

2.4 Ephesus-Mimas Route

Aim: to blend many different types of tourism at the birthplace of European civilizations and share them with hiking and cyclists.

Walking Time: Possible to walk in all seasons except for the summer times.

General Information: The route consists of 773 kilometers with 15 tracks. The 450 km hiking route consists of 49 hiking trails. The route starts from the Temple of Artemis, one of the seven wonders of the ancient world in Ephesus, ends at Karaburun, whose ancient name is Mimas. This route includes the blue route, the vineyard route and the olive route. Six of the 12 cities of the Ionian Civilization, the most powerful and most influential in the ancient world, are located on this route. The region, which is called Ionia from the Gulf of Izmir to the Gulf of Güllük, forms the basis of today's European Civilization. Famous names like Herodotus, Homer, Thales, Pythagoras, Diogenes lived in Ion Civilization. The 12 Ionian cities estimated to be founded in 1000 BC are on the route; Ephesus (Ephesus-Selcuk), Colophon (Değirmendere-Menderes), Lebedos (Ürkmez-Seferihisar), Teos (Sığacık-Seferihisar), Klazomenai (İskele-Urla) and Erythrai (Ildırı-Çeşme). The 190 km route between Cesme and the Temple of Artemis meets the standards of the European Cycling Network (EuroVelo). Since the route consists of places where olive and viticulture activities are carried out intensively, the "Olive Road" is used as Güzelbahçe, Çeşme, Seferihisar, Menderes and Selçuk. The Vineyard Road is 151 km length covering Selçuk, Seferihisar, Menderes, Urla, Karaburun and Çeşme trails. The blue route is a sea route that consists of sea routes and extends from the first known port to the most developed (Rotayarimadaizmir.com, 2019).

2.5 Evliya Çelebi Way

Aim: To bring together the Ottoman heritage with horseback riders, cyclists and hikers and to create alternative rural routes for the urban dodging citizens. It is marked with detailed descriptions of Evliya Çelebi.

Walking Time: Possible to walk in all seasons depending on weather conditions.

General Information: Evliya Çelebi route which is the first horse riding route in Turkey was opened to the visitors in 2011 declared by UNESCO as the 400 year of Evliya Çelebi's birth. Evliya, calls himself as the world traveler, the friend of mankind, and it is also claimed that he wore a ring bear "Evliya, the world traveler".mark on it. Evliya Çelebi, who has many features, collected his 10-volume travelogue from most of his official places for 40 years. Although the credibility of the travelogue is still open to debate, it is the most important travel guide from the 17th century to the present. The route, which consists of the first stages of Evliya Çelebi's pilgrimage journey, stretches from the Gulf of Izmit to Simav. It consists of a 330-kilometer hiking route and a 650-kilometer equestrian riding route (Finkel & Clow, 2011). The route can be walked in 22 days. Many places that witnessed the establishment of the Ottoman Empire on the route can be walked through rivers, footpaths, forest roads, paved roads from the Roman and Ottoman periods.

2.6 Fethiye Hiking Trails

Aim: To take an advantage of the natural and cultural diversity in Fethiye to create alternative walking trails.

Walking Time: Possible to walk in springs and winters.

General Information: The route, which is complementary to the 79 kilometers of the Lycian Way passing through Fethiye, is 320 kilometers long. Routes consisting of different regions of Fethiye have been implemented under the sponsorship of Fethiye Chamber of Industry and Commerce. The route is marked with red-yellow lines at 50 m intervals on paths and 200 m on dirt roads. In addition, a 689-kilometer bicycle route, jeep-safari, botanical and bird-watching routes have also been identified (Cultural Routes Society, 2019).

2.7 Hittite Trails

Aim: To revive the historical route that connects the three most important cities of the Hittite civilization.

Walking Time: Possible to walk in all seasons depending on weather conditions.

General Information: The Hittite Civilization, which was destroyed in 1200 BC, came to Anatolia in the early 2000s, is one of the most important civilizations that left its mark in history with its religion, writing, laws and urbanism. Boğazköy-Alacahöyük National Park, registered by UNESCO in 1986, is the main point of the Hittite Road. Hittite Road Hiking and Cycling Trails work was completed in 2010 by the Çorum Governorship. The length of the six bike routes is around 406 km. The route consists of 236 kilometers of walking paths and includes 17 tracks. This distance reaches 385 kilometers with connection roads (Demirel, 2012). Marking and mapping have been completed.

2.8 Gastronomy Route

Aim: To combine gastronomy and eco-tourism in a basin.

Walking Time: Possible to walk in the period from April to November.

General Information: Kizilirmak Basin Corum Gastronomy and Hiking Route consists of 25 hiking trails of 190 kilometers. The cycling part of the 305 km hiking route with alternative routes is 606 km long. 7 different bicycle routes are planned. The driving route is 702 km. Starting from Kırşehir, thc motorway passes through Kırıkkale, Ankara, Çankırı, Çorum, Sinop and Samsun. This road extends to where Kızılırmak flows into the sea. The 359-kilometer-long Hittite Road, full of historical beauty of Çorum, is another walkable route. It is also suitable for jeep-safari tours of 1186 km, which includes a cruising road and all bike routes. Foods offered for walkers with gastronomic motivation include different kinds of local foods (Gastronomiyolu.com, 2019).

2.9 Idyma Way

Aim: To protect and promote to the natural and historical beauties of Gökova Nature Conservation Area.

Walking Time: Possible to walk in all seasons depending on weather conditions.

General Information: In 1988 Gokova region was proclaimed as Turkey's first "Special Environmental Protection Area". Akyaka is one of the most important ornithology areas in Turkey. Idyma was historical name of the region. It consists of 10 routes with a length of 150 kilometers.

2.10. Independence Way

Aim: To remind the spirit of national struggle is to create an alternative route for hikers.

Walking Time: Possible to walk from April to November.

General Information: It is the land road created by the efforts of volunteers and inmates during the War of Independence to bring the ammunition from Istanbul and Russia,. Mustafa Kemal's words, " I keep my eyes on Sakarya, my ears on İnebolu", emphasize the importance of Independence Road. The entire road consists of 105 km in the direction of İnebolu-Kastamonu-Çankırı-Ankara. The most common part is Ersizlerdere, Karacehennem Strait, Ecevithan, Çataltepe-Ödemiş and Çuhadaroğlu. Catak Pond and Doganlar Castle are other places to see on the road. The other part of the trail is 10 km and consists of the Kağni road (Kaynaş, 2019). It was declared a National Park in 2019. In 2007, the marking was completed and put into service.

2.11 Kaçkarlar

Aim: To create a route to nature for hikers and climbers in winter.

Walking Time: Possible to walk from August to September and from February to March.

General Information: Kaçkar Mountains separating the Black Sea which is the extension of the Caucasus from Anatolia are located between Rize and Hopa. The highest peaks are Altıparmak, Kavran and Verçenik. All 8 hiking trails are marked and signposted. The degree of difficulty is moderate. Hostels are available near the glacial lakes. The most preferred hike is climbing the Kaçkar Mountain at 4000 meters. It has three glacier masses, numerous glacial lakes, sharp basalt peaks, high plateaus, and a biodiversity that has been shown to be numbered in the world (Clow and Gardner, 2012).

2.12 Küre Mountains

Aim: To contribute to ecotourism.

Walking Time: Possible to walk between May and November.

General Information: The Kure Mountains is one of the 9 hot spots in Turkey that are chosen by World Wide Fund for Nature (WWF) as one of the 100 hot spots that should be immediately protected.. The Küre Mountains, which have the status of a National Park, have borders to both Kastamonu and Bartın. The hot spot is defined as regions with an endemic species of more than 1500, which lost 70% of their original vegetation WW (WWF, 2019). The hiking trails created as an eco tourism project prepared by the Küre Mountains National Park Directorate were put into service in 2010. The route is 482 km long which has 36 tracks along its length, reaches 762 kilometers along with alternative routes (KDMP, 2019). There is also a 828-kilometer mountain bike route. Each level has a level of difficulty and is suitable for all hikers. The route is also connected to Yenice and Independence Road.

2.13 Lycian Way

Aim: To mark the paths connecting the ancient cities of Lycian Civilization and to create a walking route for nature lovers with the traces of Lycian Civilization. Turkey is the first track in the country made to identify and preserve the ancient paths.

Walking Time: Possible to walk from February to May and from September to November.

General Information: Lycian civilization, which means "Land of Light, lived in the Teke Peninsula. It is a civilization which reached at the peak of its power in the 6th century. It was a democratically independent, artistically developed sea tribe. The Lycian Way was created by discovering and then marking a historical trade route linking 25 ancient Lycian cities. It is the first cultural route in Turkey.

It takes 1 month to walk 540 km. The route consists of old Roman roads, paths and mule roads. All are marked in international standards. It addresses many hikers with different levels of difficulty (Clow and Richerson, 2000; Clow, 2018).

2.14 Mount Ararat

Aim: To bring together nature lovers and climbers with high peaks.

Walking Time: Possible to walk in all seasons depending on weather conditions.

General Information: With peaks over 3000 meters and is considered one of Europe's highest peaks Mount Ararat was also a part of Urartu Kingdom a thousands years ago. The reason it is called Turkey's roof, is to have four of the country's highest peak. Height in order; Great Ağrı (5137 m), Süphan Mountain (4058 m), Küçük Ağrı (3896 m), Tendürek Mountain (3533 m) peaks are the most preferred routes for both national and international hikers and mountaineers. Although the summit of Mount Süphan is in Bitlis, its northern slopes are within the boundaries of Ağrı. Apart from these peaks, Mount Köse, Balık and Küp Lakes are the other trails preferred by hikers. On the route, it is possible to see the works of the Urartian Kingdom and the Ishak Pasha Palace, the only palace structure in Anatolia that was built in 1784 and survived from the Ottoman Empire. Furthermore, the rock structure, lakes, numerous bird species, flora and fauna formed as a result of volcanic eruptions have been preserved as a national park since 2004. Turkey's most massive extinct volcano with a glacier of Mount Ararat, the Ark is one of the legend of Christian pilgrimage routes. For this reason, it also hosts the hikers who come within the scope of faith tourism (Cultural Routes Society, 2019).

2.15 Mysia Trails

Aim: To be highlighted from Misi to Gölyazı, Nilüfer's historical, touristic and natural values, to ensure the diversification of tourism.

Walking Time: Possible to walk in all seasons depending on weather conditions.

General Information: Mysia Roads, which has horseback, bicycle and hiking trails, takes its name from the name used in the region in Antiquity. There are also areas such as Aviation Facility of Nilüfer Municipality, Misi Camping Area, Equestrian Hiking Trail. Route of Trans Mysia is organized for hiking, cycling and horse-hiking trails. All are marked. The 310 km long route appeals to extreme sports enthusiasts, hikers of all levels of difficulty, and those who are interested in Bursa's history, culture, flora, fauna and rural life (Nilüfer Municipality, 2019).

2.16 Peninsula Trails

These excursion routes include İzmir/Efes-Mimas, Olive Way, Vineyard Route and Blue (Sea) Route. Details are provided on the Ephesus-Mimas Route.

2.17 Phrygian Trail

Aim: To present nature lovers with the artifacts from the Phrygian Civilization.

Walking Time: Possible to walk in spring.

General Information: The road is named after the Phrygian Civilization, which lived in an area covering Afyonkarahisar, Ankara, Eskisehir, Kutahya, Konya, Isparta and Burdur between 1200-700 BC. Gordion (Yassıhöyük), Pessinus (Ballıhisar) and Yazılıkaya (Midas City) are important centers of the Phrygian Kingdom. It is possible to see the rock monuments that easily form because of the fact that the Phrygian cities are composed of lava of Turkmen Mountain on the Phrygian road which consists of ancient ways connecting the cities of this civilization. The 506-kilometer route passes through 60 villages. Directional signs are placed at the start and end points marked in red and white in international standards. It was put into service in 2012. There is no water problem on the route. Most of the route is suitable for cycling (Cultural Routes Society, 2019; Sarı, 2011).

2.18 Sarıkamış Hiking Trails

Aim: To bring nature lovers together with alternative tourism activities.

Walking Time: The appropriate walking time varies according to the expectation of the hikers. Sarikamis, which is usually snowy for 5 months, is preferred by nature hikers in spring.

General Information: It consists of hiking, cycling and track vehicle trails. The difficulty level of the entire marked route is different. There are 256 kilometers of hiking trails, 356 kilometers of cycling routes and 428 kilometers of driving routes on 21 different routes. Sarıkamış is a route where many different tourism activities can be done with its historical importance, flora, fauna and ski resort (Demirel, 2012b).

2.19 St. Paul Way

Aim: To contribute to rural tourism through the interaction of nature and faith tourism by raising awareness about St. Paul, a very important character in the history of Christianity. It was created to give an idea of where St. Paul's first travels in Asia Minor took place.

Walking Time: Possible to walk in spring.

General Information: It is a route that follows the journeys of Saint Paul, one of the most influential figures in the history of Christianity, who is not one of the apostles of Jesus Christ, who never even met him, to spread the doctrine of Christianity. There are many camping areas along the 500 km route. The duration of the route is 27 days. It is located between Perge and Yalvaç district, northeast of Lake Eğridir. Since the route hosts many different civilizations, there are many historical elements that can be seen on the route. It is a convenient route for cycling as it crosses Roman roads, forest roads and paths. It was opened in 2008. It is an advantage for walkers that the starting point is close to Antalya Airport. Eğridir is the basic transportation and camping area. Village houses and pensions are also located on the route (Clow, 2018).

2.20 Sufi Trail

Aim: To revive the spaces of Sufis, scholars and poets who lived in different centuries in Anatolia and to raise awareness on the themes of culture and belief to the new generation and to convey the spiritual values possessed.

Walking Time: Possible to walk in all seasons depending on weather conditions.

General Information: Built in 2014 and 801 km in length, the Sufi Road is a route designed for hiking and cyclists marked by international standards. It is possible to see and experience places of worship of different religions on the route extending from Istanbul to Konya (https://sufitrail.com).

2.21 Sultan's Trail

Aim: To create a nature and history walking route in long distance by following the routes of the I and II Vienna Campaigns performed by Suleiman the Magnificent. It also aims to remind the concepts of hospitality, respect, justice and peace.

Walking Time: Possible to walk in all seasons depending on weather conditions.

General Information: It is planned as an international route. In 1529, the Habsburg dynasty built a castle in the village of Simmering near Vienna, where Sultan Suleiman the Magnificent established his car on the Vienna Expedition. The route starts from Austria, Hungary, Croatia, Serbia and Bulgaria and ends at the Topkapi Palace in Istanbul after visiting the Süleymaniye Complex where the Kanuni mausoleum is visited. In Turkey, which is part of the 390, a total of 2100 kilometers of roads (Cultural Routes Society, 2019).

2.22 Trakia Trails

Aim: To create a single portal by planning destinations from tourist products in the Thrace region.

Walking Time: Possible to walk in all seasons depending on weather conditions.

General Information: The Thrace tourism route is known by its name and combines with the titles of nature, faith, culture, gastronomy and flavor. Turkey's first vineyard trail is located in this region (Ergüven, 2015).

2.23 Tolerance Way

Aim: To see the cultural and natural beauties by creating walking paths.

Walking Time: Possible to walk in all seasons depending on weather conditions.

General Information: In the region between Izmit (Nicomedia) and Iznik (Nikaia), which is dominated by the Romans. In addition, it is aimed to highlight the historical fabric of the city and the development of rural areas. The establishment of the Nicomedia-Nikaia relationship as well as the preservation of the recent heritage is one of the aims. Starting from Izmit, the Tolerance Road crosses the Samanli Mountains, Izmit Bay, Iznik Lake and ends in front of the Hagia Sophia Mosque, meeting with Evliya Celebi and Sufi Roads for the last 8 kilometers. The name Tolerance for the road was given by the Roman Emperor Galerius in 311. This decree gave Christians the freedom to establish churches and worship. In 313, the rights were extended with the Edict of Milan and the damages suffered by Christians were tried to be compensated (Çalık, 2017). The length of the route in this region, which is very important in the history of Christianity, is 126 km.

2.24 Troy Culture Route

Aim: To create a route consisting of places where the Trojan War takes place and to share the path with cyclists, hikers and nature lovers within the scope of sustainable tourism.

Walking Time: Possible to walk in all seasons depending on weather conditions.

General Information: From Canakkale to Assos ancient harbor is an area of 120 kilometers covering an extensive area. This is only the first stage. The project started in 2018 is still ongoing. This area's historical name was Troas. Troas has never lost its significance over the course of history. It carries the traces of the Trojan Wars, the Hellens, Lidians, Persians, of Alexander the Great, Rome, Byzantium and the Ottoman Empire. Also, the Dardanelles Campaign was an indication of how the region played a key role in history in the way that it changed the course of world history as well as the First World War itself.

2.25 Via Egnatia

Aim: To revive as a way of communication it aims to be a route of friendship and peace in which people, goods, ideas can be exchanged.

Walking Time: Possible to walk in all seasons depending on weather conditions.

General Information: It is planned as an international route. Via Egnatia was a passageway between East and West, with Via Appia in Italy. Albania, Macedonia, Greece and Turkey in Thrace region is passing through (Ergüven, 2015). The part passing through the region is 475 km. The best part of the road is in Albania.

2.26 Yenice Forests

Aim: To create an ecotourism activity that can take place in the region.

Walking Time: Possible to walk between April and November.

General Information: As in Kure Mountains National Park, one of the 9 hot spots in Turkey is Yenice Forests. It is regulated by the Governorship of Karabük and Yenice District and opened in 2009. The route, which consists of 21 tracks in 210 km length, reaches 396 kilometers with alternative routes. The courses are divided into daily, short and long runs. The entire route is marked and marked (Cultural Routes Society, 2019). In addition, the most important part of the Yenice Forests is the Natural Conservation Areas and the Arboretum section, which also includes monumental trees.

Places to visit are not included in the walking periods mentioned in the above section.

Table 3. Cultural Routes in Turkey

	Route	Distance(km)	Location	Content	Relevant Institution	Classification and Type
1	Abraham's Path	170	Between Şanlıurfa, Harran and Göbeklitepe	History, religion	CRS	International, A1
2	Between Two Seas	60	İstanbul Karadeniz and Marmara	Panaroma	CRS	National, B3, B4
3	Carian Trail	820	Between Büyük Menderes River and Dalaman Stream	Water, history, nature	CRS	National, B2
4	Ephesus-Mimas Way	773	Selçuk, Menderes, Seferihisar, Güzelbahçe, Urla, Çeşme and Karaburun	History, nature, cycling horse riding	CRS	Regional, B2, A1
5	Fethiye Walking Tracks	320	Fethiye	Nature, history, cycling	CRS	Local, B2, B3
6	Gastronomy Route	305	Kızılırmak Delta	Cycling, gastronomy	CRS	National, A1
7	Hittite Trail	236	Çorum	History, cycling	CRS	National, B2
8	Idyma Way	150	Akyaka and Gökova	Nature	CRS	Local, B3
9	Independence Trail	105	İnebolu-Kastamonu	History, nature, cycling	CRS	National, A1
10	Küre Mountains Trail	482	Kastamonu (Azdavay, Cide, Pınarbaşı, Şenpazar) ve Bartın (Kurucaşile, Ulus)	Nature, mountain trekking	CRS	National, B3
11	Lycian Way	540	Teke Peninsula	History, water, camping	CRS	National, B2
12	Mount Ararat	-	Ağrı	Nature, nature walking, history and religion	CRS*	National, B3
13	Mysia Tracks	310	Between Misi and Gölyazı Arası	Nature, cycling, camping, history	Bursa Nilüfer City Hall	Local, B2
14	Peninsula Trails	-	İzmir	History, nature, cycling, olive trees, vineyards	İzmir Metropolitan Municipality	Local, A1
15	Phyrgian Trail	506	Afyonkarahisar, Ankara, Eskişehir, Kütahya, Konya, Isparta and Burdur	History, cycling	CRS	National, B2
16	Sarıkamış Trails	256	Kars	Nature, cycling	CRS	National, B3
17	St. Paul Way	500	Between Perge-Eğridir	Religion, history, nature, horse riding	CRS	National, A2
18	Sufi Trail	801	Between İstanbul and Konya	Religion, history, nature	CRS	National, A1
19	Sultan's Trail	390	Between Vienne and İstanbul	Nature, history	CRS	International, A1
20	The Evliya Çelebi Way	330	İzmit Gulf-Simav	History, cycling, horse riding	CRS	National, A2
21	The Kaçkar	-	Between Trabzon and Erzurum	Nature, nature walking	CRS	National, B3
22	Tolerance Way	126	İzmit-İznik	Nature, history, religion	CRS	National, B1, B2
23	Trakya Routes	-	Trakya	History, nature, religion, vineyards	Multiple shareholders	Regional, B1
24	Troy Culture Route	120	Between Canakkale-Assos	History, nature	Multiple shareholders	Regional, B2
25	Via Egnatia	475	Albania, Macadonia Greece and Turkey	Culture, history	CRS	International, A1
26	Yenice Forest Trails	210	Karabük	Nature	CRS	National, B3

The alphabetically listed table is prepared by the author

* CRS: Cultural Routes Society

*** Above chapter has been translated into English from the Cultural Routes Specialization chapter of Tourist Guides' Specialization book (Tanrısever, 2019; edt: Koroglu, O., Guzel, O, 2019).*

Figure 1. Map of Cultural Routes in Turkey
**Map is prepared by the author*

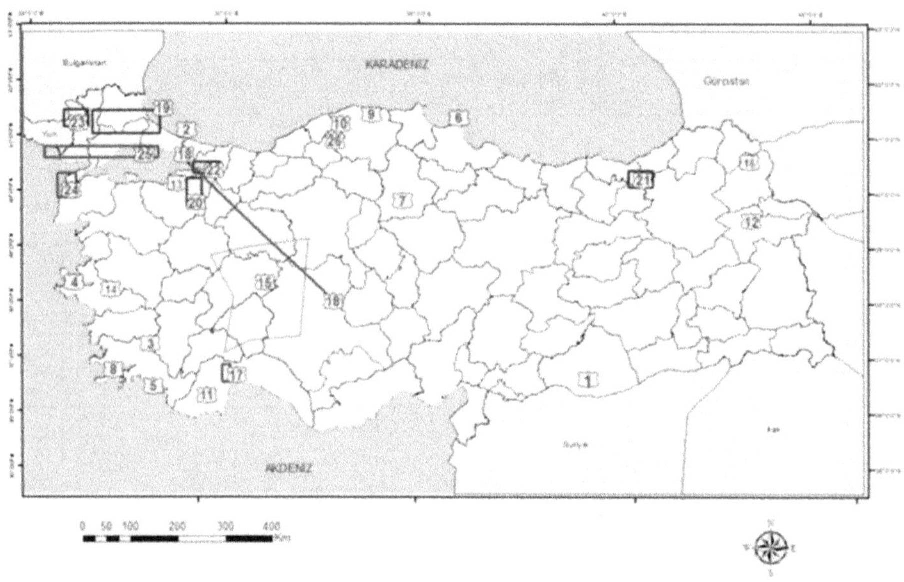

3. METHODOLOGY

There is no respectably source about cultural routes in Turkey. Information on some of the cultural routes is available on the Cultural Routes Society website. However, there is no data in the literature that contains information such as walking time, length, and appropriate walking time. In this context, this study will contribute to the literature by providing comprehensive source related to cultural routes in Turkey, classification and grouping of the routes. Secondly, it does not provide classes on this topic in tourism education institutions in Turkey, and this issue is not included in the seminars held in the tourist guidance profession. Unfortunately, the mountaineering federations provide trainings on this subject, which concerns the tourist guidance profession, only with a mountaineering perspective. Within the scope of this study, this deficit in practice has been resolved and the module of the class has been proposed.

The study is a heuristic research. In the studies carried out with the aim of exploratory research, it is tried to determine with the main lines what a problem is actually felt and which variables occur under the influence, what are the most appropriate approaches to illuminate the situation. The techniques used with the research method followed are as flexible as possible; what, why and how to collect and evaluate is not well known in advance. Based on this information, although exploratory research does not provide general recommendations in a study, it can shed light on important decisions about how and in what detail the problem should be addressed. In the study, interview technique, which is one of the qualitative data collection methods, was used. In the study, the opinions of the stakeholders were discussed face to face with the help of structured interview form. The interview form used in the research was formed by utilizing expert opinions on qualitative data collection methods based on literature review. The interview form consists of two sections and a total of 15 questions.

In the study, 5 different persons from different backgrounds were interviewed: an expert from the Cultural Routes Society board; a licensed tour guide with 15 years of trekking experience; a mountain climber and also instructor from Turkey Mountaineering Federation; a mountain climber who is also academic; and an academic who is a also tourist guide. On October 29-31, the average half-hour interviews were recorded with a voice recorder. In addition, the date, time and location of each interview were noted. Research findings were obtained from observation, interview and document analysis. Therefore, the data obtained from three different sources were examined comparatively. In addition, the information obtained through literature review is presented as tables in the previous sections. In the lights of the findings obtained, routes in Turkey, according to aim, walking time and general information are listed above.

4. FINDINGS

4.1 Answers of the Participants

Demographical information of the participants is given in Table 4.

Table 4. Demographical Information of Participants

No	Gender	Age	Job	Route Experience	City
1	Female	70	Route trekking expert and route creative	50 years	Antalya
2	Male	45	Tourist guide	15 years	Muğla
3	Male	38	Mountain Climber/Instructor	13 years	Gebze
4	Male	34	Mountain Climber/Academic	14 years	Kastamonu
5	Female	38	Tourist guide/Academic	15 years	Istanbul

According to nine questions, answers of the participants are given below. Questions 4 and 5 were combined because of the similarity of the content and were handled together.

Table 5. Answers of Q1

P1	Flora and fauna, direction finding, trekking tricks, camping
P2	Communication skills, flora, survival in nature, camping, nutrition, finding water sources in nature, climate, drama
P3	Definition of routes, selection of suitable routes, determination of alternative routes, access to the route in emergencies, geographical features of determined routes, seasonal evaluation of routes, structure of land, slope of land, water sources, crossing points of wild animals, sheltering areas, flora and fauna effects of selected routes, identifying the visitors, creating human resources to serve.
P4	Direction finding, flora and fauna, access to the route in emergencies, crossing points of wild animals, sheltering areas on the route, proper nutrition, social skills, access to water sources, route knowledge, digital skills.
P5	Proper clothes and material choice, knowledge of climate and land structure of the region, direction and mapping, healthy food choices, camping, survival in nature, crossing points of wild animals, finding water sources, flora and fauna knowledge of the region.

Q1. *"What must be in the Cultural Routes Course?"*

Q2. *"What are the features that a tourist guide should have in order to specialize on the Cultural Route?"*

Common points of all participants are the love of nature and the environment and the desire to spend time in nature. The first and second participants emphasized that the tourist guide must have sufficient walking experience. The third participant pointed that tourist guides must be social and innovative in human relationships and must have a desire to explore new places.

Q3. *"If training on Cultural Routes is prepared, how should it be?"*

Participant 1,2,4 and 5 said that training must be both theoretical and practical. Participant 3 emphasized that there should be an educational content suitable for the target audience to always prioritize protecting the environment and nature.

Q4. *"Which courses should be given in this training?"*

Table 6. Answers of Q4

P1	Proper material knowledge, nature psychology, knowledge and leadership in nature, nutrition and fluid intake, mapping, good communication with local people, sustainability, tourist types.
P2	Communication skills, survival in nature, camping, finding water sources in nature, climate, drama.
P3	Psychology, communication, foreign language, knowledge and leadership in nature, sports education, nutrition and fluid intake.
P4	Regional knowledge, nutrition, nature protection
P5	Cultural and geographical knowledge of the route, concepts of activities, communication skills, wild life information.
Common points of all	Basic first aid training, finding direction in nature, camping, shelter and rescue knowledge in nature, flora and fauna, knowledge of geography.

Q5. *"How long should the training be? (Every day program) What is your expected time for the theoretical and practical part of the training?"*

Table 7. Answers of Q5

P	Theoretical	Practical
1	1 month	6 months
2	2 weeks	1 week
3	3 months	1 year
4	1 month	1 year
5	1 month	1 year

Q6. *"Who and what organization should organize this training? In short, who should be the stakeholders of this training?"*

All participants expressed the same opinion and said that governorships, universities, tourism directorates, municipalities, health directorates, non-governmental organizations, professional chambers, extreme sports clubs, youth sports directorates should be stakeholders.

Q7. "*Should it be a prerequisite for Cultural Route training? If yes, what should be*?"

Except for the first participant the others said "yes". Participants 2,4 and 5 pointed at university degree. The third participant said that "*The feasibility studies of the routes should be done with precision and target groups should be determined according to the feasibility studies.*"

Q8. "*What are your recommendations for guides that specialize in cultural routes*?"

Table 8. Answer of Q8

P1	This does not make for the money. Firstly, they should be fall in love the nature. So they must belong to nature. They should listen to the nature. After all nature learning will start.
P2	Everywhere is a learning place if they want to learn. Before they start to the course, they should start to learn the nature from everywhere.
P3	They need to get to know their geography and social structure very well, especially to travel and to evaluate their time in order to get to know different geographies and cultures.
P4	If they do not have communication and survival skills, they should not to take the course.
P5	Practice is the most important detail in this specialize area. The more time in nature the more experience they will have. They should learn every detail of the region especially from local people. Local people will be the most important reference source for them. So they need to have best communication skills ever.

4.2 Tourist Guide Education in Turkey

Tourist profile, expectations, travel motivations of tourists have changed with the developing world and differentiating needs. Cultural tourism emerges after this differentiation. The need for individuals to discover new places, new things and sense of curiosity is one of the main motivations of cultural tourism. Culture tourists have differentiated in the process and some cultural tourists wanted to design culture and nature. Culture tourists can move on with a package tour or they can move independently. Studies by Grenier (2005), Butler, Hannam (2012) and Bui (2009) show that the main sources of motivation for independent tourists are to increase cultural capital in order to experience wildlife and adventure, to see natural beauties in the countryside and to develop personal development. Although cultural route tourists are generally independent tourists, they are also preferred by the following groups.

- Individual walkers,
- Organized groups,
- Self-guided groups,
- University-school groups,
- Nature lovers, botanists, bird watchers,
- Excursions,

- Horse riders,
- Experienced walkers,
- Pilgrim walkers (Tanrısever, 2019).

Tourist guides can serve all these groups above. Cultural routes have different degrees of difficulty besides tourist types. Some are also within the national park. Therefore, it is obligatory to have a tourist guide. However, this guidance service is provided by mountain counselors or field guides instead of licensed tourist guides. Cultural Routes are one of the professionalization fields in the area of Tourist Guides. However, the curriculum of Tourism departments in Turkish universities are lack of such subject. According to the results of the interview, the main features that should be found outside the requirements of the law in order to provide guidance are determined as follows.

- Physical endurance,
- Knowledge of flora and fauna,
- Ability to find direction,
- Map reading,
- First aid information,
- Basic camping knowledge,
- Trekking leadership,
- General knowledge about the region,
- Material knowledge.

Turkey has four different types of tourist guide education systems.

- Two year university degree,
- Four year university degree,
- Master degree,
- Culture and Tourism Ministry (Certificate Program)

After having license, tourist guides start to specialize on specific areas like diving, museums, art history and so on. One of these specializing areas should be cultural routes. While this study contributes to the realization of this expertise in practice, it will contribute to the theory with the suggestion of training module content. The training module is designed in two stages as theoretical and practical. Each given theoretical training will be practically applied in the field.

Table 9. Education Module Suggestion

Technical Education	Flora and Fauna	Trekking Leadership	Regional Education	Sustainability	Communication	Nutrition
Camping	Flora values of the route.	Trekking leadership and ethic	Meteorological forecasts	Nature and environment protection	Communication skills and effective communication techniques	Water and nutrition
Mapping, finding direction, GPS usage	Fauna values of the route	Trekking planning, alternative plan creating	General regional geography	Sustainable nature	Field visit rules	Minerals and nutrition
Security	Crossing points of wild animals	Emergency management	General regional history and periods	Sustainable tourism	Local people and culture	Energy metabolism
Rope landing	Wild life of the region	Creating route information and ratings	Cultural heritage	Climate changing and global warming	Collective living rules	Proteins, lipids, carbohydrate
First aid	Finding water sources in nature	Survival in nature	Tangible heritage		Nature physiology	Nutrition in nature
Proper material choice					Tourist types	
Search and Rescue					Drama	

CONCLUSION

Culture and hiking trails are known in most European countries for many years and in the development stage in Turkey in the last 20 years. For this reason, specialization in the field of tourism and the establishment of the infrastructure has been realized in recent years. Tourist guides are the most important building blocks in this area. In the study, a collective resource was prepared for tourist guides and guidance students who will specialize in this field and a training module was developed. According to the results of this research, this module is divided into seven main topics. Since the duration of the training will vary according to the geographical area where the training will be given, no agreement has been reached. In Turkey, people who specialize in this area is a few. Therefore, the interviewees are the key names in this field. Consensus of all participants; it is not possible to specialize in this field of people who do not have interest in nature, experience and desire to stay in nature. Tourist guidance education differs according to each country as it affects countries' own geography and identity. This education in Turkey is designed as two-year college, bachelor, graduate and certification. Specialization in the fields is provided either by the certificates obtained after these trainings/education or individually. Specialization training/education in cultural routes has not been given before and this study is important since it will be the first study both theoretical and practical in this field. Subsequent researchers may develop this program according to their country and tourist needs.

REFERENCES

Ban, Z., Chen, F., Ban, T., & Ban, C. (2016). Cultural routes: theoretic system and its enlightenment to china heritage practice. *Proceedings of the 2016 International Forum on Management, Education and Information Technology Application*. 10.2991/ifmeita-16.2016.175

Bastemur, C. (2009). *Likya Yolu ve Çevresinin Turizm ve Rekreasyon Potansiyelinin Araştırılması*. Ankara Üniversitesi Fen Bilimleri Enstitüsü Yayınlanmamış Yüksek Lisans Tezi.

Bastemur, C. (2011). A new destination for alternative tourism; Lycian Way. *International Conference on Tourism (ICOT)*. http://iatour.net/pdf/Proceedings%20ICOT%202011%20.pdf

Bastemur, C. (2013a). Mimar Sinan rotasının doğal ve kültürel peyzaj izleri. Ankara Üniversitesi Fen Bilimleri Enstitüsü Yayınlanmamış Doktora Tezi, Ankara.

Bastemur, C. (2013b). Mimar Sinan eserlerinin özel ilgi kapsamında değerlendirilmesi. VIII. Uluslararası Sinan Sempozyumu, 39-44.

Bui, D. T. (2009). *Tourism industry responses to the rise of sustainable tourism and related environmental policy initiatives: The case of Hue city, Vietnam* (Doctoral dissertation). Auckland University of Technology.

Butler, G., & Hannam, K. (2012). Independent tourist's automobilities in Norway. *Journal of Tourism and Cultural Change*, *10*(4), 285–300. doi:10.1080/14766825.2012.718441

Çalık Ross, A. (2017). Hoşgörü başkenti Nikomedia ve Galerius'un Hoşgörü Fermanı'nın çeviri denemesi. *Sosyal Bilimler Araştırma Dergisi*, *6*, 135–145.

Campolo, D., Bombino, G., & Meduri, T. (2016). Cultural landscape and cultural routes: Infrastructure role and ındigenous knowledge for a sustainable development of inland areas. *Procedia: Social and Behavioral Sciences*, *223*, 576–582. doi:10.1016/j.sbspro.2016.05.350

ÇEKÜL. (2019). *Çevre ve Kültür Değerlerini Koruma ve Tanıtma Vakfı*. https://www.cekulvakfi.org.tr/haber/kultur-rotalari-planlama-rehberi-cikti

Clow, K. (2011). *Compiled from the presentation made at the Open Museum Summit*. Academic Press.

Clow, K. (2017). *Likya Yolu; Türkiye'nin İlk Uzun Mesafe Yürüyüş Rotası*. BKM Yayıncılık.

Clow, K. (2018). *Aziz Paul Yolu*. Hil Yayınları.

Clow, K., & Gardner, C. (2012). *The Kackar: Trekking in Turkey's Black Sea Mountains*. Upcountry Yayıncılık.

Clow, K., & Richerson, N. (2000). *Likya Yolu*. Okyanus Yayınları.

COE. (2019). *Avrupa Konseyi*. https://www.coe.int/en/web/cultural-routes/by-theme

Demirel, E. (2012a). *Sarıkamış Yürüyüş Parkurları, Serhat Kalkınma Ajansı*. Academic Press.

Demirel, E. (2012b). Hitit Yolu. İstanbul: Hil Yayınları.

Ducassi, S. I., & Rosa, M. (2005). *A new category of heritage for understanding, cooperation and sustainable development. Their significance within the macrostructure of cultural heritage. The role of the CIIC of ICOMOS: Principles and methodology*. Icomos.

Ergüven, M. H. (2015). Gastronomy and wine tourism as a variety of special interest tourism: Thracian Vineyard Route. *Turkish Studies, 10*(10), 449–464.

Finkel, C., & Clow, K. (2011). *The Evliya Çelebi Way*. Upcountry Yayınevi.

Görmüş, S. (2017). Kültürel peyzaj değerlerini korumak için yeni bir yaklaşım: kültürel rotalar. *Plant Peyzaj ve Süs Bitkiciliği Dergisi,*7-9.

Grenier, K. H. (2005). *Creating Caledonia; Tourism and Identity in Scotland, 1770–1914*. Imprint Routledge.

ICOMOS. (2019). *Uluslararası Anıtlar ve Sitler Konseyi*. https://www.google.com/search?q=%C4%B1comos+cultural+routes&oq=%C4%B1comos+cultural+routes&aqs=chrome.69i57j0.4623j0j4&sourceid=chrome&ie=UTF-8

Karataş, E. (2015). *Kültür Rotası Planlama Rehberi, Çevre ve Kültürel Değerlerini Koruma ve Tanıtma Vakfı ve Tarihi Kentler Birliği Ortak Yayını*. Stil Matbaa.

Kervankıran, İ., & Çuhadar, M. (2014). Turizm rotalarının oluşturulmasında coğrafi bilgi sistemleri kullanımı. III.Disiplinlerarası Turizm Araştırmaları Kongresi.

Kültür Rotaları Derneği. (2019). http://cultureroutesinturkey.com/c/

Merriam-Webster. (2019). *Sözlük*. https://www.merriam-webster.com/dictionary/trek

Nillson, M., & Mekonnen, T. (2016). Performing the "post-secular" in Santiago de Compestela. *Annals of Tourism Research, 57*, 18–30. doi:10.1016/j.annals.2015.11.001

Nilüfer Belediyesi. (2019). *Mysia Yolları*. http://www.nilufer.bel.tr/dosya_yoneticisi/mysia.pdf

Robert, H., and Eryurt, H. (2013). Culture Routes in Turkey. *e-dialogos, 3*, 6-7.

Sarı, İ. (2011). Frig Yürüyüş Yolu, 23-24 Aralık 2011. Kültür Turizmi Zirvesi Sunumu, Kültür Turizm Bakanlığı, Ankara.

Sugio, K. (2005). A consideration on the definition of the setting and management/protection measures for cultural routes. *15th ICOMOS General Assembly and International Symposium: 'Monuments and sites in their setting - conserving cultural heritage in changing townscapes and landscapes'*, 17 – 21.

Tanrisever, C. (2016). The Necessity of cultural routes education for tourist guides in Turkey. *Bartın Üniversitesi İktisadi ve İdari Bilimler Fakültesi Dergisi*. http://iibfdergi.bartin.edu.tr/

Tanrısever, C. (2019). *Turist Rehberliğinde Uzmanlaşma Özel İlgi Turundan Özel Konulara (edt: Güzel, Ö, Köroğlu, Ö.)*. Nobel Yayınevi.

WWF. (2019). *Dünya Doğayı Koruma Vakfı*. https://www.worldwildlife.org/

Chapter 13
Ayvalik an as Art Route for Tourism Guides Within the Frame of Art Tourism

Sabriye Çelik Uğuz
https://orcid.org/0000-0002-2424-649X
Balıkesir University, Turkey

EXECUTIVE SUMMARY

Art is the most important element that directs people to a destination and gives them an original experience in a developing and changing understanding of tourism. Art tourism motivates people to travel, facilitate communication with different cultures, enhance social pride, and contribute significantly to the region's economy. Regions with artistic and touristic attractions are active with art tourism, and they have an art destination brand. In this study, Ayvalik's potential for art tourism is investigated. At the same time, art events, art venues, and artist atelier/studio in Ayvalik are presented with numerical and visual examples. Ayvalik is developed as a new art route within the framework of art tourism and offers various suggestions for its tourist guides.

INTRODUCTION

Art is an expression that awakes emotions and thoughts and a language that tells about societies. The best way to know and learn about society is art. Another way is tourism. Along with fulfilling the needs of people such as rest, entertainment, culture, and art, tourism is a phenomenon that brings nations closer together and provides interaction. When we look at the relationship between art and tourism, both are experiences and as an extension of these experiences, art tourism is becoming more and more common. Art and artists make it easy to experience the destination with an authentic and unique sense of place. Art, creating a site-specific reality, plays an active role in the image of the destination (Jolliffe and Cave, 2012: 15). It is also among the most important reasons for traveling and it shapes the expectations of tourists and contributes to the development of tourism. While the artistic activities that are increasing

DOI: 10.4018/978-1-7998-3725-1.ch013

in recent years have led to the development of tourism, the interest in art has also increased with the increase in tourism activities (Bhatta, 2017: 180).

When culture is defined as maintaining throughout a person's life and shaping his lifestyle, and its relationship with tourism is created, it appears as cultural tourism. Cultural tourism essentially focuses on cultural values and lifestyles on the one hand and social environments where cultural and artistic activities are carried out on the other hand (Andrew and Andrew-Essien, 2013: 276). While cultural tourism in the past has been mostly associated with high culture and cultured people, today it contains many popular cultural attractions that include cultural tourism, sports, living heritage, recent nostalgia and the daily life of local communities. Cultural tourism moving from the concrete heritage of the past to the abstract products of contemporary culture includes heritage tourism (Richards, 2007:2), due to visiting historical artifacts and art tourism as it benefits from current art activities and expands towards creative tourism (Bonet, 2003).

Art tourism, which develops in cultural tourism and takes part as a new touristic product in tourism literature, is a source of motivation for travel and has the power to increase the tourism demand. Many cities and countries attract an influx of tourists by creating an art-based image with its artists, artworks, art activities and programs. The main factor underlying the development of art tourism in cities such as Paris, London, Milan and New York is the understanding and experiencing the contribution of art and culture to urban development by local administrators and politicians. At this point, art becomes an income-generating factor locally and nationally through tourism (Currid, 2007: 2). Art tourism, along with contributing to the urban and national economy, renews the physical and social character of a destination. It brings local people and tourists together, increases community pride, enhances art-related organizations and creates a livelier place (Mandala, 2015: 130; Watson, 2013).

While countries/regions/cities identified with art and art tourism are increasing, it is seen that art has become an institution in Turkey and there has been a movement in the artistic activities in recent years and there have been developments in art tourism. Istanbul, which is an international art center for renewing the image of the country and organizing contemporary art activities, is the focal point of art tourists thanks to its eye-catching museums and galleries, huge exhibitions, fairs, festivals, and biennials. At the same time, Ayvalık district of Balıkesir province, which draws attention with its natural, historical and cultural structure and touristic facilities, hosts art and art activities. In Ayvalık, which is rapidly approaching to become an art destination, art is a part of urban identity and the district renewed with art and came alive with artist workshops. In this study, the potential of being an art destination is revealed by researching the art and tourism environment of Ayvalık. In the context of the study, it has been determined that the district has a liberating and dynamic art environment thanks to its art galleries, projects, events, artists and art workshops and has turned into a city known for art in terms of tourism. Ayvalık, which strengthens the city image with art and is a vibrant city with art tourism, offers an alternative to tourist guides as an art route. In line with these findings, suggestions were made to tourist guides for the development of Ayvalık as an art destination and for more tourists to visit.

THE RELATION OF ART AND TOURISM: THE CONCEPT OF ART TOURISM

The relationship between travel, tourism and art forms art, artworks, artists, the imagination of local people and tourists, and ideas about tourist places. Art tourism, which is a new tourism product and growing rapidly, provides the promotion of travel. Since the desire to see, experience and recognize art is directly related to tourism, art tourism should be supported to attract artists and tourists to destinations (Bennett, 1995; Foster, 2015; Jackson, 2003; Jagodzinska, 2017; Markusen and Schrock, 2006; Smith, 2012). Art tourism includes tours, holidays or other trips away from home, as well as those who travel exclusively to see art elsewhere, including traveling to see art (Franklin, 2018: 399).

In Hughes's (2000) art tourism model, the level of tourists' interest in art is categorized. It defines art tourists in two ways as the arts-core and the arts-peripheral. For the arts-core tourists, the arts are the main attraction for the visit. These are people who want to see artistic productions, watch a higher standard of art performance, experience the art-artist and be in a better environment with art. In the category of art-peripheral, those who have other reasons for travel and make choice before leaving the destination to attend an artistic event are art-peripheral tourists. Art-peripheral tourists, who have travel purposes other than art, will experience special interest and awareness towards art as a result of deciding to participate in artistic events. In the report of "Art Country" in 2015 prepared by The Australia Council for the Arts, it is stated that art tourism is important for Australia and approximately 2.4 million foreign art tourists visited Australia between 2013 and 2014. It is also determined that 38% of foreign tourists attended at least one art event (Australia Concil for the Arts, 2015).

Whether a place is a center due to art or a secondary attraction is a confusion. Although it is known that Broadway (New York) and West End (London), the center of entertainment and art, attract great numbers of tourists for musical, theater and concert, it is not clear how many people travel due to art. Even if art is not the main reason for travel for most of the people, artistic events are an important secondary source of attraction and attract arts-core tourists. In addition to the major tourism centers, the seaside resorts in the UK are visited for sun, sea, culture and entertainment purposes, but it is a good opportunity to see artists and watch artistic shows. Art festivals in Edinburgh and Glastonbury are also very successful examples of attracting tourists (Hughes, 2002: 53). Art is the most important part of the holiday experience for tourists and the strongest component of other touristic travels. In deciding to visit a holiday village, town, and city, the integrity and balance of the space are created with the distinctive feature of the art as well as the existence of art is enough.

From classical times until today, art has led people to visit neighbors and distant cities and regions. There is a lot of evidence that traveling for art and culture was distinct and even highly organized before the modern period. It is known that there are "tourist graffiti" In Egyptian tombs which date back to 1244 BC and these areas were opened to visitors (Yoyotto, 1960). Likewise, within the Laws of Plato when travel arrangements, facilities, and organizations in the 5th century in Greece are explained in detail, it is stated that art travelers were abundant. It is told that if travelers for cultural and artistic purposes were ordinary people, they were hosted by priests in temples and if they were with the high status they were hosted in the mansions of rich, wise or virtuous people. Special arrangements have been made to recognize the cultural and artistic areas and encouraging travel in ancient Greece and one of these important arrangements are people named ''proxenoi'' (proxeny) assigned officially to host guests in every state of Greece (O'Gorman, 2010: 7). In the Roman emperor, Augustus reign art and culture; enjoyable, educational, informative was the main reason for the travel. To see the magnificent monumental structures and artifacts, travelers should be welcomed by the citizens of Rome with good hospitality bonds. At the same time, there was a large tour which was official and obligatory that formed the origins of the Grand Tour: starting from Rome to Little Asia (especially Troy, the origin of Rome), then Egypt and it ended up returning to Rome. The number of guides that guided and informed art tourists on this tour for a fee was also growing (Lomine, 2005: 82). It is seen that the reasons for this premise that improve the journey reveal the Grand Tour and the place of art on the Grand Tour, a strong connection with modernity, has led to a serious development in tourism görülmektedir (Walton, 2005; Franklin, 2008; Gretzel and Jamal, 2009; Rakic and Lester, 2013; Caton, 2014; Long and Morpeth, 2016;). After the fall of the Roman Empire, the systematically progressing art tourism activities continued with the Grand Tour. The Grand Tour, which spanned between 1600 and 1750 AD and lasted four months to three years, included visits by young British noblemen to natural, historical, cultural, artistic and scientific places in Europe for educational purposes. Travel along the routes of the Grand Tour; has shown that the part of the search for a well-lived life and apart from observing art, art concept should be experienced. With industrialization, the artists and those who travel for art purposes have not only contributed to the development of art tourism but have led to the creation of new tourist attractions. Since the middle of the twentieth century, destinations identified with artists begun to form increasingly and have become desirable by tourists (Franklin, 2018). Today, art tourism; it is preferred for keeping the tight ties established for reasons such as enjoyment, creativity, intellectual and emotional (Miles, 2013).

The feeling of being in the social and cultural atmosphere of the destination, along with the urge to feel the power of art to transform personal emotions, enables the development of art tourism. Being sensitive to nature, environment, culture Modern art tourists are eager to travel to distant Japanese islands, China's wilderness regions, islands outside of Australia, high desert areas of Texas and Nevada. Founded in 1988, there are a growing number of travel agencies behind industry pioneers such as Martin Randall Travel, which enters the art tourism market. These include Cox and Kings, Tate Travels, Ciceroni Travel, ACE Cultural Tours and Holts Tours, which facilitate art-based travel (Franklin and Papastergiadis, 2017; Kirshenblatt-Gimblett, 1998).

ART BASED URBAN DEVELOPMENT

It is a fact that art tourism focuses mostly on urban areas (Pengiran-Bagul, 2013: 2). Cultural industry policies and "creative city" strategies have been developed by central and local governments, city planners and cultural policymakers for urban development and restoration, and in these strategies active role of art is adopted. Art, city, and tourism are combined in a common strategy with the focus on topics such as "cultural/urban tourism" and "art/city image and publicity" within the framework of art and urban renewal (Landry, Greene, Matarasso and Bianchini, 1996; Markusen, 2006; Markusen, Gilmore, Johnson, Levi and Martinez, 2006; Mathews, 2010; Pratt, 2008; Smith 1996; Ward, 2016).

Art, artists, artworks and art venues are considered "an opportunity" when transforming an old industrial city and/or its district. The New Jersey Performing Arts Center in Newark, USA, is an effective force in reviving a city and turning it into a tourist destination (Lowrie and Frisch, 2007: 4). Urban and artistic strategies are created in Chicago, transforming the remaining areas of its industrial background into places like stores, museums, bars, restaurants, and galleries. The best practice for the development of the city with art and tourism is happening in Wicker Park Quarter. Wicker Park gains a national reputation as a quality urban cultural area with music and art environment and takes advantage of development with art (Lloyd, 2002: 518). Along with cities such as New York and Los Angeles, the City of New Orleans is making significant progress in art tourism. New Orleans, the center of jazz and blues music, remains as the brilliance of art and tourism. Jackson Square, an art park in the triangle of tourism, art and artist identities, is a carnival with street events; turns into a workshop with artists productions and with the display of products turns into an outdoor art gallery which becomes the center of attention for tourists (Sheehan, 2014: 52). Built-in 2006 in Kentucky State/Louisville in the form of a 90-room boutique hotel and a public museum displaying contemporary artworks, the 21c Museum Hotel provides the city's renewal of art and tourism (Miller and Washington, 2006: 61).

The revival of the United Kingdom's cities and districts which were affected by the industrialization period is done via art. In London, from "Bankside Power Station" which was transformed into "Tate Modern" a art museum and gallery, "The Lowry" in Manchester/Saltford which is a art complex for visual and histrionics arts and the 20 meters high street art of Anthony Gormley' Angel or Angel of the North at Newcastle/Gateshead are most important examples of the fact (Evans and Shaw, 2004). Likewise, the urban areas near coasts that were affected by the urban transformation after Industrilazition become a new destination brand with the civic policies centered around the art. Margate coastal town in the Thanet District is reconstructed with the Guidance of Thanet District Council, Turner Contemporary Art Gallery and especially the visual artists (Comunian and Mould, 2014).

The Guggenheim museum at Bilbao city in Spain/Bask District, being one of the most important designs of Modern architecture, Transforms the city which was the center of industry and mining both economically and culturally with a single museum. The museum, while being the most important symbol and the driving force behind the rejuvenation of the city, also encourages the new cultural and entertainment industry and increases the numbers of Tourism establishments like cafes, restaurants, hotels and souvenir shops (Saunders, 2012). Since its opening in 1997, the museum brings in more than 1 million visitors of which 88% is foreigners (Gonzalez, 2011).

The capital of United Arab Emirates Abu Dabi, Declares the rapidly growing Saadiyat island the " Art Island" creating a new platform for artists and boosting tourism (Ameize, Mady and Benson, 2011: 98). As one of the biggest museums in the world, the Guggenheim, is the center of contemporary art in Asia and offers a unique experience to tourists from all over the world (Anderson, 2012). Seul, having performance arts as a tourist attraction transforms art tourism into a sustainable niche market. The attraction of Traditional and modern dance, music and theatre along with the Korean performing arts Benefits tourism a lot and increases the number of art tourists each year (Lim and Bendle, 2012: 668-669). Starting around the 1980s Singapore becomes a globalized city for art and with the scope of art tourism the slogan "Bring the world to Singapore, Bring Singapore to the world" increases the rate of development (Chang, 2008: 829). The link between tourists and art in Indonesia's favorite tourism center Bali Island is solidified by the art route between Sanur, Mas, and Ubud. The route which is used since 40 years ago and accompanied by art and artists performances starts from the Sanur beach and continues with The The Mas village which is famous with its extraordinary Wood sculptures then ends with Ubud which has a painting tradition and is known with its many galleries, museums, and workshops (Jenkins and Romanos, 2014: 297)

The single most lively city in terms of art in Turkey is Istanbul. Since the 1980s with the launch of galleries, exhibitions and in 2004 with the opening of Istanbul modern along with the Pera Museum, Central Istanbul and Sabancı Museum respectively increased the growth of the city. With the support of the Istanbul Biennial, 2010 Europe Capital of Culture: Istanbul and Istanbul Foundation for Culture and Arts many Art festivals took place and Istanbul became an international art center and favorite of tourists.

AYVALIK AS AN ART DESTINATION

Ayvalık is a seaside town in Edremit Gulf/Balıkesir which is located on the Northern Aegean coast. Ayvalık District which is surrounded by Gömeç in north, Dikili in south, Bergama districts in east and

Aegian sea at west, with its Weather, pine forests, Olive gardens, under and above water beauties, the archipelago called Ayvalık Islands, Ayvalık Islands Nature Part with its flora, fauna and endemic varieties, countless coves, shoreline accompanied by amazing beaches are a natural attraction with readily available transportation opportunities. Unique natural beauties accompanied by ancient cities, architecture, food culture, traditions, cultural-artistic centers, and touristic areas that stands out with its cultural attractions. Ayvalık, also known as "City of Islands", includes Ayvalık Islands Nature park which Covers 19 of 22 and islands except for a part of Cunda Island is under protection since 21.04.1995 with its 19.950 ha territory. The Nature Park's cornerstones are its location, the geological, aesthetical, natural, historical and cultural features.

Ayvalık which was known as Kydonies became a figure in history in the 6th century BC. It was built upon the Greek colonies on the Hekatonessos islands then the city was moved to the place it is today. The district was where rich people resided in 1653 and It is the biggest Industry and trade center of the region after Izmir. As a result of olive trading and farming, the district has industrial buildings and also olive factories, soap plants, oil mills and trading depots from which olive, olive oil and other olive products are made. With the Academy building(Gymnasium Kydonion) which was built in 1803, Ayvalık becomes the Education center of the region. Along with the growth of the Olive industry and economic development, cultural development with academy leads to the increase of intellectual growth. Also being an Independent settlement during the Ottoman Empire's reign, Ayvalık had both Muslim and Greek residents and as a result, the City's Identification and the shape are carried to today (Özel, 2011: 18; Uçar, 2014:7). The district hosting different cultures in history is affected by Girit and western Thrace because of the exchange and has a rich culture and it is the main reason why its cuisine is so famous. The cuisine of Ayvalık is consists of olive oil dishes, seafood, and appetizers made from herbs.

The urban architecture of Ayvalık with the growth of the olive industry in the 19th century is in a gradual order starting from the port on the shore and industrial buildings, leading to the hilltops and hillside which consists of more houses and churches. The monumental and symmetrical shapes of Ayvalık houses are a sign of Neo-Classical understanding of architecture with their gates similar to antic temple architecture, the pillars of doors with the vegetal arrangement, ion and composite corner beads. Leaf motifs above the windows, edges of frames below the windows, fronton acroterium and corner decorations (Akın, 2015: 130). As a member of the Union of Historical Towns and a part of UNESCO temporary heritage list with the "Ayvalık Industrial Landscaping" Ayvalık has 2006 registered cultural heritage pieces transforming the city into an outdoor museum almost mending streets together with the doors, iron fences of the balconies, window sashes, stained glasses, and door knocks. Within the historical structure, the architectural and aesthetic fabric of the city becomes a whole with the artists, art centers and activities (Figure 1).

Figure 1. Ayvalık's Urban Fabric

The basis of Ayvalık's urban fabric consists of artists who lives/lived there, makes it so that city is remembered with the artists. Greek writer, painter Iconograph Fotis Kontoğlu, Greek writer İlias Venezis, writer, reporter, translator Ahmet Yorulmaz, writer Feyza Hepçilingirler, poet Turgut Baygın, TV and series producer Güner Namlı. Also painters such as Nedim Gündür, Fikret Mualla Saygı, Fuat Mensi Dileksiz, Orhan Peker, Burhan Uygur, Mustafa Rüçhan, Muzaffer Akyol; Fotographers: Ara Güler, Nazım Timuroğlu, Önder Aksoy, Teoman Madra, Bülent Kürşad and Emine Berkan; Cinema artists: Nihat Durak, Nail Pehlivan, Uğraş Salman, Nur Akalın and many more artists who are deeply in love with Ayvalık, are gonna be inspected one by one and will be involved in other projects which will aid the city in future.

Figure 2. Ekrem Adyar and Bedri Karayağmurlar's Artist Ateliers

Having carried the Turkish Painting Art to the global arena, Orhan Peker's love for the Ayvalık's nature with his dreams of Summer Academy and the works he published for the city is especially important for Ayvalık. In 1976 he continues being an exemplary person with the "Ayvalık-76" art gallery opened in Ayvalık. The plastic artists and artist studios following the steps of the artists like Orhan Peker and many more who helped the city gain its identity are the deciding factor in the visitors present and future in Ayvalık. The Transformation of Ayvalık's architectural and aesthetic symbol stone buildings into artist workshops and helping the city gain its identity is long-lasting effort in the city (Figure 2). 38 plastic artists and 32 atelier/studios (Table 1) residing in the Ayvalık help the city reinforces the link between artistic identity and city's culture and with the displays of the artistic works it creates an interaction between residents and tourists and also Artists with the audience. The harmony between art, city, and tourism together shows that Ayvalık is an art destination.

Artist ateliers, art houses, art galleries, and art-themed tourist businesses (cafes, bars, restaurants, pensions, boutique hotels) and graffiti, which represent street art, create an atmosphere full of art for both local people and tourists. Many activities, from music to literature, from conversations to plastic arts, from theater to dance and cinema shows, from workshops to festivals, are presented to art audiences in the district. Ayvalık Municipality Orhan Peker Art Gallery, which keeps art alive in Ayvalık all year, Miço Art Gallery, which was newly established to bring art and collection enthusiasts together, Olgay Art House and Gallery that combines art and artist friendship on Cunda Island, brings vitality to the city's art environment with exhibitions. Ayvalık Karagöz Art House, an old carpentry shop, turned into an art-house suitable for Ayvalık architecture in 2005. Together with Karagöz Art Gallery, the places bring international artists together and provide activities such as art projects, events, and accommodation of artists (Figure 3). Besides, Ayvalık Muhip Özyiğit Culture and Art Center, which organizes art exhibitions and speeches, and Oktay Ekinci Cartoon House, where the art person A. Uğur Bilge started the cartoon workshop, embellish Ayvalık's art life.

Ayvalık is remarkable with art academies in addition to art houses and galleries (Figure 4). Ayvalık International Music Academy (AIMA), led by Prof. Filiz Ali, has contributed to the training of young musicians since 1998. She both introduces Ayvalık via music with music master class courses and concerts and also creates an alternative by establishing the relationship between art and tourism. Thanks to music events and Ayvalık International Music Festival (July), which has been organized for 6 years and also AIMA, which has been working for 22 years to develop classical music in Turkey and adding artistic value to Ayvalık, Ayvalık had the opportunity for music tourism. The place turned from a historical olive oil factory to an Art Factory is used for theater practices and other performing arts conducted by Ayvalık Art Association. The International Ayvalık Theater Festival (April), which has been going on for 10 years and held at the Art Factory, ensures Ayvalık to be mentioned with the theater (Figure 4). Films screened by Ayvalık Support Design Academy, and art exhibitions, seminars, and art workshops, and Turgay Gönenç Cinema Library (SinemaTek), While Art and Education Support Association, Kafedemia and art shop are run together.

Table 1. Artist who lives in Ayvalık and Artist Ateliers/Studios

	Name and Surname	Art Area /Artist Atelier/Art Studio/Gallery	Location
1	**Abdullah Şengörenoğlu**	Sculpture /Art Studio	Cunda
2	**Alev Gözonar**	Conceptual Art/Art Studio and Artist Atelier	Palabahçe-Zekibey
3	**Alp Tamer Ulukılıç**	Painting/Two Artist Atelier	Macaron-Hayrettinpaşa and Sefa Çamlık
4	**Arif Buz - Mustafa Sevinç**	Painting/ Artist Atelier	Hamdibey
5	**Ayşen Büyükyıldırım**	Painting/Artist Atelier	Macaron-Hayrettinpaşa
6	**Ayşen Yıldırım - Burhan Yıldırım**	Painting/ Art Studio and GALERİDA	Küçükköy
7	**Ayşenur Kocatopçu**	Painting/ Art Studio	Cunda
8	**Ayşın Özen**	Ceramic-Painting/ Art Studio	Sefa Çamlık
9	**Bahadır Gökay**	Conceptual Arts/ Art Studio	Şirinkent-Sahil Kent
10	**Banu Küpeli Ürüçoğlu-Yasin Ürüçoğlu**	Painting/Artist Atelier	Palabahçe-Zekibey
11	**Bedri Karayağmurlar**	Painting-Poetry/Artist Atelier	Macaron-Hayrettinpaşa
12	**Bülent Şavkın - Mehmet Sümbeli**	Ceramic - Sculpture/Artist Atelier	Palabahçe-Zekibey
13	**Dilek O'Keeffe**	Painting-Ceramic/Art Studio	Ayvalık
14	**Ekrem Aydar**	Painting/Artist Atelier	Palabahçe-Zekibey
15	**Emine Alışık**	Painting/Artist Atelier, Olgay Art House and Gallery	Cunda
16	**Eyüp Öz**	Sculpture /Artist Atelier	Sefa Çamlık
17	**Fuat Çağatay**	Painting/Art Studio	Palabahçe-Zekibey
18	**Fügen Leman**	Painting/Art Studio	Palabahçe-Zekibey
19	**Gülgün Haksal**	Painting/Art Studio	Cunda
20	**Gülseren Kayalı – Yüksel Erdoğan**	Painting/Art Studio	Şirinkent-Sahil Kent
21	**Hakan Urul**	Painting –Theatre/Artist Atelier	Palabahçe-Zekibey
22	**Hale Arpacıoğlu**	Painting/Artist Atelier	Macaron-Hayrettinpaşa
23	**Hyesung Lee - Arif Aşçı**	Conceptual Art - Painting, Photography /Art Studio	Sakarya
24	**Metin Benek**	Painting /Art Studio	Sefa Çamlık
25	**Muhittin Karakuş**	Sculpture /Artist Atelier	Sanayi Sitesi
26	**Naile Cimit**	Ceramic/Art Studio	Sakarya
27	**Rabia-Uğur Çalışkan**	Painting-Sculpture/Art Studio	Küçükköy
28	**Sibel Sümer**	Painting/Art Studio	Şirinkent-Sahil Kent
29	**Suna Tüfekçibaşı**	Painting/Artist Atelier and Art Gallery	Küçükköy
30	**Tunca Subaşı**	Painting/Artist Atelier ve Arkipel Art Shop	Macaron-Hayrettinpaşa and Palabahçe-Zekibey
31	**Teoman Manacıoğlu**	Painting /Art Studio	150 Evler
32	**Türkan Gündoğdular**	Ceramic-Painting/Art Studio	Macaron-Hayrettinpaşa

Figure 3. Miço Art Gallery and Karagöz Art House

Figure 4. AIMA and Art Factory

Figure 5. Barbara Residency, Devil's Coffehouse and Midi Cafe Bistro

At the same time, Art 10 Opinion Workshops in 2012 with Pınar Kür, who was a delightful writer of our literature, constitutes the conditions for literary tourism in the district. Today, Bonjour Pension, formerly used as a French consulate, provides tourist services in a historical building, Literature and tourism relationship is maintained by Onur Orhan and Emine Ayhan with ''Reading and Writing Workshop''. Defne Koryürek and Curator Vasıf Kortun, at Ayvalık-Mutluköy Guesthouse, allow people who produce scientific and artistic projects on ecology, gastronomy and/or agriculture, while Barbara Residency, opened by art lover and collector Şerif Kaynar from Ayvalık in 2016, it becomes a place open to the use of art/artists. An old Greek Tannery, which is disappearing in Ayvalık architecture, turned to an artist house with a new identity, and the adjacent house and garden become an artist studio and courtyard garden. This artist's house an art workshop, which finds meaning with art, makes a significant contribution to Ayvalik's art tourism. Located in Palabahçe, Devil's Coffehouse (Şeytanın Kahvesi) hosts art talks and interviews with tea, coffee, and grove water. Midi Cafe Bistro in Sakarya Quarter has been bringing the taste of world cuisines to Ayvalik since 2017 and organizes art exhibitions and art interviews (Figure 5).

ArtClan Cunda International Art Workshop has been bringing together national and international artists in Cunda since 2015. Productions made in the fields of painting, sculpture, ceramics, and photography, and plastic art, etc. and then exhibited in Ayvalık Orhan Peker Art Gallery (Figure 6). This qualified art workshop brings the artist and art lovers together, shows that Ayvalık's art environment is very valuable. Ayvalık, experiences art with Ayvalık Culture and Art Days (August-September), welcomes filmmakers from all over the world as part of Başka Sinema: Ayvalık Film Festival and films screened. Panels, short interviews take place with the film's director, actors or crew. This international festival motivates cinema tourism and makes a great contribution to the city brand of Ayvalık.

Figure 6. ArtClan 2019, Ayvalık Orhan Peker Art Gallery

Besides the center of Ayvalık (Kydonia, Ayvali) and Cunda Island ("Moshonisia" meaning "fragrant"), Küçükköy (Yeniçarohori) is another important art place. Küçükköy, a village of janissaries, founded in 1462 when Fatih Sultan Mehmet planned to take Midilli, continues this feature by calling the Greeks of Yeniçarohori between 1800-1900. Between 1893 and 1913, the identity of the city shaped by the Bosnian people coming from the Balkans. Beginning with the Janissaries, the village where the Greeks, refugees, and Bosniak people live respectively; It reflects the history with its old people, stone houses and streets. This village, having a hard time trying to keep its past alive, supported by artists and entrepreneurs who value culture and art. The village, where museums, artist workshops, art galleries, art/science projects, and touristic places renewed, revived thanks to the existence of the art and the identity of the artist and is rapidly moving towards an art village. Barkhan Museum/Gallery, Artura Vintega Gallery, Artura Art Craft, Yeniçarohori Antique Art, Atelier Art Center and Gallery Ida support art and artistic activities in Küçükköy, a route within the art destination. On the other hand, Narlı Kız Mansion welcomes tourists with painting art and art exhibitions organized in the mansion in cooperation with Atelier Gallery.

FUTURE RESEARCH DIRECTION

In the study, as an alternative tourism type for tourist guides, art tourism was examined and presented with the example of Ayvalık. Art route which is a new touristic product, should be expanded by the application of tourist guides by including travel agencies within the scope of tour routes. In future studies, publications related to art tourism should be increased. As in the case of Ayvalık, it is suggested to reveal examples of art destinations.

CONCLUSION

With the developing and changing understanding of tourism, tourists want to see not only a new place but also to experience the people and therefore their culture and art. One of the most important attraction forces a tourist to a new destination and an experience is an art. Whether it is a tourist or an art tourist, those who travel towards art venues and events participate in the art environment without any discrimination. Cities that give priority to art tourism as a product of tourism increase their social pride while obtaining a significant amount of tourism income. While art is important for the destination, it can expand the art and tourism industry. Museums, galleries, workshops, art houses, become contact points for tourists. Similarly, artistic events such as exhibitions, fairs, carnivals, festivals, and auctions provide tourist attractions. At this point, art tourism is effective in the following ways: It encourages people to travel, works of art reflect society and culture, local people-artists-tourists understand each other better, and this whole process contributes to the economy of the city.

The city identity of Ayvalık forms integrity with the artist's profile. Works of art, interviews, and festivals attract visitors to Ayvalık by creating a sense of place. Art tourists are both fascinated by the architecture of the city and experience unique experiences with art venues. The rising culture level through art, new identity stone buildings, and exciting activities offer a better quality of life for residents while revealing the sense of art tourists discovering. Art is a force that shapes and enriches Ayvalık's life.

As an art destination, Ayvalık is an alternative for tourist guides. As it is a new and remarkable art route, Ayvalık is the center of attention for artists and tourists, and this interest is increasing day by day. Reshaping under the leadership of art, Ayvalık is an example of many local governments. The number of artists, art institutions and events, and art-based enterprises increased. However, art tourists experience different and special experiences and an image with art content for the destination created by explaining these experiences. The art route is important for the development of Ayvalık and its surroundings. The number of art tourists and tourism revenues should be increased by working with travel agencies and tourist guides.

In Ayvalık, art tourism should develop rapidly and a tourism strategy with art content formed. Since Ayvalık is an art destination, it should be a candidate to become the art center and even the art capital of the Aegean-Marmara Region. After entering the UNESCO World Heritage List, the buildings that remain idle should be transformed into museums, galleries, theaters, cabaret, symphony and opera buildings, and touristic establishments, and they should be kept alive with art. To become a brand and a destination in art tourism with all its values, Ayvalık should be supported by central and local administrators, civil society organizations, businesses and investors associated with art and tourism.

REFERENCES

Akın, B. (2015). Ayvalık evleri'nin cephe karakterinin oluşumuna etki eden faktörlerin değerlendirilmesi. *Sanat Tarihi Dergisi, 24*(2), 121–138. doi:10.29135td.273892

Amaize, O., Mady, T., & Benson, E. (2011). The need for academic programs in tourism, arts & culture, sports, and healthcare management in the UAE. *World Journal of Management, 3*(1), 96–108.

Anderson, C. E. (2012). *Art Tourism, Guggenheim Effect and Abu Dhabi*. Retrieved October, 24, 2020, from https://confluence.cornell.edu/display/tam2011/Art+Tourism%2C+Guggenheim+Effect+and+Abu+Dhabi

Andrew, E. H., & Andrew-Essien, E. (2013, July). Arts, tourism and the environment: An overview of critical intersections in Nigeria. *Mediterranean Journal of Social Sciences, 4*(6), 275–284.

Australia Council for the Arts. (2015). *Arts nation: An overview of Australian arts*. Retrieved October, 24, 2020, from http://www.australiacouncil.gov.au/research/arts-nation-an-overview-of-australian-arts/

Bennett, T. (1995). *The birth of the museum*. Routledge.

Bhatta, C. (2017). The role of arts in promoting tourism: A case of tourism development in Thamel area. *Journal of Advanced Academic Research, 3*(1), 177–184. doi:10.3126/jaar.v3i1.16627

Bonet, L. (2003). Cultural tourism. In R. Towse (Ed.), *A Handbook of Cultural Economics* (pp. 187–193). Edward Elga Publishing Inc. doi:10.4337/9781781008003.00029

Caton, K. (2014). Underdisciplinarity: Where are the humanities in tourism education? *Journal of Hospitality, Leisure, Sport and Tourism Education, 15*, 24–33. doi:10.1016/j.jhlste.2014.03.003

Chang, T. C. (2000). Renaissance revisited: Singapore as a 'Global City for the Arts'. *International Journal of Urban and Regional Research, 24*(4), 818–831. doi:10.1111/1468-2427.00280

Comunian, R., & Mould, O. (2014). The Weakest link: Creative industries, flagship cultural projects and regeneration. *City. Cultura e Scuola, 5*(2), 65–74.

Currid, E. (2007). *The Warhol economy: How fashion, art, and music drive New York city*. Princeton University Press.

Evans, G., & Shaw, P. (2004). *The contribution of culture to regeneration in the UK: A review of the evidence*. DCMS.

Foster, H. (2015). *Bad new days: Art, criticism, emergency*. Verso Books.

Franklin, A. (2018). Art tourism: A new field for tourist studies. *Tourist Studies*, *18*(4), 399–416. doi:10.1177/1468797618815025

Franklin, A. S. (2008). The tourism ordering: Taking tourism more seriously as an ordering of a global world. *Civilisations*, *57*(1-2), 25–39. doi:10.4000/civilisations.1288

Franklin, A. S., & Papastergiadis, N. (2017). Engaging with the anti-museum? Visitors to the museum of old and new art. *Journal of Sociology (Melbourne, Vic.)*, *53*(3), 670–686. doi:10.1177/1440783317712866

Gonzalez, S. (2011). Bilbao and Barcelona in motion." How urban regeneration "models" travel and mutate in the global flows of policy tourism. *Urban Studies (Edinburgh, Scotland)*, *48*(7), 1397–1418. doi:10.1177/0042098010374510

Gretzel, U., & Jamal, T. (2009). Conceptualizing the creative tourist class: Technology, mobility, and tourist experiences. *Tourism Analysis*, *14*(3), 471–481. doi:10.3727/108354209X12596287114219

Hughes, H. L. (2000). *Arts, entertainment and tourism*. Butterworth-Heinemann.

Hughes, H. L. (2002). The role of theatre and music in urban tourism. In D. A. Jelincic (Ed.), Culture: A driving force for urban tourism (pp.51-55). Academic Press.

Jackson, M. R. (2003). Investing in creativity: A study of the support structure for U.S. artists. *The Journal of Arts Management, Law, and Society*, *34*(1), 43–58. doi:10.3200/JAML.34.1.43-58

Jagodzinska, K. (2017). *Art museums in Australian*. Jagiellonian University Press.

Jenkins, L. D., & Romanos, M. (2014). The Art of tourism-driven development: Economic and artistic well-being of artists in three Balinese communities. *Journal of Tourism and Cultural Change*, *12*(4), 293–306. doi:10.1080/14766825.2014.934377

Jolliffe, L., & Cave, J. (2012). Arts tourism. In P. Robinson (Ed.), *Tourism: The key concepts* (pp. 13–15). Routledge.

Kirshenblatt-Gimblett, B. (1998). *Destination culture: Tourism, museums, and heritage*. University of California Press.

Landry, C., Greene, L., Matarasso, F., & Biancini, F. (1996). *The Art of urban regeneration: Urban renewal through cultural activity*. Comedia.

Lim, C. C., & Bendle, L. J. (2012). Arts tourism in Seoul: Tourist orientated performing arts as a sustainable niche market. *Journal of Sustainable Tourism*, *20*(5), 667–682. doi:10.1080/09669582.2011.636817

Lloyd, R. (2002). Neo-bohemia: Art and neighborhood redevelopment in Chicago. *Journal of Urban Affairs*, *24*(5), 517–532. doi:10.1111/1467-9906.00141

Lomine, L. (2005). Tourism in Augustan society (44 BC–AD 69). In J. K. Walton (Ed.), *Histories of tourism: Representation, identity and conflict* (pp. 69–87). Channel View Publications. doi:10.21832/9781845410339-006

Long, P. E., & Morpeth, N. (Eds.). (2016). *Tourism and the creative industries: Theories, policies and practice*. Routledge. doi:10.4324/9781315735870

Lowrie, K., & Frisch, M. (2007). *The arts community, arts village development and promotion of arts in woodbridge township*. National Center For Neighborhood And Brownfields Redevelopment.

Mandala, L. (2015). Arts, tourism & cultural diplomacy. In C. Lord (Ed.), Arts and America: Arts, culture, and the future of America's communities (pp.129-140). ABD: Americans for the Arts.

Markusen, A. (2006). Urban development and the politics of a creative class: Evidence from a study of artists. *Environment & Planning A*, *38*(10), 1921–1940. doi:10.1068/a38179

Markusen, A., Gilmore, S., Johnson, A., Levi, T., & Martinez, A. (2006). *Crossover, how artists build careers across commercial, nonprofit and community work*. The Arts Economy Initiative, Project on Regional and Industrial Economics, Humphrey Institute of Public Affairs, University of Minnesota.

Markusen, A., & Schrock, G. (2006). The Artistic Dividend: Urban Artistic Specialisation and Economic Development Implications. *Urban Studies (Edinburgh, Scotland)*, *43*(10), 1661–1686. doi:10.1080/00420980600888478

Mathews, V. (2010). Aestheticizing space: Art, gentrification and the city. *Geography Compass*, *4*(6), 660–675. doi:10.1111/j.1749-8198.2010.00331.x

Miles, M. (2013). Art and culture: The global turn. In E. Grierson & K. Sharp (Eds.), *Re-imagining the city: Art, Globalization and Urban Spaces* (pp. 19–38). University of Chicago Press.

Miller, R. K., & Washington, K. D. (2006). *The 2007 travel & tourism market research handbook* (10th ed.). Richard K. Miller & Associates.

O'Gorman, K. (2010). Historical giants forefathers of modern hospitality and tourism. In K. O'Gorman (Ed.), *Tourism Giants* (pp. 3–17). CABI. doi:10.1079/9781845936525.0003

Özel, M. K. (2011). Kent strüktürü ile tapınma yapıları arasındaki ilişki bağlamında Ayvalık Hamidiye Camisi. *MSGSÜ Mimarlık Fakültesi Dergisi*, *7*(11-12), 15–39.

Pengiran-Bagul, A. H. B. (2013). *Art tourism: An exploratory study in Sabah, Malaysia*. Retrieved October, 24, 2020, from http://www.ijbts-journal.com/images/main_1366796758/0083-Awangku.pdf

Pratt, A. C. (2008). Creative cities: The cultural industries and the creative class. *Geografiska Annaler. Series B, Human Geography*, *90*(2), 107–117. doi:10.1111/j.1468-0467.2008.00281.x

Rakic, T., & Lester, J. A. (2013). *Travel, tourism and art*. Ashgate.

Richards, G. (2007). Global Trends in cultural tourism. In G. Richards (Ed.), *Cultural tourism: global and local perspectives* (pp. 1–24). Haworth Hospitality Press.

Saunders, C. E. (2012). *Art tourism, Guggenheim effect*. Retrieved October, 24, 2020, from https://confluence.cornell.edu/display/tam2011/Day+16+-+Charles+Saunders

Sheehan, R. (2014). Artists' positioning in and creating authentic identities through place and tourism in New Orleans's Jackson Square. *Journal of Tourism and Cultural Change*, *12*(1), 50–67. doi:10.1080/14766825.2013.869228

Smith, N. (1996). *The new urban frontier: Gentrification and revanchist city*. Routledge.

Smith, T. (2012). *Thinking contemporary curating*. Independent Curators International.

Uçar, H. (2014). Ayvalık'ta Hagia Triada Kilisesi mimari analizi. *Trakya University Journal of Engineering Sciences*, *15*(1), 7–18.

Walton, J. (Ed.). (2005). *Histories of tourism: Representation, identity and conflict*. Channel View Publications. doi:10.21832/9781845410339

Ward, J. (2016). Down by the sea: Visual arts, artists and coastal regeneration. *International Journal of Cultural Policy*, *24*(4), 1–18.

Watson, C. (2013, May). *How the arts and culture tourism spur economic development*. Retrieved October, 24, 2020, from https://www.westerncity.com/article/how-arts-and-cultural-tourism-spur-economic-development

Yoyotte, J. (1960). Les Pelerinages Dans I'Egypte Ancienne. In J. Yoyotte (Ed.), Les Pelerinages, 'Sources Orientales' iii (pp. 17–74). Academic Press.

ADDITIONAL READING

Bonet, L. (2003). Cultural Tourism. In R. Towse (Ed.), *A handbook of cultural economics*. Edward Elga Publishing Inc. doi:10.4337/9781781008003.00029

Harcup, T. (2000). Re-imaging a post-industrial city: The leeds St. Valentine's fair as a civic spectacle. *City*, *4*(2), 215–231. doi:10.1080/13604810050147839

Löschburg, W. (1997). Und Goethe war nie in griechenland kleine kulturgeschichte des reisens. German: Kiepenheuer

Richards, G. (2001). *Cultural attractions and European tourism*. CAB International Publishing. doi:10.1079/9780851994406.0000

KEYS TERMS AND DEFINITIONS

AIMA: Ayvalık International Music Academy that was established in 1998 enables young people to meet master musicians. Combining music and urban identity, the academy contributes to Ayvalık's art tourism.

Art Destinastion: It is a geographical area designed to meet the needs of tourists traveling for art and offers touristic services.

Art Factory: It is an art center in Ayvalık. Because its main activity is theater, it is primarily a theater stage. In fact, spreading art is the main goal.

Art Route: It is a route developed for touristic purposes with artistic heritage, dynamic artists and fun artistic activities.

Art Tourism: It is a type of tourism that includes traveling to see, get to know and experience art-artist and artistic activities.

Art Tourist: It represents those who travel very exclusively to see art elsewhere and /or those who participate in artistic activities during their travels.

Devil's Coffehouse: It is a private coffehouse with exterior appearance reflecting the urban identity (architecture and stained-glass decorations) and interior place living urban history (library, paintings and photos on the walls). It is a gathering center where cultural and artistic activities are carried out.

ENDNOTES

1 Artist living in Ayvalık and the artist atelier/studio there are very important aspects of the identity and presentation of the city Selçuk Kaltalıoğlu, who presented his research in the detection of art workshops and gratitude to Arif Buz and Hakan Urul who provided their full support and knowledge

2 Author of this study is Sabriye Çelik Uğuz, being a photograph artist all photographs used in the study belongs to her work.

Chapter 14
Technology-Supported Tourist Guiding in Aviation Tourism

Azizul Hassan
The Tourism Society, UK

EXECUTIVE SUMMARY

Augmented reality (AR) offers an interactive experience of the real-world environment when an object of the real-world is augmented by computer-generated perceptual information and relevant artefacts. This is a conceptual chapter based on the review of available literature. Also, resources on the internet have also been accessed and reviewed. On the context of the Diffusion of Innovation theory, this research aims to outline AR guiding for in an airport used for tourist aviation. Biman Bangladesh Airlines, the national flag carrier of the country, is the example where this study also explains the possible challenges and benefits that AR guiding facilities can possibly have. This research outlines two specific areas of management and marketing issues are analysis on the way to implement such guiding. Findings show that from the understanding of the Diffusion of Innovation, AR guiding in these days is adopted by an 'Early Majority' who are followers and engages in reading those reviews given by the previous adopters of new services or products.

INTRODUCTION

Today's travellers are more conscious about the services or products they are accessing, where, airliners face difficulties when they think about strong competitiveness. The effective management and marketing strategy formulation and adoption are important to place positive image on the consumer mindset. In addition, the prompt responses to consumers' demands are also vital to fulfil their demands more satisfactorily. The growth of aviation market is turning as an obvious challenge in terms of fiercer competitions over the years with the introduction of more airliners across the world. However, the pace and rhythm are still to maintain a steady flow through out as the market tend to demand diversities with the development of updated marketing models. The global communities supported the move to quantify and broaden the extent of this market along with other potential segments and niches. The scopes are not limited and yet deserve more attention from all of the parties concerned to initiate and expand the plat-

DOI: 10.4018/978-1-7998-3725-1.ch014

form. Both of the management and marketing strategists demand valid and updated data and information and the existing authority of Biman Bangladesh Airlines in many instances lack the wider availability of such data and information. The market oriented and strategically definable data and information are demands of the policy planners to validate the claim that the global aviation industry demands more attention for increasing marketing viabilities. Technology is normally universal and follows free flow over the countries and nations. Still, in certain cases, the application of technology can create a form of dichotomy and imbalance between the developed and developing that outlines the gap between these two forms. This study has a set objective that is clear and distinct by nature. On the base of the Diffusion of Innovation (Rogers, 1962), the aim of this research is focused on implementing AR guiding for aviation tourism with Biman Bangladesh Airlines as the example. Also, two specific areas of management and marketing issues are analysis on the way to implement such guiding. Based on findings this study offers some suggestions to overcome the explored issues of both management and marketing.

BACKGROUND

Augmented Reality Guiding: Features and Facts

AR reality is one of the most common aspects of gradual technological changes (Hassan & Dadwal, 2016). This is very often viewed as the later stage of virtual technology that is commonly applied in entertainment and gaming industry. Characteristically, this technology is a sort of reality simulated by computer generated visuals where a user can experience a virtually created reality (Chang & Cheung, 2001; Kounavis, et al., 2012). According to the Augmented reality blog (2020), the popularity of this technology as a tool of guiding has been increasing.

AR is a digital technology. This technology when viewed through a specific device, makes changes a person's perception of their physical surroundings. AR as an innovative technology has similarity with Virtual Reality (VR) (Hassan, 2018). However, as a key feature of AR, the real-world environment cannot be replaced but can be augmented by overlaying digital components. Till date, most probably the most mentionable example of AR app is Pokemon Go. AR actually has surpassed the gaming world and moved into the world of marketing. Tourism marketers normally use AR for adding useful information of graphics to an environment that is viewed through a compatible device. In general, AR is experienced through computing devices as Smartphones, tablet and similar. Thus, the technology stays as relatively cheaper for the consumers as most of the VR enabled headsets or devices. In almost all cases, AR functions hand-in-hand with other mobile technologies as GPS tracking and cameras.

In the last few years, AR has turned as a popular technology in the tourism industry. One of the key reasons for such popularity is that AR enables tourism businesses to operate in tourism for enhancing the physical environments that they actually try to encourage customers for visiting of the local sights and relevant facilities.

Like other activities, the application of technology in tourism requires to be well researched due to the demand of information that customers demand prior to make a travel plan. Also, one of the main reasons to adopt AR in the tourism industry is the recent lifestyle changes of the tourists over the last decade (Shabani & Hassan, 2017). Tourists are more information demanding to plan a tour. Their information search tends to continue even after completing the tour. The use of AR in the tourism industry is still a relatively recent development that results the emergence of new users, all the time. AR can serve for ensuring much of this information is available to them. Such information is made available for all time at times that is most relevant. Tourists in these days are in a habit of using Smartphones than ever before even at the time of travelling. Thus, the use of AR guiding app on these Smartphones turns as a great help for them.

Arguably the most innovative and effective uses of AR is its application for travel and tour guiding. AR application in a relatively bigger or unknown tourist destination can be beneficial. AR has importance as tour guiding. The key purpose of any trip by a tourist is to explore new sights and destinations. This can well challenging to travel in places or destinations that are unfamiliar places. Thus, AR supported Maps or AR apps for navigation can make this type travel relatively well managed and easier. Away from the comfortable and lavish hotel environments, AR apps are developed by some travel companies and experts within the tourism industry (Hassan & Shabani, 2017). These allow tourists for enhancing tourist attractions and physical locations that possibly can allow tourists to point their Smartphone at a landmark or building and thus to learn more about it, in the real-time environment. As an example, tourists as users can point their Smartphone at a big sized tourist attraction or airport and instantly they will be provided with details of that place, or even they can aim their tablet compute at a historic building or landmark and will be presented with detailed information and routes to navigate. This offers the ability for greatly enhancing the entire travel experience by allowing the tourists for taking information on the go. AR guiding and the Diffusion of Innovation can be interrelated.

Augmented Reality Guiding and the Diffusion of Innovation

In a simple sense, this theory aims to explain why, how and in which rhythm a newer form of technology or idea will extend within a culture. This theory is concerned with technologies, artefacts and ideas that normally pass from conception to use (Spencer et al., 2012). According to Rogers (2003), the sequences that this theory follows are knowledge, persuasion, decision, implementation, and confirmation. The adoption categories of this theory are innovators, early adopters, early majority, late majority, and laggards. The Diffusion of Innovations Theory relies on the innovation-decision process and possibly can be valid in diverse fields and cultures. This theory follows that the technological innovations are connected through specific channels of the social system and acts over a period of time.

Figure 1. The Diffusion of Innovations Theory (Source: semacraft.com, 2020).

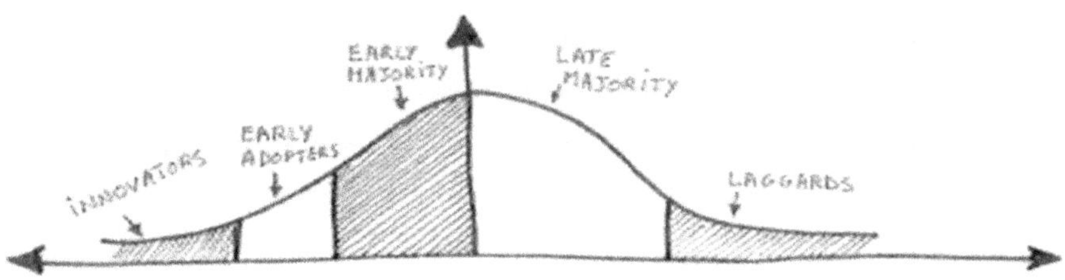

The stages or sequences within which this theory performs are mainly related to five steps. The adopters of this theory are categorized as having some key characteristics. The first group of adopters is the "Innovators" who are the groups of those peoples responsible for exploring newer forms of technologies or ideas. The second stage is the ''Early Adopters'' and this group can be identified as the ''Opinion Leaders'' that is people who are responsible for offering affirmative testimonials of new services or products followed by searching ways for improvements or competencies. The involvement of this group of adopters necessitates some sort of encouragements means for changing. They also offer the ways for using the services or products. The third group is the ''Early Majority'' that is termed as Followers and engages is reading those reviews were given by the previous adopters on new services or products. The fourth group of adopters is the ''Late Majority''. This group of adopters can be termed as sceptics, in a sense that these peoples are not eager on changing and adopt a specific product or service only when they strongly believe that they are missing or left behind for not using or adopting them. This group can get involved with evidence, reviews, case studies or accessing marketing materials in order to explore the ways of working. The fifth and the last group of adopters are the ''Laggards''. This group is a form of a descriptor. This group likes conventional means of communications and mainly adopts a certain type of product or service only when they think that there is no alternative available for them to adopt. This group relies on reviews, comments by others, and available statistics on specific products or services.

The diverse literature on diffusion and dissemination highlighted an important area of contestation in paradigms of diffusion. Most diffusion research has addressed proactively developed innovations whose main mechanism of spread are centrally driven and controlled. But many innovations in service delivery and organization occur as ''Good Ideas'' in local services, which spread informally and in a largely uncontrolled way (diffusion) (Greenhalgh et al., 2004). Smerecnik and Andersen (2011) believe that, the Diffusion of Innovation Theory can also lead the tourism and hospitality organisations into sustainability which is a very hot and competitive topic for all hotels to put themselves into challenge with. In addition, being green and saving resources for the next generations is something that can come with diffusion of innovation; thus managers can correlate adaptation of innovation with the general innovativeness of hotels and resorts in order to stay sustainable.

Based on Rogers's overview (1995), a number of components related to the Diffusion of Innovation (DOI) Theory can be identified: first, ''Network Structure'' that is the adoption of innovations by individuals as powerfully influenced by the structure and quality of their social networks (Fennell & Warnecke, 1988; Valente, 1996; and West et al., 1990). Second, ''Homophily'' that is the adoption of innovations by individuals as more likely if they are 'homophilous' (i.e. have similar socioeconomic, educational, professional, and cultural backgrounds with current users of the innovation) (Fennell & Warnecke, 1988; Fitzgerald et al., 2002; and West et al., 1999). Third, ''Opinion Leaders'' that are when some persons pose a particular influence on the beliefs and actions of their colleagues (Becker, 1970). Expert opinion leaders normally exert influence through their status and authority when peer opinion leaders exert influence through their credibility and representativeness (Fitzgerald et al., 2002; Locock et al., 2001). Opinion leaders can have either a negative or positive influence. In case a project becomes insufficiently appealing (i.e. in terms of goals, organisation and resources clarity), it will less likely to attract the support of key opinion leaders (Locock et al., 2001; Rogers, 1995). This is also important to harness the opinion leader's influence. In naturalistic settings, even though the powerful impact of social influence (such as that of opinion leaders) is well established, attempts to engage such individuals in planned change efforts have often had disappointing results. Fourth, ''Champions'' when the adoption of an innovation by individuals in an organisation becomes more likely if key individuals in their social networks are willing to support the innovation (for strong indirect and moderate direct evidence (Backer & Rogers, 1998; Markham, 1998; Schon, 1963). Fifth, ''Boundary Spanners'' when an organisation is more likely to adopt an innovation if those people who have significant social ties both inside and outside the organisation become able and willing to link the organisation to the outside world relating this specific innovation. Such individuals can play a vital role to capture the ideas that can become organisational innovations (Rogers, 1995; Kimberly & Evanisko, 1981). AR guiding can have several likely impacts.

Augmented Reality Guiding: Possible Impacts

To gauze the possible impacts of AR guiding can be tricky. This research selected AR guiding in aviation tourism with Biman Bangladesh Airlines as the case that is the national flag carrier of the country. This renders importance in terms of the application of AR guiding for aviation tourism and the national airliner setting as the topic to analyse. However, from the general viewpoint, the impacts of AR guiding should be general as activities of the airliner are maintained within a set legislative and administrative framework. AR guiding can expect to get attention from the parties involved and the results should hopefully benefit the tour planners to come to an exact conclusion to make it competitive and profitable. The results also tactically serve demands of both general and special interest tourist segments. However, the extent of acceptation of suggestions exerted from the tourists those used AR guiding can be nominal because of the direct influence and involvement of an innovative and sophisticated technology as AR. However, AR guiding would definitely become useful for tourists where strong commitment is attached.

Initially, a significant weakness of AR guiding would be to create personal interaction of the tourists with AR to feel them comfortable and attract them to use it. This would be a key challenge of typical tourist personality, as well. However, an initial pilot test can possibly be done before finally making AR guiding into the operation. This can help to identify the areas on which the tourists need to stress on and work more. The future of AR guiding can be promising.

Suggestions for the Success of Augmented Reality Guiding in Bangladesh

The implementation of AR guiding can be possible to benefit Biman Bangladesh Airlines. The five key identified issues of management that the airliner is currently facing are typical for a country where the private sector has yet to flourish. The five issues are: first, mismanagement; second, political party supported management body; third, unfit person in the unfit place; fourth, lack of expertise; and fifth, lack of strict chain of command. However, to implement AR guiding, some very typical management issues are required to be overcome as analysed below:

From the management context:

Mismanagement: This should be treated as the most crucial challenge to overcome. An autonomous management body should be formed of relevant experts and thus to ensure strict management. Management professionals are realizing the importance of understanding the local cultures and situations while coping with working in diverse environments across the world. For example, some specific word may have negative meaning in specific countries and this kind of situation can destroy brand image and thus to fully jeopardize the management potentials. But this kind of issues is typical and the caliber of an expert management body depends on tackling this kind of situation in a truly professional way.

Political party supported management body: The managerial body should remain free from political interventions or influences. The formation of a certain type of management body can have both positive and negative impacts on different cultures and countries. There should be a clear interrelationship between the theoretical framework and the application. However, the entire management can fail in cases when political parties get involved and pressurise to form in accordance with their preferences.

Unfit person in the unfit place: The selection of a dedicated management body should be formed to fit the right person for the right place. These are important when the religious facts appear as priority concerns. The similar type of products or services may not have similar responses and acceptability in both Asia and Europe. This becomes particularly true involved with historical conflict between two countries or nations.

Lack of expertise: The demand to meet fundamentals of management as concepts is also turning as important. Experts with vast experience should be offered place and should be settled in the management body. Management strategies have moved away from traditional form that used to engage more people than technologies. In recent years, the general management strategies and approaches involve the use of more technologies and equipment than peoples that certainly rely on expertise.

Lack of strict chain of command: Management policies planners very often tend to fail capitalize the full benefits where the differences in marketing are acute. This is particularly important in areas of cultural dissimilarities and very particularly when language turns as a barrier for business growth. The chain of command should follow a clear organogram with details of the assigned duties and responsibilities. AR guiding can incorporate in the airliner industry of Bangladesh. However, there are always some prospects and some challenges.

Augmented Reality Guiding for Bangladesh Biman Airlines: Prospects and Some Challenges

National airliners are the representation of a nation and advocates for national pride internationally and Biman Bangladesh Airlines is an example. The airliners are managed and marketed by both public and private sector agencies that helps to create and retain the betterment of national identity and pride. Non-conventional markets are more focused to advertisements and thus to promote the airliner reputations. In many cases, airliners adopt experimental marketing strategies that are mostly creative by nature. The marketing approaches are getting more technology based with the application and adoption of numerous platforms. The technological know-how with proper knowledge in relevant areas turn as crucial and the trend is particularly important for airliner marketing. Marketers in the modern world are normally equipped with technologically updated tools in addition to social media and mobile phone supported devices. These are relatively easier to offer the consumer more freedom to select the products or services from wider number of choices. The online world is expandable and is vast in terms of number and features. Some factors are relevant, where individuals or organisations involved with marketing can access to developed elements for marketing. These are responsible for the up gradation of marketing approaches and strategies. Campaigns and advertisements to promote sales and increase performances are also part of the entire process of marketing. For aviation industry, the message is simple that peoples have to fly to perform their day to day duties. This is also important in a marketplace, where the businesses are expanding with the intervention and injection of huge capital investments.

This study has offered a clear focus on a national airliner with Biman Bangladesh Airlines as a case, other than having scattered goals. For the first objective, the findings are clear and precise. This study identifies the key areas of management and marketing failure and offers detailed discussions from the grass-root levels. The basic finding that this study revealed is that at the current situation, clearly that this airliner is unable to offer the desired services to the consumers. The management and marketing initiatives are mostly ineffective and the passengers are less likely satisfied with the ways this airliner dealt with them. The move is clear that both of the traditional management and marketing contexts are changed, where consumers enjoy more freedom than ever before to product or service selection and purchase. This is also true for the circumstance that generates substantial interrelation between the buyer and seller. The decisions to manipulate the market situation are very often influenced by the set policies and strategies. Still, the intense competition helps to bring excellence in service and products. The local meaning of a brand may be different what it originally meant to. This normally creates complexities in diverse market places with language and cultural differences. This is thus obvious that Biman Bangladesh Airlines can less aptly have increased number of products or services consumers with impacts on management and marketing. This study suggests the adoption and application of technology induced management and marketing that are not years old, rather this is the outcome of this study. The origin of this kind of marketing appeared as opposed to traditional types that relatively outline less useful paradigms useful for approaches in business.

Biman Bangladesh Airlines visibly has a strong management body. The management body is led by the Chairman, the Chief Executive Officer (CEO) and the Board of Directors that is led by the Executive Director. The Chairman is typically nominated by the Bangladesh Government. This position is normally designated by the ex-air force chief of Bangladesh. The Chief Executive Officer (CEO) is appointed by mainly by the Board of Directors with the intervention of the responsible ministry, the Ministry of Civil Aviation and Tourism (MOCAT). The appointment of directors and formation of the Board of Directors, the government's involvement become very clear. At any stage, this airliner cannot ignore the direct or indirect influence and engagement of the government. The existing management body of Biman Bangladesh Airlines is formed of 11 positions. Out of these, 9 positions are for the Directors, while, the other 2 are the Chairman and the CEO cum Management Director. The Board of Directors are mainly pioneered by the Bangladesh Air Force officials with at least 2 Secretaries of the Bangladesh Government.

The management strategy adoption and operational activities are mainly designed and supported by the Chief Executive Officer (CEO). This person also acts as the Managing Director. This position is mainly filled by a foreigner with expertise in relevant areas. This person mainly draws the marketing strategies and chalks out the ways for their implementations. The main two areas that this airliner focuses in terms of marketing are concentrating on service improvement and offering better customer services. The current Chief Executive Officer (CEO) cum Managing Director appears as keen to listen to customer feedback and responses. This shows the inclination of this person to develop customer services. However, being a foreigner, the person cannot become always able to design, get supported and implement the marketing strategies and policies according to his pattern that should normally become parallel to the current local and international economic situations. In most cases, the independent decisions taken by this person cannot be fully implemented due to the lack of necessary supports from responsible persons and officials. Successful marketing on the other side, depends on sufficient resources and the Bangladesh Biman Airlines in many cases, cannot take desired marketing strategies. This is a common fact that turns the entire marketing system ambiguous.

Augmented Reality Guiding: Future Research Directions

A tourist has to find and turn a tour as interesting and rewarding. This should be both theoretically and practically sound, effective and set objective oriented. Designing a tour study cannot merely be viewed as recreational work. This normally requires soundness and expertise in conducting independent travel arrangements where a travel plan has to pass through several thoughts and corrections for ensuring perfection and accuracy. The final tour plan also follows the same. Being a tourist, a person should be initially confident that he or she would be able to complete the tour successfully. Prior experience of the tourist normally turns as the main strength coupled with the known environment of the travel site. However, the future travel areas that a tourist supports would be to highlight the importance of effective tour guiding and that can obviously be AR supported. This can essentially become a typical feature of any travel destination and hence this does render usability for the tourists, parties, interest groups or beneficiaries. AR guiding can have influences in a developing country like Bangladesh.

CONCLUSION

On the base of the Diffusion of Innovation (Rogers, 1962) and the basis of findings, this study proposed some suggestions to overcome the existing situations. Findings of this research establishes that AR guiding in the most recent times is adopted by the ''Early Majority'' who are the followers and read the reviews given by the previous adopters on new services or products. On the other side, AR guiding is also represented by the ''Boundary Spanners''. The suggestions are rooted from the opinion engagement. First, understanding the management and marketing failure of Biman Bangladesh Airlines. Findings of this study have identified five basic issues from the management perspective: the five issues are: mismanagement, political party supported management body, unfit person in the unfit place, lack of expertise, and lack of strict chain of command. Again, the five issues related to marketing as corruption, ineffective marketing plans and branding, poor customer service, lack of adequate resources and coordination, and limited number of destinations. The main limitation of this study was the corporate environment that acted as barrier to get support on data and information collection. In most cases, these were treated as highly confidential and non-transferable. Also, the permission seeking to participate with the research and to offer interviews were also challenges. These acted as key barriers and thus squeezed the breadth of this study and the information. The existing employees mostly hesitated to offer any form of correspondence. Rather, discouraged the researcher to continue the research activities possibly having a kind of fear of losing their jobs. There supports could definitely have enriched the contents of this study. Other than this, the limitation of resources for research activities was the other major constraint of this study. Future research studies will need to consider these limitations in order to produce a comprehensive study.

REFERENCES

Augmented Reality Blog. (2020). *How augmented reality can revolutionize the hospitality industry.* Retrieved March 10, 2020, from: http://www.augmentedrealitytrends.com/augmented-reality/hospitality-industry.html

Backer, T. E., & Rogers, E. M. (1998). Diffusion of Innovations Theory and Work-Site AIDS Programs. *Journal of Health Communication, 3*(1), 17–28. doi:10.1080/108107398127481 PMID:10947372

Becker, H. S. (1970). *Sociological work: Method and substance.* Aldine.

Chang, M. K., & Cheung, W. (2001). Determinants of the intention to use Internet/ WWW at work: A confirmatory study. *Information & Management, 39*(1), 1–14. doi:10.1016/S0378-7206(01)00075-1

Fennell, M. L., & Warnecke, R. B. (1988). *The Diffusion of Medical Innovations: An Applied Network Analysis*. Plenum. doi:10.1007/978-1-4684-5436-9

Fitzgerald, L., Ferlie, E., Wood, M., & Hawkins, C. (2002). Interlocking Interactions, the Diffusion of Innovations in Health Care. *Human Relations*, *55*(12), 1429–1449. doi:10.1177/001872602128782213

Greenhalgh, T., Robert, G., Macfarlane, F., Bate, P., & Kyriakidou, O. (2004). Diffusion of Innovations in Service Organizations: Systematic Review and Recommendations. *The Milbank Quarterly*, *82*(4), 581–629. doi:10.1111/j.0887-378X.2004.00325.x PMID:15595944

Hassan, A. (2018). Destination Image Formation: The Function Analysis of Augmented Reality Application. In M. Khosrow-Pour (Ed.), *The Encyclopaedia of Information Science and Technology* (4th ed., pp. 4031–4037). IGI Global.

Hassan, A., & Dadwal, S. (2016). Search Engine Marketing – An Outlining of Conceptualization and Strategic Application. In W. Ozuem & G. Bowen (Eds.), *Competitive Social Media Marketing Strategies* (pp. 219–234). IGI Global. doi:10.4018/978-1-4666-9776-8.ch011

Hassan, A., & Shabani, N. (2017). Usability Analysis of Augmented Reality for Tourism Destination Promotion. In S. K. Dixit (Ed.), *Routledge Handbook of Consumer Behaviour in Hospitality and Tourism* (pp. 434–441). Routledge. doi:10.4324/9781315659657-49

Kimberly, J. R., & Evanisko, M. J. (1981). Organizational innovation: The influence of individual, organizational, and contextual factors on hospital adoption of technological and administrative innovations. *Academy of Management Journal*, *24*, 689–713. PMID:10253688

Kounavis, C. D., Kasimati, A. E., & Zamani, E. D. (2012). Enhancing the tourism experience through mobile augmented reality: Challenges and prospects. *International Journal of Engineering Business Management*, *4*, 1–6. doi:10.5772/51644

Locock, L., Dopson, S., Chambers, D., & Gabbay, J. (2001). Understanding the role of opinion leaders in improving clinical effectiveness. *Social Science & Medicine*, *53*(6), 745–757. doi:10.1016/S0277-9536(00)00387-7 PMID:11511050

Markham, A. N. (1998). Life online: Researching real experiences in virtual space. In C. Ellis & A. P. Bochner (Eds.), *Ethnographic Alternatives Book Series* (pp. 244–264). Altamira Press.

Meyer, M., Johnson, D., & Ethington, C. (1997). Contrasting attributes of preventive health innovations. *Journal of Communication*, *47*(2), 112–131. doi:10.1111/j.1460-2466.1997.tb02709.x

Rogers, E. M. (1962). *Diffusion of innovations*. The Free Press.

Rogers, E. M. (1995). *Diffusion of Innovations* (3rd ed.). Free Press.

Rogers, E. M. (2003). *Diffusion of innovations* (5th ed.). Free Press.

Schon, D. A. (1963). *Reflection of concepts*. Humanities Press.

Semacraft. (2020). *Diffusion of Innovation Theory*. Retrieved March 12, 2020, from: http://semacraft.com/blog/wp-content/uploads/2010/11/blogDiffusionImage.jpg

Shabani, N., & Hassan, A. (2017). Augmented reality for tourism service promotion in Iran as an emerging market. In V. Nadda, S. S. Dadwal, & R. Rahimi (Eds.), *Promotional strategies and new service opportunities in emerging economies* (pp. 116–129). IGI Global. doi:10.4018/978-1-5225-2206-5.ch005

Smerecnik, K., & Andersen, P. (2011). The Diffusion of Environmental Sustainability Innovations in North American Hotels and Ski Resorts. *Journal of Sustainable Tourism*, *19*(2), 171–196. doi:10.1080/09669582.2010.517316

Spencer, A. J., Buhalis, D., & Moital, M. (2012). A hierarchical model of technology adoption for small owner-managed travel firms: An organizational decision-making and leadership perspective. *Tourism Management*, *33*(5), 1195–1208. doi:10.1016/j.tourman.2011.11.011

Valente, T. W. (1996). Social Network Thresholds in the Diffusion of Innovations. *Social Networks*, *18*(1), 69–89. doi:10.1016/0378-8733(95)00256-1

West, M. A., & Farr, J. L. (1990). Innovation at work. In M. A. West & J. L. Farr (Eds.), *Innovation and creativity at work: Psychological and organizational strategies* (pp. 309–333). Wiley.

ADDITIONAL READING

Hassan, A., & Ramkissoon, H. (2017). Augmented Reality for Visitor Experiences. In J. N. Albrecht (Ed.), *Visitor Management* (pp. 117–130). CABI.

Hassan, A., & Shabani, N. (2017). Usability Analysis of Augmented Reality for Tourism Destination Promotion. In S. K. Dixit (Ed.), *Routledge Handbook of Consumer Behaviour in Hospitality and Tourism* (pp. 434–441). Routledge. doi:10.4324/9781315659657-49

Rahimi, R., Hassan, A., & Tekin, O. (2017). Augmented Reality Apps for Tourism Destination Promotion. In S. Rezaei (Ed.), *Apps Management and E-Commerce Transactions in Real-Time* (pp. 236–251). IGI Global. doi:10.4018/978-1-5225-2449-6.ch011

KEY TERMS AND DEFINITIONS

Aviation: Aviation in a simple meaning is flying with an aircraft like an airplane. This also involves the activities and industries attached to flight like air traffic control.

Bangladesh: Bangladesh is a South Asian county marked by many waterways and lush greeneries. The country has tremendous potentials for tourism development.

Biman Bangladesh Airlines: The national airliner of Bangladesh.

Hazrat Shahjalal International Airport: This is most prominent and largest international airport in Bangladesh. This airport is located in Kurmitola, the northern part of the capital city Dhaka.

Technology: Technology is the scientific knowledge application for practical purposes, especially in industry.

Tour Guide: A tour guide is the person who guides visitors in the language of their choice and interprets an area's natural and cultural heritage which person generally possesses an area-specific qualification usually issued and/or recognised by the appropriate authorities.

Tourism: Tourism is the peoples' activities travelling to and staying in places outside their normal environment for leisure, business and other purposes and not for more than a consecutive year.

Chapter 15
Technological Innovation Research With Guided Tours:
Recent Trends and Future Directions

Abdullah Tarinc
https://orcid.org/0000-0001-5824-4882
Manavgat Tourism Faculty, Akdeniz University, Turkey

Abdullah Karaman
https://orcid.org/0000-0001-7934-0451
Tourism Faculty, Selcuk University, Turkey

Halil Sunar
Tirebolu Mehmet Bayrak Vocational High School, Giresun University, Turkey

Kürşad Sayin
Silifke Tasucu Vocational High School, Selcuk University, Turkey

EXECUTIVE SUMMARY

Researchers and academicians have shown many studies on how technological developments can be used in tour guidance. The common objectives of these studies are to provide tourists with a better tour experience, to assist tour guides, to help people who cannot participate in the tour, and to develop tour guide robots. The technologies used in tour guidance vary according to the type and content of the tours. In this study, technological innovations and applications used in tour guiding in the world will be investigated and explained in detail. In addition, information on technological innovations and trends that are foreseen and planned to be used in touristic tours in the future will be given. With this study, it is aimed to reveal the technological applications used today in tour guidance and technological developments that will guide the tour guidance in the future.

DOI: 10.4018/978-1-7998-3725-1.ch015

INTRODUCTION

Due to the continuous technological invention and improvement, society has changed the way of work, communication, and living (Madsen et al. 2016). Societies inevitably keep up with technological developments in every field. Information and communication technology (ICT)-based technological developments have effects on human life such as trade, production, health, transportation, tourism, and education. It can be said that the demands and expectations of the tourists participating in tourism activities differ, especially considering the transition to the information society and the social integration of technological developments that are realized rapidly. It is seen that tourists have been using information communication technology applications in the process of holidays and post-purchase behavior since the purchasing decision process. To meet the wishes and desires of tourists who develop themselves in the use of technology, it is necessary to follow and apply the information communication technologies and innovations in the tourism industry.

While it is seen that tourism-related enterprises make intensive technological investments in the tourism industry where intense competition is experienced, it is seen that technological guidance is influenced and benefited from tourist guidance, which has an important place in tourism. Today, many technological research and development activities are planned to meet the expectations of the tour experiences of the tourists and tour guides to be able to make a more effective tour. Therefore, the fact that tourists have the opportunity to further develop themselves in terms of information and the necessity of integrating tourist guides to technology increases the importance of ICT.

With the intensive use of information and communication technologies by countries, regions, tourist destinations, historical sites, museums, tourists and tourist guides, these applications are expected to evolve in terms of number and quality in the future. The main purpose of this chapter is to give information about the ICT currently used in tour guidance and to give a preliminary introduction of the applications that are intended to be used in tour guidance in the future.

RECENT TRENDS IN TOUR GUIDANCE IN TERMS OF TECHNOLOGICAL INNOVATION

Tour guiding practice has, in many respects, matured over the past 50 years, and in contributing to and responding to this maturation, there has been a burgeoning of tour guiding research (Weiler & Black, 2015). Buhalis and Law (2008) stated that information and communication technologies changed tourism since the 1980s. Besides, according to Martin and Herrero (2011), many researchers of new information and communication technologies have directed in-depth and comprehensive research on information communication technologies and tourism.

In terms of tourism, the use of information technologies should not be considered only in terms of tourism enterprises. The use of information technologies has entered into daily life activities, and the efficiency of consumers using information technologies is increasing day by day. In this context, it can be said that tourist consumers need and use information technologies at every moment of their travels. At the same time, it is seen that tour guides, which undertake the promotion of tourist destinations to tourists and assist tourists from pre-touristic to post-touristic travel, benefit from the information technology. On the other hand, the provision of information-based services by tourist destinations to both tourists and guides will help to make the tour activities more successful. Today, there are many technological applications used by tourists, tour guides, and tourist destinations. In this section, the necessary information about recent trends in terms of tour guidance will be given through literature research.

Technology Used by Tourists

The use of smartphones with many applications that can be downloaded and run has become increasingly common. Pence (2010) compares the smartphone to the Swiss Army knife as a modern communication tool with many features such as a global positioning system (GPS), camera, compass, wifi support, web browser, and accelerometer. "Mobile tourists" can go through a new way of traveling as "virtual tourists" for which the traveling experience is empowered by the information and entertainment coming "also" (but of course not exclusively) from their new "travel buddy": the mobile technology device (Palumbo et. al., 2014). With the help of the applications installed on the smartphones, tourists can make comments about the touristic destinations in advance, to obtain various information during the tour and to inform other tourists about the tour after the tour. As a broad and interconnected process from the beginning of the tour to the end of the tour, such as accessing information, obtaining information about the guide, security, destination information, and transportation services, the element of technology emerges sensibly (Çakmak and Demirkol, 2017).

The development of information and communication technologies, and especially the Internet, has created a new, more experienced, demanding tourist type that is difficult to please. At the same time, Buhalis and Law (2008) also emphasized the development of ICT and the internet. They said that the internet is important for new tourists who develop themselves, seek information, and seek value for money and time. In their study, they summarized the desires of tourists for ICT, including (Wang et. Al., 2016):

(1) Pursuing personal traveling preferences and schedules;
(2) Value for time and less willingness to wait or put up with delays;
(3) Searching for travel-related information through the Internet;
(4) Booking online tickets and making room reservations;
(5) Making online purchases;
(6) Conducting price comparisons on different travel websites;
(7) Communicating in the virtual travel communities;
(8) Offering an e-complaint handling system;
(9) Asking for multimedia service;
(10) Providing mobile facilities and applications such as Wið, short-messaging service, and multimedia-messaging service.

Buhalis and Law (2008) stated that many tourists prefer to share their travel experiences and recommendations with others, and the Virtual Travel Community (VTC) has become one of the favorite areas for publishing travel experiences. In addition, Wang et. al. (2002) said in their study that online travelers are eager to meet other travelers with similar attitudes, interests, and lifestyles. Customer satisfaction is an important factor in the evaluation of travel effectiveness. In their study, Buhalis and Law (2008) stated that travel agencies could analyze the content of VTCs, understand the satisfaction and behavior of their customers, take corrective actions to improve their offers and increase brand awareness with the help of VTCs.

Nowadays, some tourists want to visit a tourist destination by participating in group tours, museums or historical sites with a guide and a certain tour program, as well as tourists who want to visit their touristic travel freely without being bound to any program. Especially in small groups, since it is thought that organizing a trip with a tour guide will be costly, these people prefer to travel on their own and benefit from point guidance services (Ahipaşaoğlu, 2006). The ability to explain the preferred language and allow people to organize their trips is an important feature of mobile tourist guides.

Technology Used by Guides

Tourist guide, according to the World Federation of Tourist Guide Associations, is a person who guides visitors in the language of their choice and interprets the cultural and natural heritage of an area which person normally possesses an area-specific qualification usually issued and/or recognized by the appropriate authority (www.wftga.org). As a leader, the tour guide has to organize (i.e. provide direction, access, and control) and build the team (i.e. ensure group cohesion and morale). As a mentor, the guide acts as an interactional (or 'middleman') and is an educator for the group (Howard et. al., 2001).

The importance of using technology in the tourist guidance profession is increasing day by day. The applications developed, especially with the widespread use of smartphone technology, allow communication and information transfer between tour guides and tourists to be made faster and more effectively. As an example of these applications, the pocket guide system is an aid in finding directions and allows tour guides to move around the area more easily (Tekin et. Al., 2017).

The headset system implemented in museums and historical sites, GPS-navigation technologies, smartphones, and developments in transportation systems affect the guides on their professional practices. On the other hand, it also affects tourists participating in the tour very positively, and it is thought that it increases participation or efficiency (Çakmak and Demirkol, 2017).

Another convenience that technology provides to guides is that they can carry hundreds of books about the places they visit in a single device using smartphones or tablet computers instead of sourcebooks that they have to carry along with themselves.

Another application that tourist guides use when performing their narrations is a mobile microphone and headset system. These applications allow tourist guides to provide visitors with more comprehensive information about the artifacts. An example of these applications is the Tourist Guide Wireless Headset System. Together with these systems, visitors can easily listen to tourist guides even if they are in a very noisy environment. As a result, there is no disconnect between the tourist guides and the visitors, but the visitors are prevented from moving away from the subject.

It enables the tour program to be carried out more successfully thanks to the communication groups formed by the tour guides with the help of various communication programs on social media with the tourists participating in the tour or with other tour guides. For example, with the communication groups formed with the tourists participating in the tour, the tourists are in constant communication before, during, and after the tour, and both instantaneous information flow is made, and possible problems are solved quickly. Besides, in contact with other guides working in the region, information is obtained about the most appropriate route selection in terms of the instant density and tour route of the region.

With the new generation of technologies, it is almost perfect nowadays for translating speech into text automatically. Therefore, utilizing very fast processor devices/applications, it will be possible to provide guidance services in different languages to tourists with different languages in the same group by converting them to alternative languages in a very short time (Bahar et. al., 2019).

Technology Used by Tourist Destinations

In terms of tourism regions, the intensity of the use of ICT will enable the region to exist in the market and create a competitive advantage in international tourism in the coming years (Buhalis, 1998). Therefore, in the future, tourism regions can be divided into developed and underdeveloped tourism regions depending on the rate of usage of information and communication technologies (Rimmington and Kozak 1997). In this case, what should be done should first be to analyze what kind of technology a region is located in and whether the current applications meet the expectations and requirements of the consumer and whether future studies should be conducted should be shed light (Sarı and Kozak, 2005).

Many ICTs are used in tourism destinations. One of them is the tour guide robot application used in museums. Rhino was the first tour guide robot proved in a real museum that was equipped with a laser, sonar, infrared, and stereo cameras (Bueno et. al., 2011). With collision avoidance, mapping, positioning, route planning, task planning, and user interface, Rhino served as a guide robot at the Deutsches Museum in Bonn, Germany (Yıldız, 2018). Later on, the same research team developed Minerva, an improved tour-guide robot. All their efforts where mainly focus on making the interaction more sophisticated instead of modifying the system control aspects (Bueno et. al., 2011). After Rhino and Minerva, many countries have produced a large number of tour-guide robots. Perhaps the most advanced of these is the tour guide robot, Pepper, used in the Smithsonian museum by the United States, which has been implemented in recent years. Pepper's features include answering questions, telling stories, dancing, playing games, and posing for selfies. Although Pepper cannot speak naturally, it uses sound and gestures to create interactions, and also images, text and video can be viewed with the help of the tablet located on the robot's chest (www.si.edu). The robots mentioned above generally provide visitors with basic information about the museum, while also directing them to several exhibitions (Stricker et. al., 2012). It is known that tour guide robots are used not only in museums but also in exhibitions, universities, and open-air areas. Mona/Oskar guide robots used at the Opel Sales Center at Berlin are examples of guide robots used in the exhibitions. Mona / Oskar is a robot based on the information and entertainment. The information and guide robot Mona / Oskar can recognize and talk to visitors with individually structured speech output. It also displays information on the screen and offers guided tours (www.ipa.fraunhofer.de).

Another area where tour guide robots are used is universities. The purpose of using tour guide robots in universities are safe navigation, self-safety, power, and portability constraints. The tour guide robot Roslyn is an autonomous system that can carry out promotional tours of the Department of Computer Science and Engineering at Nevada University, Reno. The robot is not only able to navigate the halls, but navigate them in a socially acceptable manner (Sweet, 2014).

The fact that robot tourist guides only serve indoors suggests that robotic tourist guides will be limited to indoor spaces only, and therefore, guidance services will be continued by people in open spaces. However, it can be said that Robohon, who provides guidance in Kyoto, Japan, is the pioneer of robot tourist guides that provide outdoor guidance. Being very small and light in size, Robohon requires visitors to carry it, making it suitable only for individual or couple tourists rather than crowded tourist groups (Yıldız, 2019).

In the future, it is expected that the hotel businesses will provide pioneering guidance services with the help of data from customers' previous experiences. A visitor who has entered the hotel with a passport number has also taken into consideration the length of tourist's stay in the previous destinations or with the help of credit card expenditures (using big data analysis and algorithms), it will be possible to serve as a guide service on the room television screen of the areas of interest to be seen in the destination (Bahar et. al., 2019).

Advantages and Disadvantages

According to Ap and Wong (2001), economic problems in the tourism industry, strict competition, and high service standards expected by tourists complicate the work and role of tour guides. From this point of view, it can be said that technological developments offer various advantages and disadvantages to tourist guidance.

Now tourists are qualified for guidance services and have expectations that make them feel special. Tourists learn about descriptions via the Internet and ask for more specific information from the guides and improve them to be more equipped. At this point, guides can develop themselves more easily by using technology effectively, and they can get ready for tours.

The guides are obliged to make the holiday experiences of the customers unique, and in this direction, they can benefit from technological tools during the tour. However, advances in technology also reveal new challenges for guidelines (Çolakoğlu et. al., 2017). New developments in technology have, in many cases, fulfilled the functions of the guidelines. For example, in some historical sites, visitors can rent portable audio players and navigate without the need for a guide. Therefore, the technology also starts to restrict the work area of guides, and can provide guided tours without any group or individuals who enjoy traveling without being bound to the guide (Çakmak and Demirkol, 2017). Considering the developments in robot tourist guides from past to present; It can be said that the role of robot tourist guides in the profession of tourist guidance will increase in direct proportion with the developments in the field of technology and therefore, in the medium or long term, robotic tourist guides will pose the risk of replacing human tourist guides in the tourist guidance profession (Yıldız, 2019).

Based on Çakmak and Demirkol's (2017) study to measure the impact of technological developments on the tourist guidance profession, we can list the advantages and disadvantages of technological developments for tourist guidance as follows.

- Thanks to the communication technologies, before and after the tour, the guide-tourist relationship becomes stronger than before,
- Navigation technology to eliminate the problems experienced in the direction,
- Thanks to the resources that the guides carry with them, they can go to touristic destinations more fully equipped.
- To be able to become more knowledgeable in cultural and geographical terms in a shorter period than the previous ones,
- Thanks to advanced communication networks and social media, guides have the opportunity to communicate directly with customers and take additional tours,
- Technology is a driving force in preparing guides for the region in a shorter time
- The risk that technology-based guidance practices can replace guides in the future,
- The lack of technology-integrated guidelines lags in the competition

FUTURE DIRECTIONS IN TOUR GUIDANCE IN TERMS OF TECHNOLOGICAL INNOVATION

The applications that are used today, developing and foreseen to be offered to tourists in the future, are explained below in the frame of technological innovation in tour guidance.

Mobile Tourist Guide

The most used technologies in the planning stage of travel are the Internet and e-commerce. Recently, the intensive use of the Internet and smartphones has caused many changes in the tourism industry. The development of information communication technologies, mobile communication, and personal navigation technologies enabled the use of mobile tour guide applications. Applications that provide information about the destination have been developed in various ways to be used in many attractions, such as museums, historical sites, and natural beauty areas.

With the new generation technologies, the need for a visual guide instead of a physical guide will be formed. The resulting need can be rendered interactive in 3D by hologram guides. In recent years, the first humanoid robot Sophia'ya citizenship by Saudi Arabia, as well as a very close time to the virtual guidance of humanoid robots by the guidance rooms "guidance badge," will arise in need (Bahar et. al., 2019).

Interactive Tour Guide Robot

The development of information and communication technologies has recently increased the interest in interactive speech applications. There are many types of research and studies on interactive guidance application and applicability in museums and ruins (Trahanias et. al., 2000; Greeff and Lalioti, 2001; Fino et. al., 2013; Almeida and Yokoi, 2002; Burgard et. al., 1998). In their study, Almeida and Shigeo (2002) stated that interactive guidance applications not only guide users, give instructions, advise and perform tasks, but also improve communication with visitors through verbal and nonverbal interactions such as facial expressions, hand gestures and gaze gestures. Fino et al. (2013) developed an interactive tour guide application that includes historical places within the framework of cultural tourism using 3D animation, AR, and QR code. Users can visit the touristic routes and symbols of the city with an augmented reality application with the created website or QR code.

In the study of Tung and Law's (2017) research on human-robot interactions in the tourism and hospitality industry, it was suggested that the transfer of tourist experiences to the robots with cloud robot technology and past tourist experiences with robots could be presented to other tourists in the future. The TOURBOT system developed by Roussou et al. (2001) on interactive robot tour guide in museums, proposed a model that is personalized, interactive and provides remote access with its web interface.

On the basis of all these interactive tour-guide robot applications, the interaction of the tourists according to their culture will increase the value of the tours. Although it is quite difficult to incorporate the cultural features of the visitors into the robot cloud application, it will be an important start in terms of human-robot interaction in the future. It is important to develop an interactive tour-guide robot technology to increase the attractiveness of museums, increase the effectiveness of museum guidance, and provide a comprehensive experience to the visitors.

Virtual Reality and Tour Guidance

İlhan and Çeltek (2016) pointed out that technology is constantly improving in today's world and stated that new technology is introduced to individuals every day. They also said that these technologies make people's lives easier and that they are new tools or applications that are easy to use and have many functions. One of them is the VR technology developed by information communication technologies and adapted to the tourism sector. VR is defined as the use of a computer-generated 3D environment – called a 'virtual environment' (VE) – that one can navigate and possibly interact with, resulting in real-time simulation of one or more of the user's five senses (Gutentag, 2010). Augmented reality (AR) – the projection of computer-generated images onto a real world view as a type of VR. AR and VR are related, and it is quite valid to consider the two concepts together (Gutentag, 2010). It is possible to develop Mobile AR applications in multidisciplinary fields such as tourism, travel, games, educational entertainment, cultural heritage, navigation, medicine, defense industry, engineering.

It is almost impossible to make a definitive judgment on how virtual reality will be in the future. However, you can learn about the future of virtual reality by looking at current research. For example, according to researchers at the University of Cambridge, the new VR technology could play a key role in diagnosing and monitoring Alzheimer's in the future – also assisting with clinical trials of medications aimed at slowing down the disease's progression. As another example, the 5G technology presented in the Qualcomm wireless event in San Diego is that VR applications can turn into a whole new dimension. With the use of virtual reality, users feel like there, and at the same time gain interaction with those using the same platform to provide experience. According to Adhani & Awang (2012), users have become able to feel objects in their virtual tours with the new ICT. Besides, users can both improve the learning experience and rebuild the old object and building with mobile AR in cultural heritage sites.

Examples of e-museum applications are related to AR activities under the guidance of tourists. When a visitor enters this next-generation museum, he/she will be identified by a location tag and, from that point on, guided by a human-faced robot capable of interacting with the person through the use of voice recognition and synthesis. However, user experience can also be improved by using personal voice-operated audio-guides and Augmented Reality (AR) devices (Bueno et. al., 2011). It was concluded that Soe et al. (2010) supported the personalized tour guides on the sites by using contextual information such as the location and profile of a tourist in their studies on virtual tour guidance with AR application (Soe et. al., 2010). The museum guide developed by Damala et al. (2008) with AG technology at the Museum of Fine Arts in Rennes, France, is another example. Looking at the research, it can be seen that the developed AR system can offer an interesting alternative to navigation and interaction features. It also provides a more natural and different way of interacting with the displayed objects. Designed by the Smithsonian National Museum of Natural History, the "Skins & Bones" application is one of the best examples of augmented reality applications in terms of museum participation and training. This application, launched in 2015, was developed for an exhibition called one Bone Hall, which was established in the 1960s and had almost no renovation in the following 50 years. This application aims to share the untold stories behind the museum's most iconic collections (Ding, 2017).

Another area where AR applications are used in tourism is cultural heritage. Many institutions and organizations authorized in the field of cultural heritage develop augmented reality applications that enrich historical sites with virtual information to be used in these fields (Haugstvedt & Krogstie, 2012). For example, the augmented reality application used in the Cathedral of Valencia was developed to show scenes that have been partially or completely lost in the past and present scenes. With this application, users can compare the baroque style vault in the cathedral and the church mihrab backrest in the past and present (Portales et. al., 2009).

CONCLUSION

The tourism and travel industry is undoubtedly affected by ICT development. Providing unique experiences to the tourists and ensuring customer satisfaction is the primary purpose of tourist destinations. With ICT, various and many convenience and comforts are offered to people's lives. It is seen that ICT applications such as web, AR, VR, mobile tour guide, tour guide robot are examined by researchers interested in tour guidance today.

In the literature research, it was determined that these technological applications add value to touristic experiences, increase tourist satisfaction, and contribute to promotional activities for destinations. However, in the future, both tourists and tourist destinations have high expectations of ICT. While mobile tour guides, augmented reality and interactive tour robot guidance applications provide many advantages for tourists, there are various disadvantages in terms of tour guidance profession. Although the tour guiding profession is under threat for tourists to give importance to interaction and educational trips, it is thought that the tour guides can overcome this problem by adapting themselves to technology.

REFERENCES

Adhani, N. I., & Awang, R. D. R. (2012, December). A survey of mobile augmented reality applications. *1st International Conference on Future Trends in Computing and Communication Technologies*, 89-96.

Almeida, P., & Shigeo, Y. (2002, December). Interactive conversational character as a virtual tour guide to an online museum exhibition. *International Conference on Computers in Education, Proceedings*, 215-216. 10.1109/CIE.2002.1185907

Ap, J., & Wong, K. K. (2001). Case study on tour guiding: Professionalism, issues and problems. *Tourism Management, 22*(5), 551–563. doi:10.1016/S0261-5177(01)00013-9

Bahar, M., Yüzbaşıoğlu, N., & Topsaka, L. Y. (2019). Akıllı turizm ve süper akıllı turist kavramları ışığında geleceğin turizm rehberliğine bakış. *Journal of Travel and Tourism Research, 14*, 72–93.

Bueno, D. R., Viruete, E., & Montano, L. (2011, December). *An autonomous tour guide robot in a next generation smart museum.* Paper presented at the 5th International Symposium on Ubiquitous Computing and Ambient Intelligence (UCAmI 2011), Riviera Maya, Mexico.

Buhalis, D. (1998). Strategic use of information technologies in the tourism industry. *Tourism Management, 19*(5), 409–421. doi:10.1016/S0261-5177(98)00038-7

Buhalis, D., & Law, R. (2008). Progress in information technology and tourism management: 20 years on and 10 years after the Internet—The state of eTourism research. *Tourism Management, 29*(4), 609–623. doi:10.1016/j.tourman.2008.01.005

Buhalis, D., & Licata, M. C. (2002). The future e-tourism intermediaries. *Tourism Management, 23*(3), 207–220. doi:10.1016/S0261-5177(01)00085-1

Burgard, W., Cremers, A. B., Fox, D., Hähnel, D., Lakemeyer, G., Schulz, D., . . . Thrun, S. (1998, July). The interactive museum tour-guide robot. *Proceedings of the Fifteenth National Conference on Artificial Intelligence and Tenth Innovative Applications of Artificial Intelligence Conference*, 11-18.

Çakmak, T. F., & Demirkol, Ş. (2017). Teknolojik gelişmelerin turist rehberliği mesleğine etkileri üzerine bir swot analizi. *Bingöl Üniversitesi Sosyal Bilimler Enstitüsü Dergisi, 7*, 221–235. doi:10.29029/busbed.317656

Çolakoğlu, O. E., Efendi, E., & Epik, F. (2017). *Tur yönetimi ve turist rehberliği*. Detay Yayıncılık.

Damala, A., Cubaud, P., Bationo, A., Houlier, P., & Marchal, I. (2008, September). Bridging the gap between the digital and the physical: Design and evaluation of a mobile augmented reality guide for the museum visit. *Proceedings of the 3rd International Conference on Digital Interactive Media in Entertainment and Arts (DIMEA '08)*, 120-127. 10.1145/1413634.1413660

Ding, M. (2017). *Augmented reality in museums*. Retrieved October 15, 2019, from https://static1.squarespace.com/static/51d98be2e4b05a25fc200cbc/t/5908d019f5e2314ab790c269/1493749785593/Augmented+Reality+in+Museums.pdf

Fino, E. R., Martín-Gutiérrez, J., Fernández, M. D. M., & Davara, E. A. (2013). Interactive tourist guide: Connecting web 2.0, augmented reality and QR codes. *Procedia Computer Science, 25*, 338–344. doi:10.1016/j.procs.2013.11.040

Fraunhofer Institute. (n.d.). *Robot Mona/Oskar*. Retrieved October 12, 2019, from https://www.ipa.fraunhofer.de/en/expertise/robot-and-assistive-systems/household-and-assistive-robots/entertainment.html

Greeff, M., & Lalioti, V. (2001, September). Interactive cultural experiences using virtual identities. *International Cultural Heritage Informatics Meeting*, 455-465.

Guttentag, D. A. (2010). Virtual reality: Applications and implications for tourism. *Tourism Management, 31*(5), 637–651. doi:10.1016/j.tourman.2009.07.003

Haugstved, A.-C., & Krogstie, J. (2012, January). Mobile augmented reality for cultural heritage: A technology acceptance study. *2012 IEEE International Symposium on Mixed and Augmented Reality (ISMAR)*, 247-255. 10.1109/ISMAR.2012.6402563

Howard, J., Smith, B., & Thwaites, R. (2001). Investigating the role of the indigenous tour guide. *Journal of Tourism Studies, 12*(2), 32.

İlhan, İ., & Çeltek, E. (2016). Mobile marketing: Usage of augmented reality in tourism. *Gaziantep University Journal of Social Sciences, 15*(2), 581–599. doi:10.21547/jss.256721

Madsen, E. S., Bilberg, A., & Grube Hansen, D. (2016, September). Industry 4.0 and digitalization call for vocational skills, applied industrial engineering, and less for pure academics. *Proceedings of the 5th P&OM World Conference P&OM*.

Martín, H. S., & Herrero, A. (2011). Influence of the user's psychological factors on the online purchase intention in rural tourism: Integrating innovativeness to the UTAUT framework. *Tourism Management*, *33*(2), 341–350. doi:10.1016/j.tourman.2011.04.003

Palumbo, F., Dominici, G., & Basile, G. (2014). The culture on the palm of your hand: How to design a user oriented mobile app for museums. In L. Aiello (Ed.), Management of cultural products: e-relationship marketing and accessibility perspectives (pp. 224-243). IGI Global.

Portalés, C., Lerma, J. L., & Pérez, C. (2009). Photogrammetry and augmented reality for cultural heritage applications. *The Photogrammetric Record*, *24*(128), 316–331. doi:10.1111/j.1477-9730.2009.00549.x

Rimmington, M., & Kozak, M. (1997). Developments in information technology: Imlications for the tourism industry and tourism marketing. *Anatolia: An International Journal of Tourism Hospitality Research*, *8*(3), 59–80. doi:10.1080/13032917.1997.9687121

Roussou, M., Trahanias, P., Giannoulis, G., Kamarinos, G., Argyros, A., Tsakiris, D., ... Schulz, D. (2001, November). Experiences from the use of a robotic avatar in a museum setting. *Proceedings of the 2001 conference on Virtual reality, archeology, and cultural heritage*, 153-160. 10.1145/584993.585017

Sarı, Y., & Kozak, M. (2005). Turizm pazarlamasina internetin etkisi: Destinasyon web siteleri için bir model önerisi. *Akdeniz İİBF Dergisi*, *9*, 248–271.

Seo, B. K., Kim, K., & Park, J. I. (2010, November). Augmented reality-based on-site tour guide: a study in Gyeongbokgung. *Asian Conference on Computer Vision*, 276-285.

Smithsonian Museum. (n.d.). *Pepper the Robot*. Retrieved October 10, 2019, from https://www.si.edu/visit/pepper

Stricker, R., Müller, S., Einhorn, E., Schröter, C., Volkhardt, M., Debes, K., & Gross, H. M. (2012). Konrad and Suse, two robots guiding visitors in a university *building*. In *Autonomous Mobile Systems 2012* (pp. 49–58). Springer. doi:10.1007/978-3-642-32217-4_6

Sweet, T. (2014*). Roslyn: The tour guide robot* (PhD Thesis). Computer Science & Engineering, University of Nevada, Reno, NV.

Tekin, Ö., Bideci, M., & Avcıkurt, C. (2017). Turist rehberliğinde yeni teknolojilerin kullanımı. In F. Ö. Güzel, V. Altıntaş, & İ. Şahin (Eds.), *Turist rehberliği araştırmaları öngörüler ve uygulamalar* (pp. 291–303). Detay Yayıncılık.

Trahanias, P., Argyros, A., Tsakiris, D., Cremers, A., Schulz, D., Burgard, W., ... Kamarinos, G. (2000, May). Tourbot-interactive museum tele-presence through robotic avatars. *Proc. of the 9th International World Wide Web Conference.*

Tung, V. W. S., & Law, R. (2017). The potential for tourism and hospitality experience research in human-robot interactions. *International Journal of Contemporary Hospitality Management*, *29*(10), 2498–2513. doi:10.1108/IJCHM-09-2016-0520

Wang, X., Li, X. R., Zhen, F., & Zhang, J. H. (2016). How smart is your tourist attraction? Measuring tourist preferences of smart tourism attractions via a FCEM-AHP and IPA approach. *Tourism Management*, *54*, 309–320. doi:10.1016/j.tourman.2015.12.003

Wang, Y., Yu, Q., & Fesenmaier, R. D. (2002). Defining the virtual tourist community: Implications for tourism marketing. *Tourism Management*, *23*(4), 407–417. doi:10.1016/S0261-5177(01)00093-0

Weiler, B., & Black, R. (2015). *Tour guiding research: Insights, issues and implications* (Vol. 62). Channel View Publications.

World Federation of Tourist Guide Associations. (n.d.). *What is a Tourist Guide?* Retrieved October 15, 2019, from http://www.wftga.org/tourist-guiding/what-tourist-guide

Yıldız, Ş. (2018). Profesyonel turist rehberlerinin teknolojik rehber algısıı belirlemeye yönelik bir araştırma: İstanbul örneği. *Journal of Social And Humanities Sciences Research*, *5*(31), 4686–4697.

Yıldız, Ş. (2019). The rise of robot guides in tourist guidance profession. *Suleyman Demirel University Visionary Journal*, *10*(23), 164–177.

KEY TERMS AND DEFINITIONS

Augmented Reality (AR): A concept is an interactive experience of a real-world environment where the objects that reside in the real world are enhanced by computer-generated perceptual information, sometimes across multiple sensory modalities, including visual, auditory, haptic, somatosensory and olfactory.[1]

Global Positioning System (GPS): Communications refers to the use of signals to transfer voice, data, image, and/or video information between locations, the main applications of which are in communications science, engineering, and technology. From the bonfires and smoke signals to early modern optical telegraphs, humankind has attempted to move communication beyond the realm of earshot.[2]

Information and Communication Technologies (ICT): Information and communications technology (ICT) is an extensional term for information technology (IT) that stresses the role of unified communications[3] and the integration of telecommunications (telephone lines and wireless signals) and computers, as well as necessary entreprise software, midware, storage, and audiovisual systems, that enable users to Access, store, transmit, and manipulate information.[4] [5]

Middleman: A person who buys goods from producers and sells them to retailers or consumers.'we aim to maintain value for money by cutting out the middleman and selling direct'.[6]

Travel Buddy: A travel buddy is another adult who travels with you.[7]

Virtual Reality (VR): A set of images and sounds, produced by a computer, that seem to represent a place or a situation that a person can take part in the VR experience.[8]

Virtual Tour: A virtual tour is a simulation of an existing location, usually composed of a sequence of videos or still images. It may also use other multimedia elements such as sound effects, music, narration, and text. It is distinguished from the use of live television to affect tele-tourism.[9]

ENDNOTES

1 https://en.wikipedia.org/wiki/Augmented_reality. Retrieved 2020-03-12.

2 https://ethw.org/Category:Communications. Retrieved 2020-03-12.

3 Murray, James (2011-12-18). "Cloud network architecture and ICT - Modern Network Architecture". TechTarget =ITKnowledgeExchange. Retrieved 2013-08-18.

4 Information and Communication Technology from. FOLDOC. 2008-09-19. Archived from the original on 2013-09-17. Retrieved 2010-09-29.

5 https://en.wikipedia.org/wiki/Information_and_communications_technology. Retrieved 2020-03-12.

6 https://www.lexico.com/en/definition/middleman. Retrieved 2020-03-12.

7 https://www.workingmother.com/momlife/13366783/what-is-a-travel-buddy/. Retrieved 2020-03-12.

8 https://dictionary.cambridge.org/dictionary/english/vr. Retrieved 2020-03-12

9 https://en.wikipedia.org/wiki/Virtual_tour. Retrieved 2020-03-12.

Chapter 16
The Future of Tourism Guidance in the Scope of Industry 4.0 and Next-Generation Technologies

Yunus Topsakal
https://orcid.org/0000-0003-3202-5539
Adana Alparslan Türkes Science and Technology University, Turkey

Mehmet Bahar
Cappadocia University, Turkey

Nedim Yüzbaşıoğlu
Akdeniz University, Turkey

EXECUTIVE SUMMARY

Next-generation technologies such as robotics, the internet of things, artificial intelligence, sensors, cognitive technologies, nanotechnology, quantum computing, wearable technologies, augmented reality, intelligent signaling, and intelligent robots have led the fourth industrial revolution, often referred to as Industry 4.0. With the rapid advance of technology, most people today rely heavily on the internet to get information while traveling anywhere, because the use of technology has deeply penetrated daily life. The internet also makes travel easier and more convenient. For instance, it is possible to plan travel using smartphones and applications and at the same time meet instant travel needs as they arise. Therefore, the aim of this study is to examine tourism guidance within the scope of the super-smart tourists of the future, to determine the usage areas of next-generation technologies in the field of tourism guidance, and to give recommendations for tourism guidance in this regard.

DOI: 10.4018/978-1-7998-3725-1.ch016

INTRODUCTION

Technology has a significant impact on human lives due to its continuous and rapid development, and therefore technology may be regarded as an unpredictable and important force that people must pay attention to (Hooijdonk, 2015). For example, while the number of mobile phone users in the world was 2.32 billion in 2017, it is expected to exceed five billion by the end of 2019 (Statista, 2018), showing the importance of mobile applications for economic industries. Smart technologies are accepted almost without notice by users and may be found in almost all areas of life today. Tourism destinations have also started to implement smart technologies in order to prevent ongoing climate change and to enrich the experience of tourists within the framework of European Union objectives (Endesa, 2008).

Next-generation technologies such as robotization, the Internet of Things (IoT), artificial intelligence, sensors, cognitive technologies, nanotechnology, Internet services, quantum computing, wearable technologies, augmented reality, intelligent signaling, intelligent robots, big data, 3D technology, and intelligent networks have led to the fourth industrial revolution, known as Industry 4.0. These Industry 4.0 technologies have started to change business environments and lifestyles due to their rapid use in business life, communication, and education. Therefore, countries, governments, local administrations, enterprises, and educational institutions have started working to adapt to Industry 4.0 technologies. The term "smart" also entered the literature with Industry 4.0. Höjer & Wangel (2015) state that technological advances are not so significant on an individual basis; rather, different technologies used in a connected, synchronized, and harmonious way impact our lives in a unified fashion. Harrison et al. (2010) define the concept of "smart" as using real-time and real-world data, integrating data, sharing data, and using analytics, modeling, optimization, and visualization to make better operational decisions. The term "smart" has become a word that describes the technological, economic, and social developments brought about by sensors, big data, open data, and new technological avenues for communication and information exchange (e.g., the IoT, GPS, and NFC) (Gretzel et al., 2015).

In recent years, intensive efforts have been made for the use of information and communications technology (ICT) in the tourism industry (Stamboulis & Skayannis, 2003). However, there are limited studies addressing how smart applications affect tourism. Considering the fact that tourism represents a significant source of income in developing countries, more attention should be paid to the adaptation of the tourism industry to next-generation technologies. Especially with the increase in the number of "Z generation" tourists, technology has found a place in the tourism sector. In this context, the concept of the super-smart tourist, also called Tourist 5.0, has been discussed in the literature. The expectations of professional tourism guides by super-smart tourists in the context of next-generation technologies have started to change. In this context, this study will first address how industrial revolutions have influenced tourism. Information is then given about next-generation technologies. Finally, the effects of next-generation technologies on professional tourism guides are discussed.

BACKGROUND

A Brief Overview of the Effects of Industrial Revolutions on Tourism

The first industrial revolution (FIR) began in England in the 1750s (Leighton, 1970). In 1781, James Watt patented the steam engine, the driving force behind the FIR. Steam engines were then used in the first train, boat, and agricultural machinery within the next 30 years (Roberts, 2015). While the steam engine was a driving force of the FIR, the textile industry was another force (Outman & Outman, 2003). The basic facilitating technologies of the FIR were the change in power supplies (steam engines running on coal and wood fuel). The FIR was focused on textiles and iron manufacturing. It aimed to increase productivity, mostly in terms of mechanical tools, and it spread a new standard of living across England from the 1750s onward (Tien, 2012). During this period, with the development of tourist transport in San Sebastian in 1820, the bath/spa facilities were frequented by many more tourists (Larrinaga, 2005). Seaside visits also started with bath facilities and spas. In these years, seaside travels became valued for health, recovery, and fashionable recognition (Beckerson & Walton, 2005).

The second industrial revolution (SIR) focused on assembly lines, as specifically designed by Henry Ford, and steel production; it aimed to increase productivity in mass production. The first effects were felt in the USA and Germany in 1860 (Tien, 2012). With the SIR, fields of business began to shift from manufacturing to services (Blinder, 2006). The main facilitating technologies of the SIR included the change in power supplies (electric power) and transportation (railways, automobiles), developments in iron and steel production, and the invention of the lightbulb. According to Janicke and Jacob (2009), the SIR enabled mass production and mass earnings. Cunard, a British cruise line, made its first transatlantic voyage in 1840. Passenger transport developed largely due to migration, but it has been used heavily for tourism purposes since World War I (Gierczak, 2011). Transatlantic travel accelerated in the 1860s and the concept of the "big tour" emerged. Hotel Ezcurra in 1870 and Hotel Bermejo in 1884 started to serve customers in Spain for tourism purposes, while in 1881 the Hotel de Inglaterra and in 1884 the Hotel Continental were introduced (Larrinaga, 2005). The Castle Line maritime company published the "Guide to South Africa" guidebook for the first time in the 1890s to guide tourists, athletes, and immigrants (Mackenzie, 2005).

The third industrial revolution (TIR) began in the last half of the 1950s and aimed at increasing the national and international operations of US companies (Leighton, 1970). The term "TIR" was first used by the economist Rifkin (2011) and the TIR was explained as the integration of Internet technology with renewable energy, aiming to increase productivity in mass customization (especially the integration of services and manufactured goods), and to use computer and communication technologies that can work in the long term and are mostly connected to brain power in enterprises (Tien, 2012). During the TIR period, after 1945, tourism entered a new phase of development with collective travel. In 1953, the Vickers Viking, a twin-engine aircraft seating 36 people, was designed to take tourists on a two-day tour of Lyon, Barcelona, Madrid, Tangiers, Casablanca, and Agadir (Gierczak, 2011).

The fourth industrial revolution was triggered by the development of ICT. Industry 4.0 can be understood as a series of technology-based smart automation and cyber-physical systems. The term "Industry 4.0" was first used at the Hannover fair in 2011 (Rojko, 2017). With Industry 4.0, digitalization has begun to change the way we do business in information management and production/service, because technologies such as light robots, tablet computers, transponders, sensors, and artificial intelligence have become more economical for enterprises (Schlund et al., 2014). ICT is now significantly affecting and changing the tourism industry. Rapidly developing smartphone applications particularly provide tourists with many benefits, such as destination guides, address finding, exchange rate calculation, and hotel or flight booking. Smartphones search for information, social network connections, navigation help, and so on. Smartphones now offer a wide range of services and can support thousands of mobile applications (Wang et al., 2012). Kramer et al. (2007) found that tourists' choices could easily be changed with smartphones.

Next-Generation Technologies of Industry 4.0

Big Data

In recent years, the tourism industry has been rapidly adapting and adopting technological developments. Therefore, enterprises and managers operating in the tourism industry should use ICT and the Internet at maximum levels. For example, visitor information, visitor needs, and visitor email addresses and phone numbers are stored in the tourism industry. In addition, businesses operating in the tourism industry can collect information from a variety of sources through social media accounts, representing a large amount of data. Tourism enterprises and actors in the tourism sector can use that "big data" for promotion, marketing, competitions, and the following of trends (Topsakal, 2019).

The benefits of big data for smart tourism and destinations are constantly becoming more obvious. From the planning of a journey to its realization and across the whole process, a large amount of data is released about accommodation, entertainment, and restaurant services, and that external information is published via various social media tools, leaving a digital mark by tourists on web-based and mobile-based services (Esen & Türkay, 2017). Big data usage is the most innovative tool to date to identify consumer behavior and improve process efficiency. The main concerns for the tourist are satisfaction with the services, keeping costs to a minimum, gaining recognition or improving their reputation, and seeing that enterprises' commitments are fulfilled (Sheoran, 2017).

Internet of Thing (IoT)

The Internet of Things (IoT) means that everything is connected to all other things over the Internet without limitations of time, space, or presence. The IoT can communicate across devices and collaborate to achieve common goals. Radio frequency identification tags, sensors, processors, chips, mobile phones, and mobile devices are some examples of the common presence of the IoT among the objects that surround us. Since these things are all connected to the Internet, they close the gap between the real world and the digital world. Therefore, the IoT facilitates the development of a variety of platforms capable of transmitting a wide range of various types of data using participating detection systems (Gretzel et al., 2015a).

The IoT provides support for information collection, analysis, automation, and control. For example, a chip embedded in entrance tickets or a smartphone application may allow tourism service providers to track tourists' locations and consumption behaviors. This also allows for location-based rescues in the event that a tourist leaves a designated route (Masseno & Santos, 2018).

Cloud Computing

Cloud Computing services are designed to provide an easy way to access online storage. Cloud computing provides access to robust web platforms and data storage through a public electronic communications network (Masseno & Santos, 2018). Cloud computing centers offer fairly centralized user information, and tourists are thus at risk of the violation of important personal privacy when using cloud resources. When cloud computing is implemented within smart tourism, a primary challenge is to ensure the security and authentication of personal information (Liu & Liu, 2016).

Blockchain

A blockchain is a constantly growing list of records, called blocks, that are linked and secured using cryptography. Blockchains provide a secure distributed database that can function without a central authority or administrator. Blockchains use a distributed peer-to-peer network to make a continuously growing list of regular records, called blocks, to create a digital ledger. Each transaction represented in a cryptographically marked block is then automatically verified by the network itself (Angraal et al., 2017). The future of the tourism industry in light of developing blockchain technology is focused on four basic points. These are simplified and safer traveler identification, improved baggage tracking, more user-friendly customer loyalty programs, and the facilitating of payments between travel agencies and airlines. Blockchains can bring advancements to the tourism industry in these four realms (Blockchain, 2017).

Mobile Applications

Specially prepared software that provides mobile services through mobile devices, especially smartphones, is called an "application" (Topsakal, 2019a). The increasing diversity of travel practices has enabled individuals to access, buy, and share knowledge anytime and anywhere (Lu et al., 2015). Tourism mobile applications help users plan and map their travels and make accommodation reservations, ticket reservations, taxi reservations, and more. Mobile applications can also improve service quality by marking the right places in tourist destinations or forecasting reservations and waiting times at various tourist sites (Amanda et al., 2018). Mobile applications can be divided into mobile applications specifically targeting tourists (e.g., Airbnb, TripAdvisor, Skyscanner) and those often used in typical travel contexts (e.g., Google maps, iMoney, Instagram) (Lu et al., 2015). Applications developed for tourism provide visitors with services such as geographical positioning, socialization, communication, security, emergency assistance (information, medical), transactions (shopping, tickets, finance, banking, etc.), information, and entertainment (Topsakal, 2019a). For example, Airbnb ("Air Bed and Breakfast") is the most popular home rental application. It was founded in San Francisco, California, in August 2008, and is now a marketplace that offers the option of listing, discovering, and renting from thousands of accommodation options online or via the smartphone app worldwide. It is possible to rent houses with different price options in more than 25,000 cities and 192 countries via Airbnb, whether it is an apartment for a night, a boat, a tent, a castle, a lighthouse for a week, or a villa for a month (Airbnb, 2018).

Near-Field Communication (NFC)

NFC technology allows tourists to browse information points about paintings, sculptures, or historical artifacts and it can be used in touristic places such as museums where the smartphone can provide information about an object in the desired language. At a train station, for example, NFC posters may be available and thus a user can be directed to an online site with information on routes, travel times, and prices using a smartphone at the station (Smith, 2015). With the use of a mobile phone, using panels placed at bus stops, NFC can provide transportation schedules, waiting times for arrival, ticket purchasing information, and so on. When a tourist approaches a work exhibited in a museum, he can get information about it, or he can pay by phone in a restaurant after eating (Çelik & Topsakal, 2019).

QR Codes (Quick Response Codes)

QR (quick response) codes as a specific form of mobile code provide solutions for connecting physical and virtual content to provide users with additional information or access to mobile services (Canadi et al., 2010). Tourists can access information about their destinations via mobile devices by scanning the QR code. They can access various contents (text, videos, links, any data or webpages) with barcode-reading software installed on their mobile devices. Differences between NFC and QR code show in Table 1.

Table 1. Differences between NFC and QR code

NFC	QR Code
The NFC interface is integrated into almost all mobile phones.	To read the QR code, a special application is needed on the mobile phone.
NFC technology has been standardized worldwide and offers a wide range of applications in areas such as industry, logistics, marketing, automotive	The QR-code can optically downgrade the customer layout
No special application required	The QR code is susceptible to dirt which reduces the first-pass read rate significantly.
NFC is based on ISO14443 standard or ISO15693 and according to the International Transmission standard, the frequency is 13.56 MHz.	QR code data cannot be changed so that they are static and cannot be changed.
Maximum distance up to 5 cm and fast transmission establishment	Long website URLs affect a very large QR code, which leads to poor read rates.
Each NFC chip has a unique ID number worldwide. Thus, each product is unique, retractable and becomes an original.	The QR Code can be processed in two steps, first opening the scanning program and second scanning the code.
A virtual 100% first pass reading rate	The cost of QR codes is low.
NFC provides the possibility to add, read and modify data on the chip at any time.	Once created and cannot be changed again

Source: "NFC-Technology". n.d.

Kiosks

Electronic touch screens are a form of smart information systems. Especially in smart destinations, these touch screens placed at points such as important tourist attractions, airports, shopping malls, and tourist information centers provide much information about the destination to tourists. Electronic touch screens are systems that can improve the experience of tourists during their stay at their destination and provide different information. An advantage of electronic touch screen technology for visitors is that it provides information 24 hours a day (Rodrigues et al., 2017).

Smart Tourist Cards

The creation of smart tourist cards standardizes experiences and makes it easier for tourists to learn about options at their destinations (transport, guides and excursions, visits to museums, performances, etc.) (Segittur, 2019). Tourist cards, which help tourists gain unique perspectives of their destinations, provide a number of benefits and advantages to users in terms of time and cost savings, and they stand out in making experiences more successful. Tourist cards are important destination marketing tools, combining various services and activities provided at the destination at an affordable price. The main facilities provided by tourist cards include free or discounted entrance to museums, churches, and historical sites; free use of public transport; map guides with tourist attractions, restaurants, shops, and amusement parks; guided tour activities; and information or discounts for car rentals, bike rentals, or city parking. Given the large number of tourist services and facilities available via a tourist card, it is an essential marketing tool for the success of destinations (Topsakal et al., 2018a). A smart card will be either a chip card or a QR code, or a mobile application.

Augmented/Virtual Reality

Augmented/virtual reality in smartphones may be used to provide information about points of interest (Sato et al., 2017). By using virtual reality and augmented reality technologies, visitors can find their destinations in virtual space. Geographical information systems, on the other hand, combine the location data of a site with its geographic location and the properties of the object and provide a wide range of information about the destination. Smart tourist guide services based on geographic information systems and mobile tour guide applications have been developed to provide digital tourism before, during, and after travel (Jwa, 2016).

Augmented/virtual reality plays an important role in tourism, especially in the promotion of cultural heritage in a colorful and realistic way. For example, in an ancient Roman theater, an environment in which users can watch a gladiator fight in the arena through their smartphones can be achieved by digital reconstructions provided by augmented reality. With augmented reality, historical places can be presented to visitors on a digital platform. Augmented reality also has the ability to guide visitors through applications such as Google Maps that offer location-based services (Smith, 2015). Before booking accommodations, tourists can visit the hotel rooms with augmented reality. Using maps created by augmented reality while navigating a destination, they can enjoy a more in-depth experience with arrows, direction signs, and digital elements. Augmented reality multimedia can personalize interactive information in order to enhance tourists' experiences (Çelik & Topsakal, 2019).

Global Positioning System (GPS)

The location-based applications have started to develop with the introduction of features such as GPS and near-field communication (NFC) on mobile devices. The best examples of location-based applications are public transport applications. SmartCity public transportation applications can use GPS to locate public transport routes passing by a location in real time. In terms of tourism, the Foursquare and Swarm applications can show location-based suggestions for restaurants, cafes, and shopping centers together with information about these places. Socialeyes, Zomato, Tinder, and Glympse are other examples of applications that offer different location-based services (Topsakal, 2019b). On the other hand, mobile tour guides provide tour information and routes to help visitors easily find and navigate the points of interest around them as they travel around their destinations (Sun et al., 2016). In this context, Shen et al. (2016) propose the IoT and big data for developing a system of smart tourism and sustainable cultural heritage.

Beacon

One of the latest technologies being used today is beacon technology, a Bluetooth-based interaction technology. Beacons are placed at certain points and emit Bluetooth radio waves. Thanks to these waves, personal discounts can be offered to visitors entering an area by the surrounding businesses in line with their preferences. In order to receive these notifications, it is necessary to download an application such as a smart tourist guide to the smartphone and enter one's preferences (Çelik & Topsakal, 2019).

Humanoid Robots

Humanoid robots started to be developed with Industry 4.0 and rapidly began being used in daily life and by enterprises. There are examples of robots in businesses. For example, a new type of robot called Sawyer has the features of automatically saving a job once it is shown the job, and unlike its predecessors, it can learn and speed up its work by extracting the results from each job that it performs. Therefore, Sawyer specializes as people perform jobs, and it shortens the time needed to perform the work while increasing the quality. Assuming that a person does a job at best in a week, Sawyer can fit it into a few hours. In addition, this robot can share experiences with its similar siblings, allowing them to start a job with excellent experience from the very beginning. From this point of view, robot guides that are being developed thanks to learning robots will update their knowledge and data day by day with interactive dialogues with tourists and the knowledge and experience they acquire, and with each passing day their work will continue to improve. It will also be possible to transfer this gained experience to robot peers providing similar services automatically and simultaneously (Sawyer, 2019).

MAIN FOCUS OF THE CHAPTER

Concept of Smart Tourism

The concept of a "smart city" encompasses many industries; in this respect, it also includes the tourism industry (Boes et al., 2015; Buhalis & Amaranggana, 2014). Boes et al. (2015) define a smart city as a connected system that connects all local organizations to provide real-time services to visitors and citizens at all times. The development of a smart city facilitates seamless access to value-added services, such as access to real-time information about public transport, enriching tourist experiences and increasing the competitiveness of destinations (Buhalis & Amaranggana, 2014).

Smart tourism includes tourist activities supported by smart technology (Gretzel et al., 2015b). Buhalis (2015) states that a smart tourism destination is a competitive tourism destination aiming to increase the social, economic, and environmental welfare of all stakeholders by successfully implementing the knowledge created by innovation supported by investments in human and social capital. According to Washburn et al. (2010), technology in smart tourism should be viewed as infrastructure rather than individual information systems, and it includes a variety of smart ICT for integrating hardware, software, and other technologies to enable people to make smarter decisions about alternatives by providing real-time awareness in the real world. Smart hotel management systems (Topsakal et al., 2018a), smart ticket (card) systems (Topsakal et al., 2018b), smart remote video monitoring systems, smart tour guide systems (Yüzbaşıoğlu et al., 2018), and smart travel agencies are used to create smart tourism (Gretzel, 2011). In short, next-generation technologies operate on the basis of efficient and effective communication in real time.

McCartney et al. (2008) state that smart tourism destinations are an important element of smart cities and Guo et al. (2014) explain that smart tourism destinations represent an important smart city strategy. Smart tourism can be created by the application of next-generation technologies in the fields of service, business, management, and governance (Yao, 2012). Smart cities may differ from smart tourism destinations because smart cities do not take tourists into account, focusing only on citizens. Smart tourism destinations can thus be seen as a derivative of smart cities. Boes et al. (2016) state that a smart city focuses on its citizens while a smart destination focuses on improving tourist experiences through ICT.

Different technologies can be found at smart tourism destinations. Technologies such as QR codes or NFC tags provide connections between the physical and digital worlds. Thus, the value of the tourist experience increases (Chillion, 2012). Access to information about nearby points of interest is made easier by the new generation of ICT (Buhalis & Amaranggana, 2014). In the case of applications, distance has no negative aspects as it did before. A smart tourism application with augmented/virtual reality can enable visitors to enjoy real-time experiences at different locations and at different times (Chillion, 2012). Smart tourism destination framework has shown in Figure 1.

Su et al. (2011) identify different main applications such as green cities, smart medical treatment, smart transportation, wireless cities, smart homes, smart city management, smart public services, and smart tourism in their study on the importance of creating smart application systems. Basically, a smart city should integrate human capital, infrastructure, and knowledge (Buhalis & Amaranggana, 2014). Since the tourism industry is one of the most suitable industries in which ICT is used extensively, the idea of smart tourism destinations has found application areas quite quickly (Koo et al., 2016). Since smart tourism destinations offer unique personalized experiences from real-time data based on ICT, each visitor can experience heterogeneous experiences even at the same destination. Therefore, the idea of a smart tourism destination can be important for destinations where attraction centers are very different and especially for locations difficult to link and market (Koo et al., 2016).

The concept of smart tourism destinations has emerged due to the development of smart cities (Buhalis & Amaranggana, 2014). While the smart city approach focuses on citizens, it may also include tourists, and the smart tourism destination approach focuses on tourists and passengers (Lamsfus & Alzua-Sorzabal, 2013). Boes et al. (2015) describe smart tourism destinations as platforms that implement ICT such as artificial intelligence, cloud computing, and the IoT to provide personalized information and advanced services created by mobile end-user tools. Lopez de Avila (2015) defines smart tourism destinations as innovative tourism destinations built on state-of-the-art technology infrastructure that is accessible to all, facilitates the sustainable development of tourist areas, and allows the interaction and integration of the visitor with the environment.

Figure 1. Smart Tourism Destination Framework
Source: Boes vd., 2016

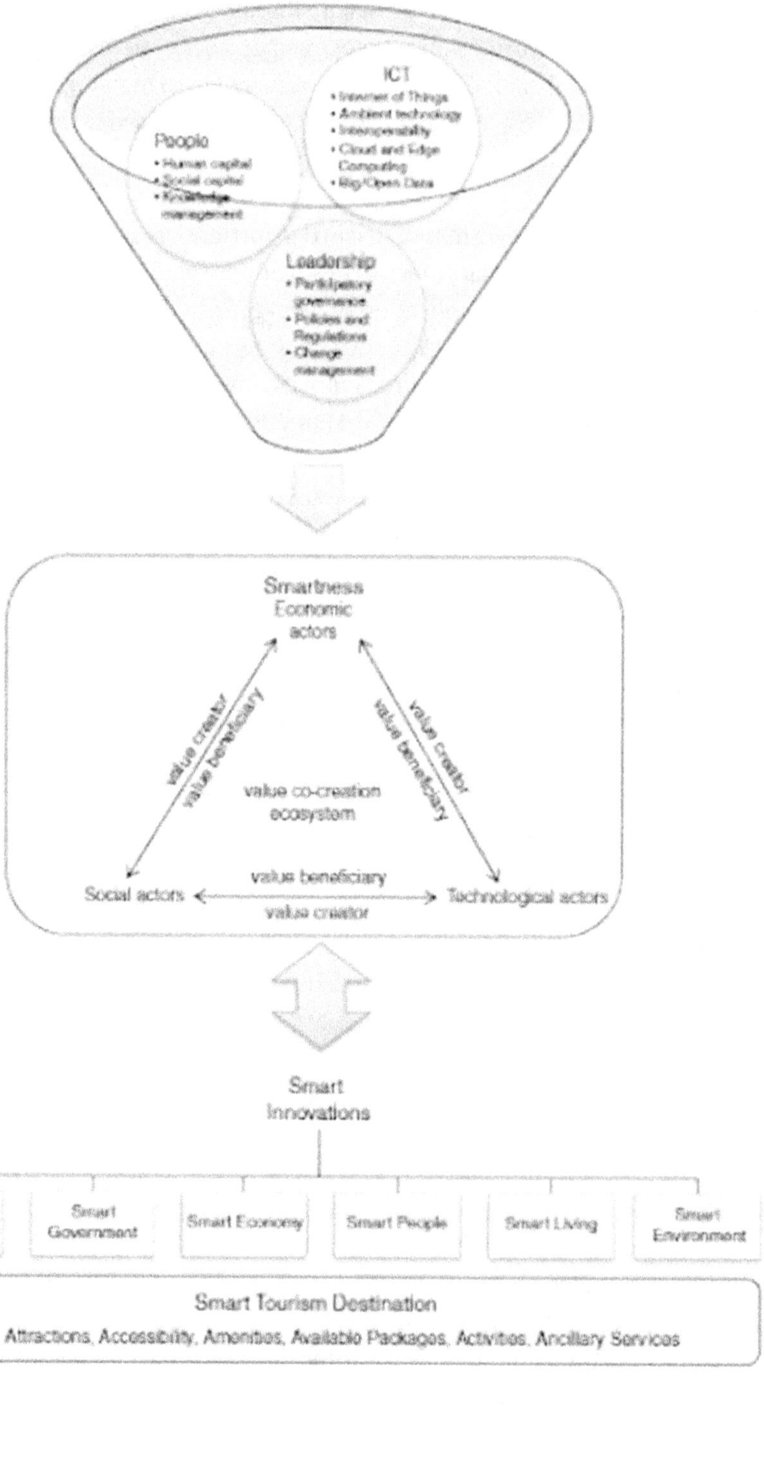

The concept of smart tourism is defined as providing open information to tourists by integrating tourism resources with ICT such as artificial intelligence, cloud computing, and the IoT (Zhang et al., 2012). Research has been conducted in many areas related to smart tourism, including the following types of studies:

- Smart tourism and smart city studies defining the application of the smart city concept and technologies in the context of tourism (Topsakal et al., 2018a);
- Studies on smart tourism destinations (Topsakal and Çelik, 2017);
- Studies on smartphone applications in tourism, focusing on the functionality and adoption of smartphone applications (Topsakal, 2018);
- Studies on smart hotels that examine the use of information technologies in the hospitality industry and digital marketing decisions of smart hotels (Topsakal et al., 2018b, Topsakal et al., 2018c);
- Studies related to smart cards that can be used in the tourism industry (Topsakal et al., 2018d);
- Studies on gamification, augmented reality, and smart tourism to use augmented reality technologies (Topsakal et al., 2018d);
- Studies on smart suggestions for tourists searching for personalized destination suggestion systems combining individual preferences and geolocation data (Yüzbaşıoğlu et al., 2018);
- Studies on smart guides defining content-based information applications for smart tourist guidance (Yüzbaşıoğlu et al., 2018).

Super-Smart Tourist (Tourist 5.0)

As part of the Fifth Science and Technology Basic Plan (FY16-FY20), announced in April 2016, the Japanese government introduced a new society, which they called Community 5.0 or Super-Smart Society. Community 5.0 provides a common community infrastructure for prosperity based on an advanced service platform. Wang et al. (2018) explain that the concept of Community 5.0 emerged in 2015 with a strategic national political initiative in Japan. Society 5.0 emerged following Industry 4.0; while Industry 4.0 focuses on production, Society 5.0 places people at the center of technology with the aim of improving quality of life and the sustainability of the results of Industry 4.0 (Costa, 2018). According to Hayashi et al. (2017), Society 5.0 aims to create new values by collaborating with several different systems of Japan. In addition, smart property development, international standardization, IoT technologies, big data analysis technologies, artificial intelligence technologies, and similar developments are expected to encourage Japan's competitiveness in "super-smart society". Smart tourism technologies have also started to change the behavior of tourists and have created a specific segment of visitors, which are considered as a new type of tourist. Basic tourist demands in the information age include the following (Wang et al., 2016):

- Keeping track of personal travel preferences and timing,
- Waiting less and experiencing fewer delays,
- Searching for travel information via the Internet,
- Online ticket booking and room booking,
- Online shopping,
- Price comparisons via different travel sites,
- Communication via virtual travel communities,

- Access to complaint management systems,
- Multimedia service queries,
- Mobile facilities and applications such as Wi-Fi, short message services, and multimedia messaging services.

The basic needs of the super-smart tourist include the following: information about the city (the state of beaches, instant alerts); emergency telephone numbers; public Wi-Fi maps; maps of tourism offices, hospitals, and local government buildings; real-time maps, schedules, and information for urban transport (metro, buses, bicycles, tourist buses, railways); real-time information about harbors, airports, and trains; traffic and road data; weather information; tourist information (hotels, restaurants, museums, tourist attractions, general interest, and entertainment); and local social networks (Mussina & Oryngazhiyeva, 2018).

According to Femenia-Serra et al. (2019), a super-smart tourist is a tourist with smart experience who interacts with stakeholders by sharing smart data. This tourist is open to innovation and is social and proactive. In this context, it can be said that the super-smart tourists will be from the Z generation, a group of fast-learning and fast-developing people born in the mid-1990s and 2000s. Considering the general characteristics of generation Z (Bahar, 2018):

- They learn fast,
- They are confident and have high self-confidence,
- They are not prone to teamwork,
- They give more importance to education and social status,
- They like to act independently,
- They have an inward-looking structure and prefer to communicate via social media,
- They cannot make friends easily,
- Technology is not a luxury but a need for this generation,
- They know what they want,
- They believe that everything is possible in life,
- They have high creativity,
- They like to access information first-hand,
- They find rules boring,
- They avoid work and social environments that do not value creativity,
- They do not like standards.

Potential Effects of Next-Generation Technologies for Tourism Guidance

The travel and tourism industries have been heavily influenced in recent years by new technological developments, such as the mobile technology revolution (Lamsfus et al., 2015). For example, visitors using smartphone applications can receive tourist information, promotions, and geolocation information; they can share their experiences instantly through applications or they can purchase tourist services through applications (INVAT.TUR, 2015). Mobile technology allows people to plan their travels and interact more with others (Gretzel, 2011). In this context, smart tourist guide applications are becoming widespread in many places. In cities like Malaga and Barcelona, local tourism authorities have created smart guide applications. These applications are installed on a smartphone by the user before traveling to the destination. In doing so, tourists can explore the area before arriving and learn about the main attractions, as well as information on local resources. The Malaga tourism guide application combines written text with sound that can be listened to in advance or in a particular touristic area (Ayuntamiento de Malaga, 2015). The iBarcelona Smartour application (Barcelona Turisme, 2015) is a mobile tourist guide application with some different features. iBarcelona provides augmented reality services that enhance the user's experience through visual effects. Other smart tourist guides that are in use like those of Malaga and iBarcelona include the following:

- The Deep Map app is a mobile system that can create personal guided tours in the city of Heidelberg, Germany. This app creates a tour by taking into account personal interests, social background (e.g., age, education, and gender), and type of transportation. It enhances the experience of tourists by reconstructing points of interest in 3D, such as virtual time travel (Malaka & Zipf, 2000).
- CRUMPET is designed to provide personalized information and service to tourists. It offers descriptions, maps, directions, and pictures for each area of interest. In the first use, tourists learn more about setting their user preferences when providing demographic information, traveling, and interacting with the system. Recommendations are based on personal interests and current location, obtained via GPS (Poslad et al., 2001).
- The MoreTourism app provides information about tourist resources based on user profile, location, schedules, and visiting times. The system consists of two parts: a mobile device using Android OS and a server that provides system-specific functions. It provides socialization and advertising as well as information. First, it allows users to interact through social networks and create activity groups in specific locations. Second, it offers different advertising campaigns according to the user profile (Rey-Lopez et al., 2011).
- MyTourGuide has been developed for collaborations with both visitors and tourism. These collaborations match visitors and businesses, and the app acts as an intermediary between tourists and tourism-related parties such as hotels, restaurants, car rental services, and attractions. The stakeholders of the app can share tour information and promote their latest activities, services, or products (Husain et al., 2012).
- GoTour is a smart tourism guide application designed for the city of Istanbul and works on Android and mobile devices. It provides information about points of interest, consulates, and city services. Based on location, it suggests tour plans for tourists while taking into account the visitor's areas of interest, distance, and time (Al-Rayes et al., 2011).

It can thus be said that super-smart tourists will join the tourism movement within the scope of Community 5.0 and Industry 4.0 technologies. When Industry 4.0 technologies and generation Z are examined together, it can be foreseen that the following changes will occur in tourism guidance in the future.

First, smart tourist guide applications, which have already been developed with the new generation of technologies, will now be able to offer location-based guidance to visitors with more advanced software. For example, the user can plan appropriate tours based on a GPS location and personal interests, entering the time that the user can allocate for the visit. In this way, the visitor will be able to save time as well as enrich his or her experience.

Taking into account the length of the stay, as a visitor enters a hotel with a passport number, he or she could obtain guidance services on the room's TV screen for the places that he or she plans to visit.

In addition, it would be possible to join existing planned trips as an additional participant in the group if information is available via the TV screen about tour programs planned in the area. In this way, while reducing costs, visitors will be able to meet new people while seeing new places. Tourist flow could also be controlled by this method in order to not exceed the carrying capacity experienced at certain times of day.

The automatic translation of speech into text is now very smooth with next-generation technologies. Therefore, by means of very fast processor devices or applications, it will be possible to provide guidance services to tourists in different languages by converting a main text into alternative languages very quickly. In this way, guidance services can become a true cultural service rather than a language-based service so that individuals who have sufficient tourism guidance certification in any language can provide service to everyone in the customer's desired language.

In addition to departments such as reception services being predicted to vanish within the scope of smart tourism, traditional tourism guidance services will also disappear completely due to reasons such as the development of next-generation technology, changing visitor profiles (particularly the participation of generation Z in tourism mobility), and the demand for personalization of services. In particular, with Apple's voice-help service via the iPhone's Siri, it is not a stretch to imagine dialogues like "Siri, show me around Topkapı Palace". The classical tourism guidance profession might therefore begin to seem outdated or antiquated, like other "old-fashioned" professions such as tinsmith and coppersmith work.

The need for a physical guide will be replaced by the need for a visual guide With next-generation technologies. That need can be met interactively in 3D by hologram guides. In recent years, the first humanoid robot, Sophia, was granted citizenship by Saudi Arabia, and virtual guidance by humanoid robots with "guidance badges" is sure to follow.

SOLUTIONS AND RECOMMENDATIONS

Online guidance services can be provided at destinations with videos prepared within the scope of Industry 4.0 for guided tours or with a live video connection. In this way, the obligation of the guide and the tourist being together in the same place is overcome. Interactive guidance services can be provided to visitors independent of time and place in any language with the applications that have been developed. When you visit Aspendos, for example, you will be able to join a tour group for a small fee after receiving a message through an application: "Do you want to participate?" This will provide both the price advantages of group membership and the freedom of individuality. Today, the consumption tendencies of visitors and the understanding of the service sector are moving in the direction of the personalization of the service provided/received. Therefore, if the visitor chooses a travel time and the amount of time to be devoted to an activity, and if the visitor also identifies areas of interest, the application will prepare a virtual tour, especially focused on the areas of his or her interest in an optimal way. Once these trends are entered into the system by the visitor, warning messages can be sent by the application along the journey and the routes that have been traveled can be recorded or explained, even if no tours or activities are planned. Thus, such applications will benefit both visitors and tourism stakeholders. Stakeholders participating in the implementation of these features will be directed to use NFC and beacon technology. In this context, the following recommendations may be made:

- Applications are now expected to offer the most suitable tour programs to visitors with new technologies. Therefore, countries should develop travel and scheduling plans in touristic regions and develop applications specific to destinations with software supported by expert guides.
- Industry 4.0 technologies and the time limits of the Z generation, demanding everything immediately, mean that hotels and guides should work together to provide guidance and tour services to visitors more quickly.
- It can now be said that the general language of the world is technology. It is inevitable that the next-generation technologies developed with Industry 4.0 will be applied in the sector; in fact, these technologies have already started to affect tourism, like all other industries. Thus, tourism guidance education curricula should be harmonized with the requirements of Industry 4.0.
- Virtual and hologram guides should be prepared. Those who cannot physically provide tourism guidance can do so virtually, particularly with the help of translation programs that will make it possible to offer such guidance in any requested language.

FUTURE RESEARCH DIRECTIONS

In this study, Industry 4.0 and the next-generation technologies in use have been examined and some suggestions have been made about the potential effects of next-generation technologies on tourism guidance. In this context, this study is a conceptual study. Although new technologies have potential effects for tourism guidance in the tourism industry, they are not being widely utilized yet. A model proposal study could be realized for tourism guidance with next-generation technologies. These technologies could be examined separately for tourism guidance education, tourism guidance standards, etc. In addition, the perceptions and expectations of tourism guidance regarding next-generation technologies can be determined by field studies or surveys with tourism guides and tourists.

CONCLUSION

Technology has emerged as both a trigger and a main force for tourist destinations (Kuflik et al., 2015). For this reason, the tourism industry is undergoing a technological transformation in order to offer its services more easily and faster because the tourism industry is growing with increasing knowledge (Benckendorff et al., 2014). Today, ICT has become an integral part of the tourist experience; this is because tourists have begun to use different technological tools to first plan their travels and then to better experience their destinations and to share them afterwards (Wang et al., 2014). Therefore, it can be said that future tourism guides will be influenced by the next-generation technologies that form the basis of the fourth industrial revolution. In this context, both the training of tourism guides and the standards of tourism guidance should be harmonized with next-generation technologies.

REFERENCES

Airbnb. (2018). *Aribnb data*. Retrieved from https://www.airbnb.com.tr

Al-Rayes, K. A., Sevkli, A., Al-Moaiqel, H. F., Al-Ajlan, H. M., Al-Salem, K., & Al-Fantoukh, N. A. (2011). Mobile tourist guide for trip planning. *The Multidisciplinary Engineering Education Magazine*, *6*, 1–6.

Amanda, R., Santosa, P., & Rizal, N. M. (2018). Analysis of tourists preferences on smart tourism in Yogyakarta (case: Vredeburg Fort Museum). *Journal of Physics: Conference Series*, *1007*(1), 1–8. doi:10.1088/1742-6596/1007/1/012040

Angraal, S., Krumholz, H. M., & Schulz, W. L. (2017). Blockchain technology: Applications in health care. *Circulation: Cardiovascular Quality and Outcomes*, *10*(9), 1–3. doi:10.1161/CIRCOUTCOMES.117.003800 PMID:28912202

Ayuntamiento de Malaga. (2015). *Listado de apps*. Retrieved from http://www.malagaturismo.com/en /apps

Bahar, M. (2018). *Attention generation Z comes*. Retrieved from https://turizmdays.com/tr/makale / dikkat-z-kusagi-geliyor-180.htm

Barcelona Turisme. (2015). *Mobile apps*. Retrieved from http://www.turismedebarcelona.net/Mobile-Apps/_1d096tFVcJUB30LvIA_sk0JzLhRr3czPe71W_IX4cDk

Beckerson, J., & Walton, J. K. (2005). Selling air: marketing the ıntangible at British resorts. In J. K. Walton (Ed.), *Histories of tourism representation, identity and conflict* (pp. 55–68). Channel View Publications. doi:10.21832/9781845410339-005

Benckendorff, P., Sheldon, P., & Fesenmaier, D. (2014). *Tourism information technology*. CABI International. doi:10.1079/9781780641850.0000

Blinder, A. S. (2006). Offshoring: The next industrial revolution? *Foreign Affairs*, *85*(2), 113–128. doi:10.2307/20031915

Blockchain. (2017). *Blockchain: harnessing it's potential in travel*. Retrieved from https://amadeus.com/en/insights/research-report/blockchain-harnessing-its-potential-in-travel

Boes, K., Borde, L., & Egger, R. (2015). The acceptance of NFC smart posters in tourism. In I. Tussyadiah & A. Inversini (Eds.), *Information and communication technologies in tourism* (pp. 435–448). Springer.

Boes, K., Buhalis, D., & Inversini, A. (2016). Smart tourism destinations: Ecosystems for tourism destination competitiveness. *International Journal of Tourism Cities*, *2*(2), 108–124. doi:10.1108/IJTC-12-2015-0032

Buhalis, D., & Amaranggana, A. (2014). Smart tourism destinations. In Z. Xiang, I. Tussyadiah, & A. Inversini (Eds.), *Information and communication technologies in tourism 2014* (pp. 553–564). Springer.

Buhalis. (2015). *Working definitions of smartness and smart tourism destination*. Retrieved from http://buhalis.blogspot.com.tr/2014/12/working-definitions-of-smartness-and.html

Canadi, M., Höpken, W., & Fuchs, M. (2010). Application of QR codes in online travel distribution. In U. Gretzel, R. Law & M. Fuch (Eds.), Information and communication technologies in tourism (pp. 137-148). Springer. doi:10.1007/978-3-211-99407-8_12

Çelik, P., & Topsakal, Y. (2019). *Industry 4.0 and smart tourism*. Detay Publication.

Chillion, P. S. (2012). *From vacation spots to smart destinations: technology and tourism: QR, apps and augmented reality for cities*. Retrieved from https://urban360.me/2012/02/08/this-place-worths-a-visit-intelligent-destinations-smart-city-technology-tourism-and-city-branding/

Costa, J. M. (2018). *Sociedade 5.0: a mudança que aí vem*. Retrieved from https://hrportugal. pt/ sociedade-5-0-a-mudanca-que-ai-vem/

Endesa. (2008). *Smartcity Malaga - a model of sustainable energy management for cities of the future report*. Spain: Direccion General de Distribucion de Endesa.

Esen, M. F., & Türkay, B. (2017). Big data applications in tourism industries. *Journal of Tourism and Gastronomy Studies*, *5*(4), 92–115. doi:10.21325/jotags.2017.140

Femenia-Serra, F., Neuhofer, B., & Ivars-Baidal, J. A. (2019). Towards a conceptualisation of smart tourists and their role within the smart destination scenario. *Service Industries Journal*, *39*(2), 109–133. doi:10.1080/02642069.2018.1508458

Gierczak, B. (2011). The history of tourist transport after the modern ındustrial revolution. *Polish Journal of Sport and Tourism*, *18*(4), 275–281. doi:10.2478/v10197-011-0022-6

Gretzel, U. (2011). Intelligent systems in tourism: A social science perspective. *Annals of Tourism Research*, *38*(3), 757–779. doi:10.1016/j.annals.2011.04.014

Gretzel, U., Reino, S., Kopera, S., & Koo, C. (2015b). Smart tourism challenges. *Journal of Tourism*, *16*(1), 41–47.

Gretzel, U., Sigala, M., Xiang, Z., & Koo, C. (2015a). Smart tourism: Foundations and developments. *Electronic Markets*, *25*(3), 179–188. doi:10.100712525-015-0196-8

Gretzel, U., Werthner, H., Koo, C., & Lamsfus, C. (2015). Conceptual foundations for understanding smart tourism ecosystems. *Computers in Human Behavior*, *50*, 558–563. doi:10.1016/j.chb.2015.03.043

Guo, Y., Liu, H., & Chai, Y. (2014). The embedding convergence of smart cities and tourism ınternet of things in China: An advance perspective. *Advances in Hospitality and Tourism Research*, *2*(1), 54–69.

Harrison, C., Eckman, B., Hamilton, R., Hartswick, P., Kalagnanam, J., Paraszczak, J., & Williams, P. (2010). Foundations for smarter cities. *IBM Journal of Research and Development*, *54*(4), 1–16. doi:10.1147/JRD.2010.2048257

Hayashi, H., Sasajima, H., Takayanagi, Y., & Kanamaru, H. (2017, September). *International standardization for smarter society in the field of measurement, control and automation*. Paper presented at the 56th Annual Conference of the Society of Instrument and Control Engineers of Japan, Kanazawa, Japan. 10.23919/SICE.2017.8105723

Höjer, M., & Wangel, J. (2015). Smart sustainable cities: definition and challenges. In L. Hilty & B. Aebischer (Eds.), *ICT ınnovations for sustainability, advances in ıntelligent systems and computing* (pp. 333–349). Springer. doi:10.1007/978-3-319-09228-7_20

Hooijdonk, R. (2015). *Technology trends 2030.* Retrieved from https://www.richardvanhooijdonk.com/en/keynote/trends2030/

Husain, W., Dih, L. Y., Yen, H. F., & Jothi, N. (2012). Mytourguide.com: a framework of a location-based services for tourism ındustry. In *Proceedings of International Conference on Computer and Information Science* (pp.184-189). 10.1109/ICCISci.2012.6297236

Invat.Tur. (2015). *Destino turistico inteligente: manual operativo para la configuracion de destinos turisticos inteligentes.* Alicante: Universidad de Alicante, Instituto Universitario de Investigaciones Turisticas.

Janicke, M., & Jacob, K. (2009). *A third industrial revolution? solutions to the crisis of resource-intensive growth.* Environmental Policy Research Centre, Freie Universitat.

Jwa, J. W. (2016). Pedestrian network models for mobile smart tour guide services. *International Journal of Internet. Broadcasting and Communication*, *8*(1), 73–78.

Koo, C., Shin, S., Gretzel, U., Hunter, W. C., & Chung, N. (2016). Conceptualization of smart tourism destination competitiveness. *Asia Pacific Journal of Information Systems*, *26*(4), 561–576. doi:10.14329/apjis.2016.26.4.561

Kramer, R., Modsching, M., Hagen, K., & Gretzel, U. (2007). Behavioural impacts of mobile tour guides. In M. Sigala, L. Mich, & J. Murphy (Eds.), *Information and communication technologies in tourism* (pp. 109–118). Springer.

Kuflik, T., Wecker, A., Lanir, J., & Stock, O. (2015). An integrative framework for extending the boundaries of the museum visit experience: Linking the pre, during, and post visit phases. *Information Technology & Tourism*, *15*(1), 7–47. doi:10.100740558-014-0018-4

Lamsfus, C., & Alzua-Sorzabal, A. (2013). Theoritical framework for a tourism internet of things: Smart destinations. *Journal of Tourism and Human Mobility*, *2*, 15–21.

Lamsfus, C., Martín, D., Alzua-Sorzabal, A., & Torres, M. E. (2015). Smart tourism destinations: an extended conception of smart cities focusing on human mobility. In I. Tussyadiah & A. Inversini (Eds.), *Information and communication technologies in tourism 2015* (pp. 363–375). Springer. doi:10.1007/978-3-319-14343-9_27

Larrinaga, C. (2005). A century of tourism in northern Spain: the development of high-quality provision between 1815 and 1914. In J. K. Walton (Ed.), *Histories of tourism representation, identity and conflict* (pp. 88–103). Channel View Publications. doi:10.21832/9781845410339-007

Leighton, D. S. R. (1970). The Internationalization of American business: The third industrial revolution. *Journal of Marketing*, *34*(3), 3–6.

Liu, P., & Liu, Y. (2016). Smart tourism via smart phone. *Proceedings of International Conference on Communications*, 129-132).

Lopez de Avila, A. (2015). Smart destinations: XXI century tourism. *Proceedings of ENTER2015 Conference on Information and Communication Technologies in Tourism.*

Lu, J., Mao, Z., Wang, M., & Hu, L. (2015). Goodbye maps, hello apps? exploring the influential determinants of travel app adoption. *Current Issues in Tourism*, *18*(11), 1059–1079. doi:10.1080/13683500.2015.1043248

Mackenzie, J. M. (2005). Empires of travel: British guide books and cultural imperialism in the 19th and 20th centuries. In J. K. Walton (Ed.), *Histories of tourism representation, identity and conflict* (pp. 19–38). Channel View Publications. doi:10.21832/9781845410339-003

Malaka, R., & Zipf, A. (2000). Deep map: challenging IT research in the framework of a tourist information system. In *Proceedings of the 7th International Conference in Information and Communication Technologies in Tourism* (pp.15-27). 10.1007/978-3-7091-6291-0_2

Masseno, M. D., & Santos, C. (2018). Smart tourism destinations priivacy risks on data protection. *Revista Eletrônica Sapere Aude*, *1*(1), 125–149.

McCartney, G., Butler, R., & Bennett, M. (2008). A strategic use of the communication mix in the destination image-formation process. *Journal of Travel Research*, *47*(2), 183–196. doi:10.1177/0047287508321201

Mussina, K. P., & Oryngazhiyeva, M. R. (2018). Smart cities as a tourist destination for smart travelers. *Innovative Economy: Perspectives of Development and Improvement*, *6*(32), 12–16.

NFC-Technology. (n.d.). Retrieved from https://www.smart-tec.com/en/auto-id-world/nfc-technology

Outman, J. L., & Outman, E. M. (2003). *Industrial revolution*. Thomson Learning.

Poslad, S., Laamanen, H., Malaka, R., Nick, A., Buckle, P., & Zipf, A. (2001). Crumpet: creation of user-friendly mobile services personalised for tourism. In *Proceedings of the 3G Mobile Communication Technologies* (pp. 28- 32). 10.1049/cp:20010006

Rey-Lopez, M., Barragans-Martinez, A., Peleteiro, A., Mikic-Fonte, F., & Burguillo, J. (2011). More tourism: mobile recommendations for tourism. In *Proceedings of the IEEE International Conference on Consumer Electronics* (pp. 347-348). Berlin, Germany: IEEE.

Rifkin, J. (2011). *The third industrial revolution: how lateral power is transforming energy, the economy, and the world*. Palgrave Macmillan.

Roberts, B. R. (2015). *The third industrial revolution: iimplications for planning cities and regions*. Urban Frontiers Working Paper 1.

Rodrigues, J. M., Ramos, C. M., Cardoso, P. J., & Henriques, C. (2017). *Handbook of research on technological developments for cultural heritage and etourism applications*. IGI Global.

Rojko, A. (2017). Industry 4.0 concept: Background and overview. *International Journal of Interactive Mobile Technologies*, *11*(5), 77–90. doi:10.3991/ijim.v11i5.7072

Sato, G., Hirakawa, G., & Shibata, Y. (2017). Push typed tourist information system based on beacon and augumented reality technologies. In *Proceedings of the IEEE 31st International Conference on Advanced Information Networking and Applications* (pp. 298-303). 10.1109/AINA.2017.167

Sawyer. (2019). *High performance collaborative robot.* Retrieved from https://www.rethinkrobotics.com/sawyer

Schlund, S., Hammerle, M., & Strolin, T. (2014). *How automation and digitization will change production*. Fraunhofer Institute for Industrial Engineering.

Segittur. (2019). Retrieved from www.segittur.es

Shen, J., Mei, T., & Gao, X. (2016). Landmark reranking for smart travel guide systems by combining and analyzing diverse media. *IEEE Transactions on Systems, Man, and Cybernetics. Systems*, *46*(11), 1492–1504. doi:10.1109/TSMC.2016.2523948

Sheoran, S. K. (2017). Big data: A big boon for tourism sector. *International Journal of Research in Advanced Engineering and Technology*, *3*, 10–13.

Smith, R. (2015). *Smart tourism tools: linking technology with the touristic resources of city destinations* (Bachelor Thesis). NHTV University of Applied Sciences, International Tourism Management and Consultancy.

Stamboulis, Y., & Skayannis, P. (2003). Innovation strategies and technology for experience based tourism. *Tourism Management*, *24*(1), 35–43. doi:10.1016/S0261-5177(02)00047-X

Statista. (2018). *Number of smartphone users worldwide from 2014 to 2020.* Retrieved from https://www.statista.com/statistics/

Su, K., Li, J., & Fu, H. (2011). Smart city and the applications. *Proceedings of the International Conference on Electronics, Communications and Control*, 1028-1031.

Sun, Y., Song, H., Jara, A. J., & Bie, R. (2016). Internet of things and big data analytics for smart and connected communities. *IEEE Access: Practical Innovations, Open Solutions*, *4*, 766–773. doi:10.1109/ACCESS.2016.2529723

Tien, J. M. (2012). The next industrial revolution: Integrated services and goods. *Journal of Systems Science and Systems Engineering*, *21*(3), 257–296. doi:10.100711518-012-5194-1

Topsakal, Y. (2018). Disabled-friendly mobile services in the context of smart tourism: Recommendations for Turkey tourism. *Journal of Tourism Intelligence and Smartness*, *1*(1), 1–13.

Topsakal, Y. (2019). Big data and tourism. In O. İçöz & M. Uysal (Eds.), *Tourism Encyclopedia - Turkey: Tourism and Hospitality Industry Basic Terms* (p. 66). Detay Publicaiton.

Topsakal, Y. (2019a). Mobile applciation in tourism. In O. İçöz & M. Uysal (Eds.), *Tourism Encyclopedia - Turkey: Tourism and Hospitality Industry Basic Terms* (p. 614). Detay Publicaiton.

Topsakal, Y. (2019b). Location based applications. In O. İçöz & M. Uysal (Eds.), *Tourism Encyclopedia - Turkey: Tourism and Hospitality Industry Basic Terms* (p. 291). Detay Publicaiton.

Topsakal, Y., & Çelik, P. (2017). A new strategy for tourism: smart destinations. In G. Haşit, Ç. Hakkı, & M. E. Merter (Eds.), *Socio-economic strategies II: management* (pp. 96–106). IJOPEC.

Topsakal, Y., Yüzbaşıoğlu, N., & Akıncı, Z. (2018c). Smart tourism: cannot tourism enterprises unable to generate smart solutions be able to compete. In *Proceedings of the 9th International Conference of Strategic Research on Scientific Studies and Education* (pp. 259-272). Antalya, Turkey: Academic Press.

Topsakal, Y., Yüzbaşıoğlu, N., & Bahar, M. (2018b). Proposal for the hotel sector on the wat to be 'smart' at the age of Industry 4.0. In *Proceedings of the Second International Congress on Future of Tourism: Innovatıon, Entrepreneurship and Sustainability* (pp. 252-256), Mersin, Turkey: Academic Press.

Topsakal, Y., Yüzbaşıoğlu, N., & Çelik, P. (2018d). New generation tourist cards: tourist card proposal for the Antalya destination. In *Proceedings of the International Antalya Congress* (pp. 1321-1333). Antalya, Turkey: Academic Press.

Topsakal, Y., Yüzbaşıoğlu, N. & Çuhadar, M. (2018a). Industrial revolutions and tourism: Turkey tourism 4.0 swot analysis and proposal for adaptation process. *Süleyman Demirel University the Journal of Faculty of Economics and Administrative Sciences, 23*(Special Issue), 1623-1638.

Wang, D., Park, S., & Fesenmaier, D. (2012). The role of smartphones in mediating the tourism experience. *Journal of Travel Research, 51*(4), 371–387. doi:10.1177/0047287511426341

Wang, D., Xiang, Z., & Fesenmaier, D. (2014). Smartphone use in everyday life and travel. *Journal of Travel Research, 55*(1), 1–12.

Wang, F., Yuan, Y., Wang, X., & Qin, R. (2018). Societies 5.0: A new paradigm for computational social systems research. *IEEE Transactions on Computational Social Systems, 5*(1), 2–8.

Wang, X., Li, X. R., Zhen, F., & Zhang, J. (2016). How smart is your tourist attraction? Measuring tourist preferences of smart tourism attractions via a FCEM-AHP and IPA approach. *Tourism Management, 54*, 309–320. doi:10.1016/j.tourman.2015.12.003

Washburn, D., Sindhu, U., Balaouras, S., Dines, R. A., Hayes, N., & Nelson, L. E. (2010). *Helping CIOs understand smart city initiatives.* Retrieved from http://www.uwforum.org/upload/board/forrester_help_cios_smart_city.pdf

Yao, G. (2012). A study on the construction framework of intelligent tourism. [Social Sciences]. *Journal of Nanjing University of Posts and Telecommunications, 14*(2), 13–16.

Yüzbaşıoğlu, N., Çelik, P., Topsakal, Y., & Bahar, M. (2018). Industry 4.0 and smart tourism: Developing a smart tourist guidance application for Antalya destinastion. *Proceedings of the Innovation and Global Issues in Social Sciences, III*, 707–718.

Zhang, L., Li, N., & Liu, M. (2012). On the basic concept of smarter tourism and its theoretical system. *Luyou Xuekan, 27*(5), 66–73.

ADDITIONAL READING

Ang, S. C., & Zaphiris, P. (2009). *Human Computer Interaction: Concepts, Methodologies, Tools, and Applications* (Vols. 1–4). IGI Global. doi:10.4018/978-1-60566-052-3

Guerra, A. G. (Ed.). (2019). *Organizational transformation and managing innovation in the Fourth Industrial Revolution*. IGI Global. doi:10.4018/978-1-5225-7074-5

Mezghani, K., & Aloulou, W. (Eds.). (2019). *Business transformations in the era of digitalization*. IGI Global. doi:10.4018/978-1-5225-7262-6

Rodrigues, J. M., Ramos, C. M., Cardoso, P. J., & Henriques, C. (Eds.). (2017). *Handbook of research on technological developments for cultural heritage and etourism applications*. IGI Global.

Ruiz, G. R., & Hernandez, M. H. (Eds.). (2018). *Augmented Reality for Enhanced Learning Environments*. IGI Global. doi:10.4018/978-1-5225-5243-7

Sabri, E. (Ed.). (2019). *Technology optimization and change management for successful digital supply chains*. IGI Global. doi:10.4018/978-1-5225-7700-3

KEY TERMS AND DEFINITIONS

Industry 4.0: The fourth industrial revolution, led by next-generation technologies such as robotization, the Internet of things, artificial intelligence, sensors, cognitive technologies, nanotechnology, Internet services, quantum computing, wearable technologies, augmented reality, smart signaling, humanoid robots, big data, 3D technology, and intelligent networks.

Smart Tourism Guide Application: Applications developed with next-generation technologies that offer guidance services based on location with more advanced software.

Smart Tourist Cards: Smart cards that combine various services and activities provided at a discounted price in an integrated manner, which can be either a chip or QR code, or a mobile application.

Super-Smart Tourist (Tourist 5.0): A tourist who provides smart data by interacting with stakeholders in the production process by using next-generation smart technologies.

Virtual/Hologram Tourism Guide: Tourism guides that provide tourist guidance in a 3D interactive way via the new generation of technologies regardless of time and place.

Compilation of References

Ababneh, A. (2017). Tourist guides and job-related problems, analysis and evidence from Jordan. *Journal of Quality Assurance in Hospitality & Tourism, 18*(2), 200–217. doi:10.1080/1528008X.2016.1189867

Abraham, S., & Ramraj, K. (2015). The Potential of Culinary Tourism in India. *ANVESHAK-International Journal of Management, 4*(2), 133–148. doi:10.15410/aijm/2015/v4i2/67724

Abuamoud, I. N., Libbin, J., Green, J., & ALRousan, R. (2014). Factors affecting the willingness of tourists to visit cultural heritage sites in Jordan. *Journal of Heritage Tourism, 9*(2), 148–165. doi:10.1080/1743873X.2013.874429

Achu, F. N. (2019). Pro-Environmental Behaviour of Attendees at a Major Sport Event in Cameroon. *Geo Journal of Tourism and Geosites, 27*(4), 1307–1320. doi:10.30892/gtg.27416-435

Adhani, N. I., & Awang, R. D. R. (2012, December). A survey of mobile augmented reality applications. *1st International Conference on Future Trends in Computing and Communication Technologies*, 89-96.

Ahipaşaoğlu, H. S. (2006). *Guidance in Tourism.* Gazi Bookstore.

Ahrens, P. (2018). Qualitative network analysis: A useful tool for investigating policy networks in transnational settings? *Methodological Innovations, 11*(1). Advance online publication. doi:10.1177/2059799118769816

Airbnb. (2018). *Aribnb data.* Retrieved from https://www.airbnb.com.tr

Akın, B. (2015). Ayvalık evleri'nin cephe karakterinin oluşumuna etki eden faktörlerin değerlendirilmesi. *Sanat Tarihi Dergisi, 24*(2), 121–138. doi:10.29135td.273892

Al Obeidat, A. A. A. (2016). The sociocultural and economic evolution of Mansaf in Hartha, Northern Jordan. *Humanities (Washington), 5*(22). Advance online publication. doi:10.3390/h5020022

Alazaizeh, M. M., Jamaliah, M. M., Mgonja, J. T., & Ababneh, A. (2019, November 2). Tour guide performance and sustainable visitor behaviour at cultural heritage sites. *Journal of Sustainable Tourism, 27*(11), 1708–1724. doi:10.1080/09669582.2019.1658766

Al-Kheder, S. A., Al-Malabeh, A. A., & Al-Momani, R. (2016). Spatial Analysis and Transportation System Review for Tourism Areas in Jordan: Ajloun City as a Case Study. *Jordan Journal of Civil Engineering, 10*(4).

Almeida, P., & Shigeo, Y. (2002, December). Interactive conversational character as a virtual tour guide to an online museum exhibition. *International Conference on Computers in Education, Proceedings*, 215-216. 10.1109/CIE.2002.1185907

Aloudat, A. Sh. (2017b). Being in the minority: The challenges of being a female tour guide in Jordan. *Networked Innovation in Guided: The 5th International Research Forum on Guided Tours.*

Aloudat, A. Sh. (2017a). *The worldview of tour guides: A grounded theory study.* Noor Publishing.

Alrawadieh, Z., Cetin, G., Dincer, M., & Dincer, F. (2019). The impact of emotional dissonance on quality of work life and life satisfaction of tour guides. *The Service Industries Journal.*

Al-Rayes, K. A., Sevkli, A., Al-Moaiqel, H. F., Al-Ajlan, H. M., Al-Salem, K., & Al-Fantoukh, N. A. (2011). Mobile tourist guide for trip planning. *The Multidisciplinary Engineering Education Magazine*, *6*, 1–6.

Al-Salameen, Z., & Falahat, H. (2009). Religious practices and beliefs in Wadi Mousa between the late 19th and early 20th centuries. *Jordan Journal of History and Archaeology*, *3*(3), 175–204.

Alserhan, B.A., Wood, B.P., Rutter, R., Halkias, D., Terzi, H., & Serhan, O.A. (2018). *The transparency of Islamic Hotels: "Nice Islam" and the "self-Orientalizing"*. Academic Press.

Alserhan, B. A. (2017). *The principles of Islamic marketing*. Routledge. doi:10.4324/9781351145688

Altıntaş, V., Yılmaz, Y., & Çetinel, F. (2011). Mobbing in Accomodation Businesses: Sample of Antalya. *E-Journal of New World Sciences Academy*, *6*(4), 439–454.

Amaize, O., Mady, T., & Benson, E. (2011). The need for academic programs in tourism, arts & culture, sports, and healthcare management in the UAE. *World Journal of Management*, *3*(1), 96–108.

Amanda, R., Santosa, P., & Rizal, N. M. (2018). Analysis of tourists preferences on smart tourism in Yogyakarta (case: Vredeburg Fort Museum). *Journal of Physics: Conference Series*, *1007*(1), 1–8. doi:10.1088/1742-6596/1007/1/012040

Amin, S. B. (2010). Tourism development and economic growth in Bangladesh: An application of the ARDL testing approach. *Bank Parikrama*, *34*, 60–82.

Amin, S. B., Kabir, F. A., Khan, F., & Rahman, S. Z. (2019). Impact of seaside tourism on host community in Bangladesh: The case of cox's bazar, north south business review. *NSBR*, *102*(10), 12.

Anderson, C. E. (2012). *Art Tourism, Guggenheim Effect and Abu Dhabi*. Retrieved October, 24, 2020, from https://confluence.cornell.edu/display/tam2011/Art+Tourism%2C+Guggenheim+Effect+and+Abu+Dhabi

Andrew, E. H., & Andrew-Essien, E. (2013, July). Arts, tourism and the environment: An overview of critical intersections in Nigeria. *Mediterranean Journal of Social Sciences*, *4*(6), 275–284.

Angraal, S., Krumholz, H. M., & Schulz, W. L. (2017). Blockchain technology: Applications in health care. *Circulation: Cardiovascular Quality and Outcomes*, *10*(9), 1–3. doi:10.1161/CIRCOUTCOMES.117.003800 PMID:28912202

ANI. (2019, September 29). Two-day apple festival concludes in Shimla. *ANI News.* Retrieved on 14 October, 2019, fromhttps://www.aninews.in/news/national/ general-news/two-day-apple-festival-concludes-in-shimla20190929155529/

Ap, J., & Wong, K. (2001). Case study on tour guiding: Professionalism issues and problems. *Tourism Management*, *22*(5), 551–563. doi:10.1016/S0261-5177(01)00013-9

Appelbaum, S. H., Semerjian, G., & Mohan, K. (2012). Workplace bullying: Consequences, causes and controls (part two). *Industrial and Commercial Training*, *44*(6), 337–344. doi:10.1108/00197851211254770

Arnott, M. L. (1975). *Gastronomy: The anthropology of food and food habits.* Mouton Publishers.

Asean Post. (2018). Could Islamic Tourism be Brunei's answer? *The Asean Post.* https://theaseanpost.com/article/could-islamic-tourism-be-bruneis-answer

Askarian, F., Shojaei, M. K., & Yavari, U. (2015). Challenges in the Development of Sports Tourism in Aras Free Zone According to Managers and Experts Opinions. *Arabian Journal of Business and Management Review.*, *3*(7), 59–69. doi:10.12816/0014521

Asutay, M. (2007). A political economy approach to Islamic economics: Systemic understanding for an alternative economic system. *Kyoto Bulletin of Islamic Area Studies, 1*(2), 3–18.

Athukorala, P. C. (2008). Export performance in the reform era: Has India regained the lost ground. *ASARC Working Papers*. Doi:10.1057/9780230228337_6

Augmented Reality Blog. (2020). *How augmented reality can revolutionize the hospitality industry*. Retrieved March 10, 2020, from: http://www.augmentedrealitytrends.com/augmented-reality/hospitality-industry.html

Au, L., & Law, R. (2002). Categorical classification of tourism dining. *Annals of Tourism Research, 29*(3), 819–833. doi:10.1016/S0160-7383(01)00078-0

Australia Council for the Arts. (2015). *Arts nation: An overview of Australian arts*. Retrieved October, 24, 2020, from http://www.australiacouncil.gov.au/research/arts-nation-an-overview-of-australian-arts/

Ayuntamiento de Malaga. (2015). *Listado de apps*. Retrieved from http://www.malagaturismo.com/en /apps

Backer, T. E., & Rogers, E. M. (1998). Diffusion of Innovations Theory and Work-Site AIDS Programs. *Journal of Health Communication, 3*(1), 17–28. doi:10.1080/108107398127481 PMID:10947372

Bahar, M. (2018). *Attention generation Z comes*. Retrieved from https://turizmdays.com/tr/makale /dikkat-z-kusagi-geliyor-180.htm

Bahar, M., Yüzbaşıoğlu, N., & Topsaka, L. Y. (2019). Akıllı turizm ve süper akıllı turist kavramları ışığında geleceğin turizm rehberliğine bakış. *Journal of Travel and Tourism Research, 14*, 72–93.

Bahçelerli, N. M., & Sucuoğlu, E. (2015). Undergraduate tourism students' opinions regarding the work conditions in the tourism industry. *Procedia Economics and Finance, 26*, 1130–1135. doi:10.1016/S2212-5671(15)00941-7

Banerjee, M. (2015). Food Tourism: An Effective Marketing Tool for Indian Tourism Industry. *International Journal of Scientific Research (Ahmedabad, India), 4*(6), 795–800.

Bangkok Post. (2019). *Thailand ready to tap busy Muslim Travel Market*. https://www.bangkokpost.com/business/1696452/thailand-ready-to-tap-busy-muslim-travel-market

Ban, Z., Chen, F., Ban, T., & Ban, C. (2016). Cultural routes: theoretic system and its enlightenment to china heritage practice. *Proceedings of the 2016 International Forum on Management, Education and Information Technology Application*. 10.2991/ifmeita-16.2016.175

Barcelona Turisme. (2015). *Mobile apps*. Retrieved from http://www.turismedebarcelona.net/Mobile-Apps/_1d096tFVcJUB30LvIA_sk0JzLhRr3czPe71W_IX4cDk

Barre, S. (2013). Wilderness and cultural tour guides, place identity and sustainable tourism in remote areas. *Journal of Sustainable Tourism, 21*(6), 825–844. doi:10.1080/09669582.2012.737798

Baruah, S. R. (2016). Promotion of Culinary Tourism as a Destination Attraction of North-East India. *IJIRSSC- Journal, 2*(1), 201-209.

Bas, E. (2011). A capital budgeting problem for preventing workplace mobbing by using analytic hierarchy process and fuzzy 0-1 bidimensional knapsack model. *Expert Systems with Applications, 38*(10), 12415–12422; Advance online publication. doi:10.1016/j.eswa.2011.04.022

Bastemur, C. (2011). A new destination for alternative tourism; Lycian Way. *International Conference on Tourism (ICOT)*. http://iatour.net/pdf/Proceedings%20ICOT%202011%20.pdf

Bastemur, C. (2013a). Mimar Sinan rotasının doğal ve kültürel peyzaj izleri. Ankara Üniversitesi Fen Bilimleri Enstitüsü Yayınlanmamış Doktora Tezi, Ankara.

Bastemur, C. (2013b). Mimar Sinan eserlerinin özel ilgi kapsamında değerlendirilmesi. VIII. Uluslararası Sinan Sempozyumu, 39-44.

Bastemur, C. (2009). *Likya Yolu ve Çevresinin Turizm ve Rekreasyon Potansiyelinin Araştırılması*. Ankara Üniversitesi Fen Bilimleri Enstitüsü Yayınlanmamış Yüksek Lisans Tezi.

Battour, M., & Ismail, M. N. (2016). Halal tourism: Concepts, practices, challenges and future. *Tourism Management Perspectives*, *19*, 150–154. doi:10.1016/j.tmp.2015.12.008

Baum, T. (2006). *Human resource management for tourism, hospitality and leisure: An international perspective*. Cengage Learning EMEA.

Baum, T. (2007). Human resources in tourism: Still waiting for change. *Tourism Management*, *50*(6), 1383–1399. doi:10.1016/j.tourman.2007.04.005

Baum, T., Cheung, C., Kong, H., Kralj, A., Mooney, S., Thanh, H., Ramachandran, S., Ruzic, M., & Siow, M. (2016). Sustainability and the tourism and hospitality workforce: A thematic analysis. *Sustainability*, *8*(8), 809–830. doi:10.3390u8080809

Bayat, A. (2007). *Making Islam democratic: Social movements and the post-Islamist turn*. Stanford University Press. doi:10.5117/9789053569832

BBC. (2012). *Olympics in Numbers*. Retrieved November 10, 2019, from https://www.bbc.com/turkce/haberler/2012/07/120723_olympics_innumbers

Becker, H. S. (1970). *Sociological work: Method and substance*. Aldine.

Beckerson, J., & Walton, J. K. (2005). Selling air: marketing the ıntangible at British resorts. In J. K. Walton (Ed.), *Histories of tourism representation, identity and conflict* (pp. 55–68). Channel View Publications. doi:10.21832/9781845410339-005

Belmonte, J. A., González-García, A. C., Rodríguez-Antón, A., & Betancor, M. A. P. (2019). Equinox in Petra: Land- and Skyscape in the Nabataean Capital. *Nexus Network Journal*, 1–23.

Benckendorff, P., Sheldon, P., & Fesenmaier, D. (2014). *Tourism information technology*. CABI International. doi:10.1079/9781780641850.0000

Bennett, T. (1995). *The birth of the museum*. Routledge.

Bentley, T. A., Catley, B., Cooper-Thomas, H., Gardner, D., O'Driscoll, M. P., Dale, A., & Trenberth, L. (2012). Perceptions of workplace bullying in the New Zealand travel industry: Prevalence and management strategies. *Tourism Management*, *33*(2), 351–360. doi:10.1016/j.tourman.2011.04.004

Berry, K. (2014). Progress and Emerging Issues in Culinary Tourism: A Study with Special Reference to Punjab. *Journal of Hospitality Application & Research*, *9*(1), 28–46.

Beswick, J., Gore, J., & Palferman, D. (2006). Bullying at Work: A Review of the Literature. Retrieved from https://www.hse.gov.uk/research/hsl_pdf/2006/hsl0630.pdf

Bhalla, S., & Bhalla, T. C. (2007). Traditional foods and beverages of Himachal Pradesh. *Indian Journal of Traditional Knowledge*, *6*(1), 17–24.

Bhatta, C. (2017). The role of arts in promoting tourism: A case of tourism development in Thamel area. *Journal of Advanced Academic Research*, *3*(1), 177–184. doi:10.3126/jaar.v3i1.16627

Bhutia, S. (2015). Sustainable tourism development in Darjeeling Hills of West Bengal, India: Issues & challenges. *Global Journal of Human-Social Science: B, 15*(3), 1-10.

Bianchi, R. (2004). *Guests of God: Pilgrimage and politics in the Islamic world*. Oxford University Press. doi:10.1093/0195171071.001.0001

Bianchi, R. (2013). *Islamic Globalization: Pilgrimage, Capitalism, Democracy, and Diplomacy*. World scientific. doi:10.1142/8810

Black, R., & Ham, S. (2005). Improving the quality of tour guiding: Towards a model for tour guide certification. *Journal of Ecotourism*, *4*(3), 178–195. doi:10.1080/14724040608668442

Black, R., Ham, S., & Weiler, B. (2001). Ecotour guide training in less developed countries: Some preliminary research findings. *Journal of Sustainable Tourism*, *9*(2), 147–156. doi:10.1080/09669580108667395

Blamey, R. K. (1997). Ecotourism: The search for an operational definition. *Journal of Sustainable Tourism*, *5*(2), 109–130. doi:10.1080/09669589708667280

Blinder, A. S. (2006). Offshoring: The next industrial revolution? *Foreign Affairs*, *85*(2), 113–128. doi:10.2307/20031915

Blockchain. (2017). *Blockchain: harnessing it's potential in travel*. Retrieved from https://amadeus.com/en/insights/research-report/blockchain-harnessing-its-potential-in-travel

Boes, K., Borde, L., & Egger, R. (2015). The acceptance of NFC smart posters in tourism. In I. Tussyadiah & A. Inversini (Eds.), *Information and communication technologies in tourism* (pp. 435–448). Springer.

Boes, K., Buhalis, D., & Inversini, A. (2016). Smart tourism destinations: Ecosystems for tourism destination competitiveness. *International Journal of Tourism Cities*, *2*(2), 108–124. doi:10.1108/IJTC-12-2015-0032

Bon, M., & Hussain, M. (2010), Tourism in the Muslim world N. Scott (Ed.), Bridging tourism theory and practice (2010), vol. 2, pp. 47-59

Bonet, L. (2003). Cultural tourism. In R. Towse (Ed.), *A Handbook of Cultural Economics* (pp. 187–193). Edward Elga Publishing Inc. doi:10.4337/9781781008003.00029

Bourdeau, L., Gravari-Barbas, M. M., & Robinson, M. (Eds.). (2016). *World heritage, tourism and identity: inscription and co-production*. Routledge. doi:10.4324/9781315546315

Bowie, D., & Chang, C. (2005). Tourist satisfaction: A view from a mixed international guided package tour. *Journal of Vacation Marketing*, *11*(4), 303–322. doi:10.1177/1356766705056628

Brandon, K. (1996). *Ecotourism and Conservation: A Review of Key Issues*. World Bank.

Briedenhann, J., & Wickens, E. (2004). Tourism routes as a tool for the economic development of rural areas—Vibrant hope or impossible dream? *Tourism Management*, *25*(1), 71–79. doi:10.1016/S0261-5177(03)00063-3

Bromley, A. E. (2018). *New Commissioners will focus on untold stories of UVa in the age of segregation*. Retrieved from https://news.virginia.edu/content/new-commissioners-will-focus-untold-stories-uva-age-segregation

Bromley, A. E. (2019). *Work on memorial to enslaved laborers gets underway*. Retrieved from https://news.virginia.edu/content/work-memorial-enslaved-laborers-gets-underway

Bromley, R. (2006). On and off campus: Colleges and universities as local stakeholders. *Planning Practice and Research, 21*(1), 1–24. doi:10.1080/02697450600901400

Bryce, D., & Carnegie, E. (2013). Exhibiting the 'Orient': Historicising theory and curatorial practice in UK museums and galleries. *Environment & Planning A, 45*(7), 1734–1752. doi:10.1068/a45359

Brymer, R. A., Perrewe, P. L., & Johns, T. R. (1991). Managerial job stress in the hotel industry. *International Journal of Hospitality Management, 10*(1), 47–58. doi:10.1016/0278-4319(91)90006-4

Bueno, D. R., Viruete, E., & Montano, L. (2011, December). *An autonomous tour guide robot in a next generation smart museum.* Paper presented at the 5th International Symposium on Ubiquitous Computing and Ambient Intelligence (UCAmI 2011), Riviera Maya, Mexico.

Buhalis. (2015). *Working definitions of smartness and smart tourism destination.* Retrieved from http://buhalis.blogspot.com.tr/2014/12/working-definitions-of-smartness-and.html

Buhalis, D. (1998). Strategic use of information technologies in the tourism industry. *Tourism Management, 19*(5), 409–421. doi:10.1016/S0261-5177(98)00038-7

Buhalis, D., & Amaranggana, A. (2014). Smart tourism destinations. In Z. Xiang, I. Tussyadiah, & A. Inversini (Eds.), *Information and communication technologies in tourism 2014* (pp. 553–564). Springer.

Buhalis, D., & Law, R. (2008). Progress in information technology and tourism management: 20 years on and 10 years after the Internet—The state of eTourism research. *Tourism Management, 29*(4), 609–623. doi:10.1016/j.tourman.2008.01.005

Buhalis, D., & Licata, M. C. (2002). The future e-tourism intermediaries. *Tourism Management, 23*(3), 207–220. doi:10.1016/S0261-5177(01)00085-1

Bui, D. T. (2009). *Tourism industry responses to the rise of sustainable tourism and related environmental policy initiatives: The case of Hue city, Vietnam* (Doctoral dissertation). Auckland University of Technology.

Burgard, W., Cremers, A. B., Fox, D., Hähnel, D., Lakemeyer, G., Schulz, D., . . . Thrun, S. (1998, July). The interactive museum tour-guide robot. *Proceedings of the Fifteenth National Conference on Artificial Intelligence and Tenth Innovative Applications of Artificial Intelligence Conference*, 11-18.

Burns, J. (2010). *Career opportunities in travel and hospitality.* Infobase.

Butler, G., & Hannam, K. (2012). Independent tourist's automobilities in Norway. *Journal of Tourism and Cultural Change, 10*(4), 285–300. doi:10.1080/14766825.2012.718441

Butler, R. (2001). Seasonality in tourism: Issues and implications. In T. Baum & S. Lundtorp (Eds.), *Seasonality in tourism* (pp. 5–21). Elsevier Science Ltd. doi:10.1016/B978-0-08-043674-6.50005-2

Çakmak, T. F., & Demirkol, Ş. (2017). Teknolojik gelişmelerin turist rehberliği mesleğine etkileri üzerine bir swot analizi. *Bingöl Üniversitesi Sosyal Bilimler Enstitüsü Dergisi, 7*, 221–235. doi:10.29029/busbed.317656

Çalık Ross, A. (2017). Hoşgörü başkenti Nikomedia ve Galerius'un Hoşgörü Fermanı'nın çeviri denemesi. *Sosyal Bilimler Araştırma Dergisi, 6*, 135–145.

Çalışkan, O., & Tepeci, M. (2008). Mobbing Behaviors at Hotels and Their Impacts on Employee Job Satisfaction and Intent to Remain. Anatolia. *Turizm Araştırmaları Dergisi, 19*(2), 135–148. doi:10.17123/atad.09003

Campolo, D., Bombino, G., & Meduri, T. (2016). Cultural landscape and cultural routes: Infrastructure role and ındigenous knowledge for a sustainable development of inland areas. *Procedia: Social and Behavioral Sciences, 223*, 576–582. doi:10.1016/j.sbspro.2016.05.350

Canaan, T. (1929). Studies in the Topography and Folklore of Petra. *Journal of the Palestine Oriental Society*, *9*, 136–218.

Canadi, M., Höpken, W., & Fuchs, M. (2010). Application of QR codes in online travel distribution. In U. Gretzel, R. Law & M. Fuch (Eds.), Information and communication technologies in tourism (pp. 137-148). Springer. doi:10.1007/978-3-211-99407-8_12

Carmody, J. (2013). Intensive tour guide training in regional Australia: An analysis of the Savannah Guides organization and professional development schools. *Journal of Sustainable Tourism*, *21*(5), 679–694. doi:10.1080/09669582.2012.744412

Carson, D., Carson, De., & Lundmark, L. (2017). *Tourism, motilities, and development in sparsely populated areas*. Routledge. doi:10.4324/9781315665665

Caton, K. (2014). Underdisciplinarity: Where are the humanities in tourism education? *Journal of Hospitality, Leisure, Sport and Tourism Education*, *15*, 24–33. doi:10.1016/j.jhlste.2014.03.003

ÇEKÜL. (2019). *Çevre ve Kültür Değerlerini Koruma ve Tanıtma Vakfı*. https://www.cekulvakfi.org.tr/haber/kultur-rotalari-planlama-rehberi-cikti

Çelik, P., & Topsakal, Y. (2019). *Industry 4.0 and smart tourism*. Detay Publication.

Census of Himachal Pradesh. (2011). *Himachal Pradesh Religious Data*. Retrieved from Census of India website on 18 October, 2019, from https://www.census2011. co.in/census/ state/himachal+pradesh.html

Chalip, L., Green, B. C., & Velden, L. V. (1998). Sources of interest in travel to the Olympic Games. *Journal of Vacation Marketing*, *4*(1), 7–22. doi:10.1177/135676679800400102

Chan, A., Hsu, C., & Baum, T. G. (2015). The Impact of Tour Service Performance on Tourist Satisfaction and Behavioural Intentions: A Study of Chinese Tourists in Hong Kong. *Journal of Travel and Tourism Marketing, 32*(1-2), 18-33. org/ doi:10.1080/10548408.2014.986010

Chandler, J. D., & Vargo, S. L. (2011). Contextualization and value-in-context: How context frames exchange. *Marketing Theory*, *11*(1), 35–49. doi:10.1177/1470593110393713

Chang, K., & Brattlof, M. (2015). *Socio-economic implications of climate change for tea producing countries*. Food and Agricultural Organisation of the United Nation Report.

Chang, K. (2014). Examining the effect of tour guide performance, tourist trust, tourist satisfaction, and flow experience on tourists' shopping behavior. *Asia Pacific Journal of Tourism Research*, *19*(2), 219–247. doi:10.1080/10941665.2012.739189

Chang, M. K., & Cheung, W. (2001). Determinants of the intention to use Internet/ WWW at work: A confirmatory study. *Information & Management*, *39*(1), 1–14. doi:10.1016/S0378-7206(01)00075-1

Chang, T. C. (2000). Renaissance revisited: Singapore as a 'Global City for the Arts'. *International Journal of Urban and Regional Research*, *24*(4), 818–831. doi:10.1111/1468-2427.00280

Charmaz, K. (2006). *Constructing grounded theory: A practical guide through qualitative analysis*. Sage.

Charmaz, K., & Belgrave, L. (2002). *Qualitative interviewing and grounded theory analysis T* (2nd ed.). Sage.

Cheng, S., Xu, F., Zhang, J., & Zhang, Y. (2010). Tourists' attitudes toward tea tourism: A case study in Xinyang, China. *Journal of Travel & Tourism Marketing*, *27*(2), 2010. doi:10.1080/10548401003590526

Cheng, Y. S., Kuo, N. T., Chang, K. C., & Chen, C. H. (2019). How a tour guide interpretation service creates intention to revisit for tourists from Mainland China: The mediating effect of perceived value. *Journal of China Tourism Research*, *15*(1), 84–104. doi:10.1080/19388160.2018.1517067

Chillion, P. S. (2012). *From vacation spots to smart destinations: technology and tourism: QR, apps and augmented reality for cities*. Retrieved from https://urban360.me/2012/02/08/this-place-worths-a-visit-intelligent-destinations-smart-city-technology-tourism-and-city-branding/

Christie, M., & Mason, P. (2003). Transformative tour guiding: Training tour guides to be critically reflective practitioners. *Journal of Ecotourism*, *2*(1), 1–16. doi:10.1080/14724040308668130

Chuang, H. T. (2009). The rise of culinary tourism and its transformation of food cultures: The national cuisine of Taiwan. *The Copenhagen Journal of Asian Studies*, *27*(2), 84–108. doi:10.22439/cjas.v27i2.2542

Chu, K. H. L., & Murrmann, S. K. (2006). Development and validation of the hospitality emotional labor scale. *Tourism Management*, *27*(6), 1181–1191. doi:10.1016/j.tourman.2005.12.011

Civil Service Bureau. (2018). *Civil Bureau Bylaw*. Available at: http://www.csb.gov.jo/csb/Legislations/bylaw

Cleveland, J. N., O'Neill, J. W., Himelright, J. L., Harrison, M. M., Crouter, A. C., & Drago, R. (2007). Work and family issues in the hospitality industry: Perspectives of entrants, mangers, and spouses. *Journal of Hospitality & Tourism Research (Washington, D.C.)*, *31*(3), 275–298. doi:10.1177/1096348007299919

Clow, K. (2011). *Compiled from the presentation made at the Open Museum Summit*. Academic Press.

Clow, K. (2017). *Likya Yolu; Türkiye'nin İlk Uzun Mesafe Yürüyüş Rotası*. BKM Yayıncılık.

Clow, K. (2018). *Aziz Paul Yolu*. Hil Yayınları.

Clow, K., & Gardner, C. (2012). *The Kackar: Trekking in Turkey's Black Sea Mountains*. Upcountry Yayıncılık.

Clow, K., & Richerson, N. (2000). *Likya Yolu*. Okyanus Yayınları.

Coakley, J. (2004). *Sports in Society: Issues and Controversies* (8th ed.). Mc Graw Hill Higher Education.

COE. (2019). *Avrupa Konseyi*. https://www.coe.int/en/web/cultural-routes/by-theme

Cohen, E. (1972). Toward a Sociology of International Tourism. *Social Research*, *39*, 163–182.

Cohen, E. (1985). The tourist guide: The origins, structure and dynamics of a role. *Annals of Tourism Research*, *12*(1), 5–29. doi:10.1016/0160-7383(85)90037-4

Cohen, E. (1985). The tourist guide: The origins, structure, and dynamics of a role. *Annals of Tourism Research*, *12*(1), 5–29. doi:10.1016/0160-7383(85)90037-4

Çolakoğlu, O. E., Efendi, E., & Epik, F. (2017). *Tur yönetimi ve turist rehberliği*. Detay Yayıncılık.

Cole, J. (2002). *Sacred place and holy war: The politics, culture and history of Shi'ite Islam*. I. B. Taubis.

Collins, V. (2000). Becoming a Tour Guide, the Principals of Guiding and Site Interpretation. Academic Press.

Commons, J., & Page, S. (2001). Managing seasonality in peripheral tourism regions: The case of Northland, New Zealand. In T. Baum & S. Lundtorp (Eds.), *Seasonality in tourism* (pp. 153–172). Pergamon. doi:10.1016/B978-0-08-043674-6.50013-1

Comunian, R., & Mould, O. (2014). The Weakest link: Creative industries, flagship cultural projects and regeneration. *City. Cultura e Scuola*, *5*(2), 65–74.

Connell, J. (1996). A study of tourism on university campus sites. *Tourism Management*, *17*(7), 541–544. doi:10.1016/0261-5177(96)89217-X

Connell, J. (2000). The Role of Tourism in the Socially Responsible University. *Current Issues in Tourism*, *3*(1), 1–19. doi:10.1080/13683500008667863

Cook, I., & Crang, P. (1996). The world on a plate: Culinary culture, displacement and geographical knowledges. *Journal of Material Culture*, *1*(1), 131–154. doi:10.1177/135918359600100201

Coon, D., & Mitterer, J. (2012). *Introduction to psychology gateways to mind and behaviour* (13th ed.). Wadsworth Cengage Learning.

Cooper, C., & Shepherd, R. (2014). The relationship between tourism education and tourism industry: Implications for tourism education. *Tourism Recreation Research*, *22*(1), 34–47. doi:10.1080/02508281.1997.11014784

Corigliano, A. (2002). The route to quality: Italian gastronomy networks in operations. In A. M. Hjalager & G. Richards (Eds.), *Tourism and Gastronomy* (pp. 166–185). Routledge.

Costa, J. M. (2018). *Sociedade 5.0: a mudança que aí vem.* Retrieved from https://hrportugal. pt/sociedade-5-0-a-mudanca-que-ai-vem/

Costa, C., Panik, E., & Buhalis, D. (2013). *Trends in European Tourism Planning and Organization.* Channel View Publication. doi:10.21832/9781845414122

Cousquer, G. O., & Beames, S. (2013). Professionalism in mountain tourism and the claims to professional status of the International Mountain Leader. *Journal of Sport & Tourism*, *18*(3), 185–215. doi:10.1080/14775085.2014.904246

Crawford, N. (1997). Bullying at work: A psychoanalytic perspective. *Journal of Community & Applied Social Psychology*, *7*(3), 219–225. doi:10.1002/(SICI)1099-1298(199706)7:3<219::AID-CASP420>3.0.CO;2-Q

Crescent. (2014). *Why Should Tour Guides Be Aware of the Special Needs of Muslims?* https://www.crescentrating.com/magazine/opinion/3656/why-should-tour-guides-be-aware-of-the-special-needs-of-muslims.html

Crescent. (2016). *Philippines accelerates plans to develop Halal Tourism.* https://www.crescentrating.com/magazine/press-releases/3832/philippines-accelerates-plans-to-develop-halal-tourism.html

Creswell, J. W. (2007). *Qualitative inquiry and research design, Choosing among Five Approaches* (2nd ed.). Sage.

Creswell, J. W. (2014). *Research Design, Qualitative, Quantitative and Mixed Methods Approaches* (4th ed.). Sage Publishing.

Crompton, J. (1979). Motivations for pleasure vacation. *Annals of Tourism Research*, *6*(4), 408–424. doi:10.1016/0160-7383(79)90004-5

Csapó, J. (2012). The Role and Importance of Cultural Tourism in Modern Tourism Industry. In M. Kasimoglu & H. Aydin (Eds.), *Strategies for Tourism Industry - Micro and Macro Perspectives* (pp. 201–232). InTech Europe. doi:10.5772/38693

Cullen, E., Kordey, N., Schmidt, L., & Gaboardi, E. (2003). *Work and family in the e- work era.* IOS Press.

Currid, E. (2007). *The Warhol economy: How fashion, art, and music drive New York city.* Princeton University Press.

Cusack, I. (2000). African cuisines: Recipes for nation Building? *Journal of African Cultural Studies*, *13*(2), 207–225. doi:10.1080/713674313

Dabney, V. (1981). *Mr Jefferson's University: A History*. University of Virginia Press.

Dahles, H. (2002). The politics of tour guiding: Image management in Indonesia. *Annals of Tourism Research*, *29*(3), 783–800. doi:10.1016/S0160-7383(01)00083-4

Damala, A., Cubaud, P., Bationo, A., Houlier, P., & Marchal, I. (2008, September). Bridging the gap between the digital and the physical: Design and evaluation of a mobile augmented reality guide for the museum visit. *Proceedings of the 3rd International Conference on Digital Interactive Media in Entertainment and Arts (DIMEA '08)*, 120-127. 10.1145/1413634.1413660

Das, D. (2012). Tourism industry in North-East Indian States: Prospects and problems. *Global Research Methodology Journal, 2*(7).

Das, R. (2017). Prospects and problems of tourism in Assam. *International Journal of Applied Research*, *3*(8), 32–35.

Datta, C. (2018). Future prospective of tea-tourism along with existing forest-tourism in Duars, West Bengal, India. *Asian Review of Social Sciences*, *7*(2), 33–36.

Davenport, L., Brockelman, W. Y., Wright, P. C., Ruf, K., & del Valle, F. B. R. (2002). Ecotourism tools for parks. In J. Terborgh, C. van Schaik, L. Davenport, & M. Rao (Eds.), *Making Parks Work: Strategies for Preserving Tropical Nature* (pp. 279–306). Island.

Davenport, N., Schwartz, R. D., & Elliott, G. P. (2005). *Mobbing: Emotional Abuse in the American Workplace* (3rd ed.). Civil Society Publications.

De Aragao, M. M. (2015). *Economic Impacts of the FIFA World Cup in Developing Countries. Honors Theses. Paper 2609. Western Michigan University*. Lee Honors College.

Deeb, L., & Harb, M. (2013). *Leisurely Islam: negotiating geography and morality in Shi 'ite South Beirut*. Princeton University Press.

Deery, M., Jago, L., & Fredline, L. (2004). Sport tourism or event tourism: Are they one and the same? *Journal of Sport & Tourism*, *9*(3), 235–245. doi:10.1080/1477508042000320250

Demirel, E. (2012a). *Sarıkamış Yürüyüş Parkurları, Serhat Kalkınma Ajansı*. Academic Press.

Demirel, E. (2012b). Hitit Yolu. İstanbul: Hil Yayınları.

Denzin, N., & Lincoln, Y. (1994). *Handbook of Qualitative Research*. SAGE Thousand Oaks.

Department of Statistics. Jordan Government. (2018). Retrieved from: http://dosweb.dos.gov.jo/product-category/jordan-in-figuers_yearbook/

Devi, K. S. & Amity University. (2017). *Cuisine Tourism in West Bengal*. Kolkata: Amity University.

Ding, M. (2017). A*ugmented reality in museums*. Retrieved October 15, 2019, from https://static1.squarespace.com/static/51d98be2e4b05a25fc200cbc/t/5908d019f5e2314ab790c269/1493749785593/Augmented+Reality+in+Museums.pdf

DiPietro, R., & Bufquin, D. (2018). Effects of work status congruence and perceived management concern for employees on turnover intentions in a fast-casual restaurant chain. *Journal of Human Resources in Hospitality & Tourism*, *17*(1), 38–59. Advance online publication. doi:10.1080/15332845.2017.1328260

Dube, M. P., & Padalia, M. (2002). *Democratic decentralization & Panchayati Raj in India*. Anamika Publishers.

Ducassi, S. I., & Rosa, M. (2005). *A new category of heritage for understanding, cooperation and sustainable development. Their significance within the macrostructure of cultural heritage. The role of the CIIC of ICOMOS: Principles and methodology*. Icomos.

Duffy, M., & Sperry, L. (2007). Workplace Mobbing: Individual and Family Health Consequences. *The Family Journal (Alexandria, Va.)*, *15*(4), 398–404. doi:10.1177/1066480707305069

Duman, T. (2010). The value of Islamic tourism: Perspectives from the Turkish experience Islam and Civilizational Renewal (ICR). Academic Press.

Duttagupta, S. & University of Waterloo. (2013). *Foreign travellers' recommendation of culinary tourism in India based on cuisine image and satisfaction with experiences at culinary establishments: An exploratory study*. Waterloo: University of Waterloo.

Eade, J., & Sallnow, M. J. (Eds.). (2000). *Contesting the sacred: the anthropology of pilgrimage*. University of Illinois Press.

Eid, R., & El-Gohary, H. (2015). The role of Islamic religiosity on the relationship between perceived value and tourist satisfaction. *Tourism Management*, *46*, 477–488. doi:10.1016/j.tourman.2014.08.003

Einarsen, S. (1999). The nature and causes of bullying at work. *International Journal of Manpower*, *20*(1/2), 16–27. doi:10.1108/01437729910268588

Einarsen, S., & Raknes, B. I. (1997). Harassment in the Workplace and the Victimization of Men. *Violence and Victims*, *12*(3), 247–263. doi:10.1891/0886-6708.12.3.247 PubMed

Einstein, M. (2007). *Brands of faith: Marketing religion in a commercial age*. Routledge. doi:10.4324/9780203938874

Elaziz, M. F., & Kurt, A. (2017). Religiosity, consumerism and halal tourism: A study of seaside tourism organizations in Turkey. *Tourism (Zagreb)*, *65*(1), 115–128.

El-Gohary, H. (2016). *Halal Tourism, is it really Halal? Tourism Management Perspective* (Vol. 19). Academic Press.

Emmons, K. (1991). Educational planning and ecotourism: Targeting school students. In J. A. Kusler (Ed.), *Ecotourism and Resource Conservation: A Collection of Papers* (Vol. 2, pp. 826–838). Association of Wetland Managers.

Endesa. (2008). *Smartcity Malaga - a model of sustainable energy management for cities of the future report*. Spain: Direccion General de Distribucion de Endesa.

Enright, M. J., & Newton, J. (2004). Tourism destination competitiveness: A quantitative approach. *Tourism Management*, *25*(6), 777–788. doi:10.1016/j.tourman.2004.06.008

Eren, R. (2019). Turkey's Food Image, Travelers' Revisit Intention and Tourist Expenditures. Anais Brasileiros de Estudos Turísticos - ABET, 9(1,2,3), 1–16.

Eren, R., & Aydin, A. (2019). Perceptions and attitudes of culinary students towards food and beverage industry in Turkey. Journal of Culinary Science & Technology, 1–25. doi:10.1080/15428052.2019.1640155

Ergüven, M. H. (2015). Gastronomy and wine tourism as a variety of special interest tourism: Thracian Vineyard Route. *Turkish Studies*, *10*(10), 449–464.

Erkol Bayram, G. (2016). The research on examination of tour guides mobbing exposure within the demographic characteristics. *The Journal of International Social Research*, *9*(47), 1185–1194. doi:10.17719/jisr.2016.1465

Esen, F. Ö., & Gülmez, M. (2017). View of Academicians and Professional Tourist Guides on Domestic Practice Tours. *International Journal of Commerce and Management*, *5*(7), 222–250.

Esen, M. F., & Türkay, B. (2017). Big data applications in tourism industries. *Journal of Tourism and Gastronomy Studies, 5*(4), 92–115. doi:10.21325/jotags.2017.140

Evans, G., & Shaw, P. (2004). *The contribution of culture to regeneration in the UK: A review of the evidence*. DCMS.

Ewert, A. W., & Shultis, J. (1997). Resource based tourism: An emerging trend in tourism experiences. *Parks & Recreation, 32*(9), 95–103.

Fairley, S. (2003). In search of relived social experience: Group-based nostalgia sport tourism. *Journal of Sport Management, 17*(3), 284–304. doi:10.1123/jsm.17.3.284

Fakhry, Piccirillo, & Saqqaf. (1996). *The Holy Sites Of Jordan*. Turab.

Farkic, J. (2018). Outdoor guiding as hospitality work. *Annals of Tourism Research, 73*(C), 197–199. doi:10.1016/j.annals.2018.04.004

Fealy, G. (2008). Consuming Islam: Commodified religion and aspirational pietism in contemporary Indonesia. In G. Fealy & S. White (Eds.), *Expressing Islam: Religious life and politics in Indonesia* (pp. 15–39). Institute of Southeast Asian Studies. doi:10.1355/9789812308528-006

FEG. (2019, Nov 11). *European Federation of Tourist Guide Association*. Retrieved from https://www.feg-touristguides.com/about.php

Femenia-Serra, F., Neuhofer, B., & Ivars-Baidal, J. A. (2019). Towards a conceptualisation of smart tourists and their role within the smart destination scenario. *Service Industries Journal, 39*(2), 109–133. doi:10.1080/02642069.2018.1508458

Fennell, M. L., & Warnecke, R. B. (1988). *The Diffusion of Medical Innovations: An Applied Network Analysis*. Plenum. doi:10.1007/978-1-4684-5436-9

Fernando, I. N. (2015). What competitive strategies way forward the regional competitiveness? A comparative economic approach to Sri Lankan tourism. *International Journal of Business and Management, 10*(4). Advance online publication. doi:10.5539/ijbm.v10n4p178

Fernando, P. I. N., Kumari, K. W. S. N., & Rajapaksha, R. M. P. D. K. (2017). Destination marketing to promote tea tourism socio-economic approach on community development. *International Review of Management and Business Research, 6*(1), 68–75.

Finkel, C., & Clow, K. (2011). *The Evliya Çelebi Way*. Upcountry Yayınevi.

Fino, E. R., Martín-Gutiérrez, J., Fernández, M. D. M., & Davara, E. A. (2013). Interactive tourist guide: Connecting web 2.0, augmented reality and QR codes. *Procedia Computer Science, 25*, 338–344. doi:10.1016/j.procs.2013.11.040

Fischer, J., Khan, I., Khemani, T., Mak, D., & Najmi, R. (2009). Jordan tourism cluster. *Report For Microeconomics of Competitiveness, 8*.

Fitzgerald, L., Ferlie, E., Wood, M., & Hawkins, C. (2002). Interlocking Interactions, the Diffusion of Innovations in Health Care. *Human Relations, 55*(12), 1429–1449. doi:10.1177/001872602128782213

Food and Agriculture Organization of the United Nations. (2015). *Global Forest Resources Assessment 2015*. Rome: Author.

Foster, H. (2015). *Bad new days: Art, criticism, emergency*. Verso Books.

Franklin, A. (2018). Art tourism: A new field for tourist studies. *Tourist Studies, 18*(4), 399–416. doi:10.1177/1468797618815025

Franklin, A. S. (2008). The tourism ordering: Taking tourism more seriously as an ordering of a global world. *Civilisations, 57*(1-2), 25–39. doi:10.4000/civilisations.1288

Franklin, A. S., & Papastergiadis, N. (2017). Engaging with the anti-museum? Visitors to the museum of old and new art. *Journal of Sociology (Melbourne, Vic.), 53*(3), 670–686. doi:10.1177/1440783317712866

Fraunhofer Institute. (n.d.). *Robot Mona/Oskar.* Retrieved October 12, 2019, from https://www.ipa.fraunhofer.de/en/expertise/robot-and-assistive-systems/household-and-assistive-robots/entertainment.html

Friedman, S., & Greenhaus, J. (2000). *Work and family, allies or enemies? What happens when business professionals confront life choices.* Oxford University Press. doi:10.1093/acprof:oso/9780195112757.001.0001

Frochot, I. (2003). An Analysis of Regional Positioning and Its Associated Food Images in French Tourism Regional Brochures. *Journal of Travel & Tourism Marketing, 14*(3-4), 77–96. doi:10.1300/J073v14n03_05

FSSAI. (2016). *Feasts Flavours and Fusion of India.* IGI Global India Pvt. Ltd.

Gamor, E., Amissah, E., Amissah, A., & Nartey, E. (2018). Factors of work-family conflict in the hospitality industry in Ghana. *Journal of Human Resources in Hospitality & Tourism, 17*(4), 482–501. doi:10.1080/15332845.2017.1328263

Gee, C. (1988). *Resort Development and Management.* Educational Institute of the American Hotel and Motel Association.

Gelbman, A., & Maoz, D. (2012). Island of peace or island of war: Tourist guiding. *Annals of Tourism Research, 39*(1), 108–133. doi:10.1016/j.annals.2011.05.008

Gharaibeh, K. M. (2012). The Developmental Role of Ajloun Forest Reserve in Jordan. *Journal of the Planner and Development, 25.*

Gibbons, R. (1992). *A Primer in Game Theory.* Coleshill: Harvester Wheatsheaf.

Gibson, H., Attle, S., & Yiannakis, A. (1997). *Segmenting the sport tourist market: a lifespan perspective.* Paper presented at Teaming for Success: A Forum on Sport Tourism, Arlington, VA.

Gibson, H. J. (1998a). Active sport tourism: Who participates? *Leisure Studies, 17*(2), 155–170. doi:10.1080/026143698375213

Gibson, H. J. (1998b). Sport tourism: A critical analysis of research. *Sport Management Review, 1*(1), 45–76. doi:10.1016/S1441-3523(98)70099-3

Gibson, H. J., Lamont, M., Kennelly, M., & Buning, R. J. (2018). Introduction to the Special Issue Active Sport Tourism. *Journal of Sport & Tourism, 22*(2), 83–91. doi:10.1080/14775085.2018.1466350

Gierczak, B. (2011). The history of tourist transport after the modern industrial revolution. *Polish Journal of Sport and Tourism, 18*(4), 275–281. doi:10.2478/v10197-011-0022-6

Glambek, M., Matthiesen, S. B., Hetland, J., & Einarsen, S. (2014). Workplace bullying as an antecedent to job insecurity and intention to leave: A 6-month prospective study. *Human Resource Management Journal, 24*(3), 255–268. doi:10.1111/1748-8583.12035

Goffman, E. (1959). *The Presentation of Every Day Life.* Penguin Books.

Gonçalves, E. C., & Maduro, A. V. (2016). Complementarity and Interaction of Tourist Services in an Excellent Wine Tourism Destination: The Douro Valley (Portugal). In M. Peris-Ortiz, M. C. D. R. Rama, & C. Rueda-Armengot (Eds.), *Wine and Tourism -A Strategic Segment for Sustainable Economic Development* (pp. 123–132). Springer. doi:10.1007/978-3-319-18857-7_9

Gonzalez, S. (2011). Bilbao and Barcelona in motion." How urban regeneration "models" travel and mutate in the global flows of policy tourism. *Urban Studies (Edinburgh, Scotland), 48*(7), 1397–1418. doi:10.1177/0042098010374510

Görmüş, S. (2017). Kültürel peyzaj değerlerini korumak için yeni bir yaklaşım: kültürel rotalar. *Plant Peyzaj ve Süs Bitkiciliği Dergisi,*7-9.

Goulding, C. (2002). *Grounded Theory: A Practical Guide for Management, Business and Market Researchers.* Sage. doi:10.4135/9781849209236

Greeff, M., & Lalioti, V. (2001, September). Interactive cultural experiences using virtual identities. *International Cultural Heritage Informatics Meeting*, 455-465.

Green, B. C. (2001). Leveraging subculture and identity to promote sport events. *Sport Management Review, 4*(1), 1–19. doi:10.1016/S1441-3523(01)70067-8

Greenhalgh, T., Robert, G., Macfarlane, F., Bate, P., & Kyriakidou, O. (2004). Diffusion of Innovations in Service Organizations: Systematic Review and Recommendations. *The Milbank Quarterly, 82*(4), 581–629. doi:10.1111/j.0887-378X.2004.00325.x PMID:15595944

Green, N. (2011). *Bombay Islam: The religious economy of the west Indian ocean, 1840–1915.* Cambridge University Press. doi:10.1017/CBO9780511975165

Green, N. (2015). *Terrains of exchange: Religious economies of global Islam.* Oxford University Press. doi:10.1093/acprof:oso/9780190222536.001.0001

Grenier, K. H. (2005). *Creating Caledonia; Tourism and Identity in Scotland, 1770–1914.* Imprint Routledge.

Gretzel, U. (2011). Intelligent systems in tourism: A social science perspective. *Annals of Tourism Research, 38*(3), 757–779. doi:10.1016/j.annals.2011.04.014

Gretzel, U., & Jamal, T. (2009). Conceptualizing the creative tourist class: Technology, mobility, and tourist experiences. *Tourism Analysis, 14*(3), 471–481. doi:10.3727/108354209X12596287114219

Gretzel, U., Reino, S., Kopera, S., & Koo, C. (2015b). Smart tourism challenges. *Journal of Tourism, 16*(1), 41–47.

Gretzel, U., Sigala, M., Xiang, Z., & Koo, C. (2015a). Smart tourism: Foundations and developments. *Electronic Markets, 25*(3), 179–188. doi:10.100712525-015-0196-8

Gretzel, U., Werthner, H., Koo, C., & Lamsfus, C. (2015). Conceptual foundations for understanding smart tourism ecosystems. *Computers in Human Behavior, 50*, 558–563. doi:10.1016/j.chb.2015.03.043

Grobelna, A. (2017). The perception of job-related motivators when choosing a career in the tourism and hospitality industry – a comparative study between Polish and Spanish students. *International Journal of Management and Economics, 53*(2), 84–106. doi:10.1515/ijme-2017-0014

Guba, E., & Lincoln, Y. (1989). *Fourth generation evaluation.* Sage.

Gumprecht, B. (2003). The American College Town. *Geographical Review, 93*(1), 51–80. doi:10.1111/j.1931-0846.2003.tb00020.x

Gumprecht, B. (2007). The campus as a public space in the American college town. *Journal of Historical Geography, 33*(1), 72–103. doi:10.1016/j.jhg.2005.12.001

Guo, Y., Liu, H., & Chai, Y. (2014). The embedding convergence of smart cities and tourism ınternet of things in China: An advance perspective. *Advances in Hospitality and Tourism Research, 2*(1), 54–69.

Guttentag, D. A. (2010). Virtual reality: Applications and implications for tourism. *Tourism Management*, *31*(5), 637–651. doi:10.1016/j.tourman.2009.07.003

H.P. Tourism Development & Reg. Act. (2002). *Himachal Pradesh Tourism Development & Registration Act-2002*. Retrieved from Himachal tourism website on 14 November, 2019 from https://himachaltourism.gov.in/wp-content/uploads/2018/ 09/58e76f7a08179_H.P.Tourism-Dev.-and-Registration-Act-2002.pdf

Hacıoğlu, N., Kaşlı, M., Şahin, S., & Tetik, N. (2008). *Tourism education in Turkey*. Detay Publishing.

Haenni, P. (2005). *L'islam de marché: l'autre révolution conservatrice*. Seuil.

Hall, C. M., & Sharples, L. (2003). The consumption of experiences or the experience of consumption? An introduction to the tourism of taste. In C. M. Hall, L. Sharples, R. Mitchell, N. Macionis, & B. Cambourne (Eds.), *Food Tourism around the World. Development, management and markets* (pp. 1–24). Butterworth-Heinemann. doi:10.1016/B978-0-7506-5503-3.50004-X

Hall, M., & Prayag, G. (Eds.), *The routledge handbook of halal hospitality and Islamic tourism*. Routledge. doi:10.4324/9781315150604

Hall, R. (1994). *Sociology of work: Perspectives, analyses, and issues*. Pine Forge Press.

Ham, S. (1992). *Environmental interpretation: A practical guide for people with big ideas and small budgets*. Fulcrum Publishing/ North American Press.

Hamza, I. M., Chouhoud, R., & Tantawi, P. (2012). Islamic tourism: Exploring perceptions and possibilities in Egypt. *African Journal of Business and Economic Research*, *7*(1), 85–98.

Handszuh. (2000). *Culinary Tourism Strategic Development: an Asia-Pacific Perspective Library*. Doi:10.1002/jtr.834

Hansen, Å. M., Hogh, A., Garde, A. H., & Persson, R. (2014). Workplace bullying and sleep difficulties: A 2-year follow-up study. *International Archives of Occupational and Environmental Health*, *87*(3), 285–294. doi:10.1007/s00420-013-0860-2 PubMed

Haq, F., & Wong, H. Y. (2010). Is spiritual tourism a new strategy for marketing Islam? *Journal of Islamic Marketing*, *1*(2), 136–148. doi:10.1108/17590831011055879

Harkin, M. (1995). Modernist anthropology and tourism of the authentic. *Annals of Tourism Research*, *22*(3), 650–670. doi:10.1016/0160-7383(95)00008-T

Harrill, R. (2005). *Guide to Best Practices in Tourism and Destination Management* (Vol. 2). The Education Institute of the American Hotel & Lodging Association.

Harrison, C., Eckman, B., Hamilton, R., Hartswick, P., Kalagnanam, J., Paraszczak, J., & Williams, P. (2010). Foundations for smarter cities. *IBM Journal of Research and Development*, *54*(4), 1–16. doi:10.1147/JRD.2010.2048257

Hassan, A. (2018). Destination Image Formation: The Function Analysis of Augmented Reality Application. In M. Khosrow-Pour (Ed.), *The Encyclopaedia of Information Science and Technology* (4th ed., pp. 4031–4037). IGI Global.

Hassan, A., & Dadwal, S. (2016). Search Engine Marketing – An Outlining of Conceptualization and Strategic Application. In W. Ozuem & G. Bowen (Eds.), *Competitive Social Media Marketing Strategies* (pp. 219–234). IGI Global. doi:10.4018/978-1-4666-9776-8.ch011

Hassan, A., & Shabani, N. (2017). Usability Analysis of Augmented Reality for Tourism Destination Promotion. In S. K. Dixit (Ed.), *Routledge Handbook of Consumer Behaviour in Hospitality and Tourism* (pp. 434–441). Routledge. doi:10.4324/9781315659657-49

Haugstved, A.-C., & Krogstie, J. (2012, January). Mobile augmented reality for cultural heritage: A technology acceptance study. *2012 IEEE International Symposium on Mixed and Augmented Reality (ISMAR)*, 247-255. 10.1109/ISMAR.2012.6402563

Haven-Tang, C., & Jones, E. (2005). Using local food and drink to differentiate tourism destinations through a sense of place: A story from Wales– dining at Monmouthshire's great table. *Journal of Culinary Science & Technology*, *4*(4), 69–86. doi:10.1300/J385v04n04_07

Hayashi, H., Sasajima, H., Takayanagi, Y., & Kanamaru, H. (2017, September). *International standardization for smarter society in the field of measurement, control and automation*. Paper presented at the 56th Annual Conference of the Society of Instrument and Control Engineers of Japan, Kanazawa, Japan. 10.23919/SICE.2017.8105723

Haynes, C. (2015). *The role of indigenous tour guides in providing authentic experience on guided tours in the Caribbean* (PhD dissertation). Pennsylvania State University.

Hazarika, K. (2011). Changing market scenaria for Indian Tea. *International Journal of Trade. Economics and Finance*, *2*(4), 285–287.

Hemsley-Brown, J., & Goonawardana, S. (2007). Brand harmonization in the international higher education market. *Journal of Business Research*, *60*(9), 942–948. doi:10.1016/j.jbusres.2007.01.019

Henderson, J. C. (2003). Managing tourism and Islam in Peninsular Malaysia. *Tourism Management, 24*(4), 447– 456.

Henderson, J. C. (2009, April 18). (20029). Food tourism reviewed. *British Food Journal*, *111*(4), 317–326. doi:10.1108/00070700910951470

Hengpoom, P., & Weerasophon, W. (2019). Sport Tourist Spending Behavior in Thailand. *The 2019 International Academic Research Conference*. https://www.icbtsconference.com/17096732/e-proceedings-amsterdam-2019

Herzberg, F., Mausner, B., & Snyderman, B. (2010). *The Motivation to Work* (12th ed.). Transaction.

Heung, V. C. S. (2008). Effects of tour leader's service quality on agency's reputation and customers'word-of-mouth. *Journal of Vacation Marketing*, *14*(4), 305–315. doi:10.1177/1356766708094752

Higham, J. (2007). *Sport tourism destinations: issues, opportunities and analysis*. Elsevier, Butterworth Heinemann. doi:10.4324/9780080474434

Higham, J., & Hinch, T. (2002). Tourism, sport and seasons: The challenges and potential of overcoming seasonality in the sport and tourism sectors. *Tourism Management*, *23*(2), 175–185. doi:10.1016/S0261-5177(01)00046-2

Hinch, T., & Higham, J. (2004). *Sport Tourism Development–Aspects of Tourism*. Channel View Publications.

Hinch, T., & Higham, J. (2019). *Sport Tourism Development*. Channel View Publications, Cromwell Press.

Hirsh, W., & Strebler, M. (1994). Defining managerial skills and competences. Gower Handbook of Management Development, 4.

Hjalager, A. (2003). Global tourism careers? Opportunities and dilemmas facing higher education in tourism. *Journal of Hospitality, Leisure, Sport and Tourism Education*, *2*(2), 1–12. doi:10.3794/johlste.22.35

Hodeck, A., & Hovemann, G. (2016). Motivation of active sport tourists in a German highland destination – a cross-seasonal comparison. *Journal of Sport & Tourism*, *20*(3–4), 335–348. doi:10.1080/14775085.2016.1235988

Hoesterey, J. B. (2015). *Rebranding Islam: Piety, prosperity, and a self-help guru*. Stanford University Press.

Hogh, A., & Dofradottir, A. (2001). Coping with bullying in the workplace. *European Journal of Work and Organizational Psychology, 10*(4), 485–495. doi:10.1080/13594320143000825

Höjer, M., & Wangel, J. (2015). Smart sustainable cities: definition and challenges. In L. Hilty & B. Aebischer (Eds.), *ICT innovations for sustainability, advances in intelligent systems and computing* (pp. 333–349). Springer. doi:10.1007/978-3-319-09228-7_20

Holloway, J. (1981). The guided tour: A sociological approach. *Annals of Tourism Research, 8*(3), 377–402. doi:10.1016/0160-7383(81)90005-0

Honey, M. (1999). *Ecotourism and Sustainable Development: Who Owns Paradise?* Island.

Hooijdonk, R. (2015). *Technology trends 2030.* Retrieved from https://www.richardvanhooijdonk.com/en/keynote/trends2030/

Hoover, E. (2009). Golden Walk" Gets a Makeover from an Auditor of Campus Visits. *The Chronicle of Higher Education, 55*(26).

Horen, G., & Dhurbojyoti, N. (2011). Problem and prospect of tea tourism in Assam - A Swot Analysis. In *International Conference on Advancements in Information Technology with Workshop of ICBMG 2011 IPCSIT* (vol. 20). IACSIT Press.

Houge-Mackenzie, S., & Kerr, J. (2013). Stress and emotions at work: An adventure tourism guide's experiences. *Tourism Management, 36*, 3–14. doi:10.1016/j.tourman.2012.10.018

Houge-Mackenzie, S., & Kerr, J. (2014). The psychological experience of river guiding: Exploring the protective frame and implications for guide well-being. *Journal of Sport & Tourism, 19*(1), 1, 5–27. doi:10.1080/14775085.2014.967796

Howard, J., Smith, B., & Thwaites, R. (2001). Investigating the role of the indigenous tour guide. *Journal of Tourism Studies, 12*(2), 32.

Huang, S., Hsu, C., & Chan, A. (2010). The tour guide performance and tourist satisfaction: A study of the package tours in Shanghai. *Journal of Hospitality & Tourism Research (Washington, D.C.), 34*(1), 3–33. doi:10.1177/1096348009349815

Hughes, H. L. (2002). The role of theatre and music in urban tourism. In D. A. Jelincic (Ed.), Culture: A driving force for urban tourism (pp.51-55). Academic Press.

Hughes, H. L. (2000). *Arts, entertainment and tourism.* Butterworth-Heinemann.

Humankinetics. (2019). *The Remarkable Growth of Tourism.* https://humankinetics.me/2019/04/04/the-remarkable-growth-of-sport-tourism/

Hurley, I. (2018). *Is there any real economic benefit to hosting a world cup?* https://psmag.com/economics/is-there-any-real-economic-benefit-to-hosting-a-world-cup

Husain, W., Dih, L. Y., Yen, H. F., & Jothi, N. (2012). Mytourguide.com: a framework of a location-based services for tourism ındustry. In *Proceedings of International Conference on Computer and Information Science* (pp.184-189). 10.1109/ICCISci.2012.6297236

Ibrahim, H. (1982). Leisure and Islam. *Leisure Studies, 1*(2), 197–210. doi:10.1080/02614368200390161

ICOMOS. (2019). *Uluslararası Anıtlar ve Sitler Konseyi.* https://www.google.com/search?q=%C4%B1comos+cultural+routes&oq=%C4%B1comos+cultural+routes&aqs=chrome.69i57j0.4623j0j4&sourceid=chrome&ie=UTF-8

IFTTA. (2019, Nov 11). *The International Forum of Travel and Tourism Advocates.* Retrieved from https://iftta.org/glossary

IITTM. (2019). *Regional Tour Guide training program.* Retrieved from IITTM website on 18 November, 2019, from https://www.iittm.ac.in

İlhan, İ., & Çeltek, E. (2016). Mobile marketing: Usage of augmented reality in tourism. *Gaziantep University Journal of Social Sciences, 15*(2), 581–599. doi:10.21547/jss.256721

Invat.Tur. (2015). *Destino turistico inteligente: manual operativo para la configuracion de destinos turisticos inteligentes.* Alicante: Universidad de Alicante, Instituto Universitario de Investigaciones Turisticas.

Iriguler, F. & Guler, M.E. (2017). Tourist Guides' Perceptions on Gastronomic Tourism in Turkey. *Journal of Tourism & Gastronomy Studies, 5*(1), 3-19.

Islamic Tourism. (2019). *Experience worry-free vacations with facilities adhering to Islamic Principles.* https://www.tourism.gov.my/niche/islamic-tourism

ITWD. (2019, June 27). International Delhi Mango Festival begins (place or time): All you need to know. *India Today.* Retrieved on 11 October, 2019, from https://www. indiatoday.in/information/story/all-you-need-to-know-about-international-delhi-mango-festival-1557301-2019-06-27

Izutsu, T. (2002). *Ethico-religious concepts in the Qur'an* (Vol. 1). McGill: McGill-Queen's Press-MQUP.

Jackson, M. R. (2003). Investing in creativity: A study of the support structure for U.S. artists. *The Journal of Arts Management, Law, and Society, 34*(1), 43–58. doi:10.3200/JAML.34.1.43-58

Jacobson, K. J., Hood, J. N., & Van Buren, H. J. III. (2014). Workplace bullying across cultures: A research agenda. *International Journal of Cross Cultural Management, 14*(1), 47–65; Advance online publication. doi:10.1177/1470595813494192

Jafari, J. (1985). Editor's page. *Annals of Tourism Research, 12*(1), 1–2. doi:10.1016/0160-7383(85)90035-0

Jagodzinska, K. (2017). *Art museums in Australian.* Jagiellonian University Press.

Jain, S. (2018, February). Tripoto: Kasol- A mini Israel of India [Web log post]. Retrieved fromhttps://www.tripoto.com/trip/kasol-a-mini-israel-of-india-5993cd64c8f9e

Jakovlev, Z., Dimitrov, N. V., Koteski, C., & Serafimova, M. (2017). Sport Tourism as the Most Important Alternative Form of Tourism - Strategies and Challenge, *Economic Development/Ekonomiski Razvoj, 19*(3), 197-207.

Jalis, M. H., Che, D., & Markwell, K. (2014). Utilising local cuisine to market Malaysia as a tourist destination. *Procedia: Social and Behavioral Sciences, 114*, 102–110.

Jamal, A., Raj, R., & Griffin, K. (2019). *Islamic tourism: Management of travel destinations.* Wallington: CABI.

Janicke, M., & Jacob, K. (2009). *A third industrial revolution? solutions to the crisis of resource-intensive growth.* Environmental Policy Research Centre, Freie Universitat.

Jawabreh, O. A. (2017). Distinction of Jordan as a destination for religious tourism. *Journal of Environmental Management and Tourism, 8*(22), 1171–1182.

Jenkins, L. D., & Romanos, M. (2014). The Art of tourism-driven development: Economic and artistic well-being of artists in three Balinese communities. *Journal of Tourism and Cultural Change, 12*(4), 293–306. doi:10.1080/14766825.2014.934377

Jennifer, D., Cowie, H., & Ananiadou, K. (2003). Perceptions and Experience of Workplace Bullying in Five Different Working Populations. *Aggressive Behavior, 29*(6), 489–496; Advance online publication. doi:10.1002/ab.10055

Jennings, G., & Weiler, B. (2006). Mediating meaning: Perspectives on brokering quality tourist experiences. In G. Jennings & N. Nickerson (Eds.), *Quality tourism experiences* (pp. 57–78). Elsevier. doi:10.1016/B978-0-7506-7811-7.50010-9

Jensen, O. (2010). Social meditation in remote developing world tourism locations–the significance of social ties between local guides and host communities in sustainable tourism development. *Journal of Sustainable Tourism, 18*(5), 615–633. doi:10.1080/09669581003615590

Jia, X., & Chaozhi, Z. (2020). "Halal tourism": Is it the same trend in non-Islamic destinations with Islamic destinations? *Asia Pacific Journal of Tourism Research, 25*(2), 189–204. doi:10.1080/10941665.2019.1687535

Jiménez-Beltrán, F. J., López-Guzmán, T., & Santa Cruz, F. G. (2016). Analysis of the Relationship between Tourism and Food Culture. *MDPI-Sustainability, 8*(5), 418. doi:10.3390u8050418

Jolliffe, L. (2007). *Tea and tourism: tourists, traditions and transformations*. Channel View Publications. doi:10.21832/9781845410582

Jolliffe, L., & Cave, J. (2012). Arts tourism. In P. Robinson (Ed.), *Tourism: The key concepts* (pp. 13–15). Routledge.

Jones, A., & Jenkins, I. (2002). A taste of wales–Blas Ar Gymru: Institutional malaise in promoting welsh food tourism products. In A. M. Hjalager & G. Richards (Eds.), *Tourism and Gastronomy* (pp. 115–131). Routledge.

Jordan Tour Guides Association. (2019). Retrieved from: http://www.tourguides.com.jo/index.php?pg=QUJPVVQgSl RHQQ==&sub=T3VyIEhpc3Rvcnk=

Jwa, J. W. (2016). Pedestrian network models for mobile smart tour guide services. *International Journal of Internet. Broadcasting and Communication, 8*(1), 73–78.

Kaddi, A. K. (2015). A Study on Creation and Development of Wine Tourism Circuits in Maharashtra. *Atna Journal of Tourism Studies, 10*(1), 1–12. doi:10.12727/ajts.13.1

Kara, D., Kim, H., & Uysal, M. (2018). The effect of manager mobbing behaviour on female employees' quality of life. *Current Issues in Tourism, 21*(13), 1453–1467. doi:10.1080/13683500.2015.1078298

Karataş, E. (2015). *Kültür Rotası Planlama Rehberi, Çevre ve Kültürel Değerlerini Koruma ve Tanıtma Vakfı ve Tarihi Kentler Birliği Ortak Yayını*. Stil Matbaa.

Karimi, S. (2008, March). *What is culinary tourism?* Retrieved from http://traveltips .usatoday.com/culinary-tourism-1910. html

Kaufman, E. (2001). *The Delirium of Praise*. Johns Hopkins University Press.

Kawashima, T. D. (2016). Travel agencies and priests as spiritual leaders: The merits of collaboration. *Tourist Studies, 16*(1), 40–56. doi:10.1177/1468797615588430

Kazaeleh, A. (2009). *Conflict between social values and organizational values in educational administration*. Daral-hamed Publication. (in Arabic)

Kervankıran, İ., & Çuhadar, M. (2014). Turizm rotalarının oluşturulmasında coğrafi bilgi sistemleri kullanımı. III. Disiplinlerarası Turizm Araştırmaları Kongresi.

Kessler, K., & Raj, R. (2018). The Rise of Heritage and Religious Tourism to Sacred Sites in Oman. In S. Yasuda, R. Raj & K. Griffin (Eds.), Religious tourism in Asia: Tradition and change through case studies and narratives (pp. 10-20). Wallington: CABI. doi:10.1079/9781786392343.0010

Kessler, K. (2015). Conceptualizing mosque tourism: A central feature of Islamic and religious tourism. *International Journal of Religious Tourism and Pilgrimage*, *3*(2), 11–32.

Khan, F., & Callanan, M. (2017). The halalification of tourism. *Journal of Islamic Marketing*, *8*(4), 558–577. doi:10.1108/JIMA-01-2016-0001

Khawas, V. (2002). *Environment and rural development in Darjeeling Himalaya: Issues and concerns.* Retrieved from https://www.academia.edu/2648241/Environment_and_rural_development_(2019, September 1)

Khawas, V. (2006). Status of tea garden labourers in eastern Himalaya: A case of Darjeeling tea industry. Council for Social Development, Sangha Rachana, New Delhi, India.

Khreis, S. (2018). Evaluation of social communication skills of the tour guide in Jordan as perceived by international tourists: an exploratory study in the archaeological site of Petra. *Dirasat, Human And Social Sciences, 45*(4).

Kimberly, J. R., & Evanisko, M. J. (1981). Organizational innovation: The influence of individual, organizational, and contextual factors on hospital adoption of technological and administrative innovations. *Academy of Management Journal*, *24*, 689–713. PMID:10253688

Kim, H. J., Chen, M. H., & Jang, S. C. S. (2006). Tourism expansion and economic development: The case of Taiwan. *Tourism Management*, *27*(5), 925–933. doi:10.1016/j.tourman.2005.05.011 PMID:32287716

Kim, H., & Park, E. J. (2013). The role of social experience in undergraduates' career perceptions through internships. *Journal of Hospitality, Leisure, Sport and Tourism Education*, *12*(1), 70–78. doi:10.1016/j.jhlste.2012.11.003

Kim, Y. H. (2013). An Assessment of Sport Event Tourists' Motivation with a Framework: A Case Study at a Southeast Conference Football Game. *Journal of Tourism Research & Hospitality*, *2*(2). Advance online publication. doi:10.4172/2324-8807.1000115

Kirkup, N., & Sutherland, M. (2015). Exploring the relationships between motivation, attachment and loyalty within sport event tourism. *Current Issues in Tourism*, *20*(1), 7–14. doi:10.1080/13683500.2015.1046819

Kirshenblatt-Gimblett, B. (1998). *Destination culture: Tourism, museums, and heritage*. University of California Press.

Kişi, N. (2019). A Strategic Approach to Sustainable Tourism Development Using the A'WOT Hybrid Method: A Case Study of Zonguldak, Turkey. *Sustainability*, *11*(4), 964. doi:10.3390u11040964

Kitiarsa, P. (2010). Toward a Sociology of Religious Commodification. In B. S. Turner (Ed.), *The new Blackwell companion to the sociology of religion* (pp. 563–579). Wiley-Blackwell. doi:10.1002/9781444320787.ch25

Kitterlin, M., Tanke, M., & Stevens, D. P. (2016). Workplace bullying in the foodservice industry. *Journal of Foodservice Business Research*, *19*(4), 413–423. doi:10.1080/15378020.2016.1185874

Kivela, J., & Crotts, J. C. (2005). Gastronomy tourism: A meaningful travel market segment. *Journal of Culinary Science & Technology*, *4*(2), 39–55. doi:10.1300/J385v04n02_03

Klein, A. (2004). Reinventing the campus tour. *University Business*, *7*(11), 51–53.

Konstantinos, D. (2001). *The World of the Nabataeans. Volume 2 of the International Conference: The World of the Herods and the Nabataeans held at the British Museum.* Academic Press.

Koo, C., Shin, S., Gretzel, U., Hunter, W. C., & Chung, N. (2016). Conceptualization of smart tourism destination competitiveness. *Asia Pacific Journal of Information Systems*, *26*(4), 561–576. doi:10.14329/apjis.2016.26.4.561

Koroglu, A., Koroglu, O., & Sarioglan, M. (2007). An Empirical Study on the Role of Professional Tourist Guides in Promoting Turkish Cuisine. *1st National Gastronomy Symposium and Artistic Activities.*

Koththagoda, K. C., & Thushara, S. C. (2016). Customer based brand equity for empowering Sri Lanka as a tourism destination: With special reference to tea tourism. *Kelaniya Journal of Management, 5*(1) January-June 2016.

Kounavis, C. D., Kasimati, A. E., & Zamani, E. D. (2012). Enhancing the tourism experience through mobile augmented reality: Challenges and prospects. *International Journal of Engineering Business Management, 4*, 1–6. doi:10.5772/51644

Kramer, R., Modsching, M., Hagen, K., & Gretzel, U. (2007). Behavioural impacts of mobile tour guides. In M. Sigala, L. Mich, & J. Murphy (Eds.), *Information and communication technologies in tourism* (pp. 109–118). Springer.

Kuflik, T., Wecker, A., Lanir, J., & Stock, O. (2015). An integrative framework for extending the boundaries of the museum visit experience: Linking the pre, during, and post visit phases. *Information Technology & Tourism, 15*(1), 7–47. doi:10.100740558-014-0018-4

Kültür Rotaları Derneği. (2019). http://cultureroutesinturkey.com/c/

Kumar, A., & Rana, V. S. (2016). Food as an Attraction for Tourism: A Study of Patiala City, Punjab. *IICMR Research Journal, 11*(1), 7–11.

Kumar, P., Badal, P. S., Singh, N. P., & Singh, R. P. (2008, January-March). Tea industry in India: Problems and prospects. *Indian Journal of Agricultural Economics, 63*(1).

Kuo, N. T., Chang, K. C., Cheng, Y. S., & Lin, J. C. (2016). Effects of tour guide interpretation and tourist satisfaction on destination loyalty in Taiwan's Kinmen battlefield tourism: Perceived playfulness and perceived flow as moderators. *Journal of Travel & Tourism Marketing, 33*(1), 103–122. doi:10.1080/10548408.2015.1008670

Kuo, N. T., Cheng, Y. S., Chang, K. C., & Chuang, L. Y. (2018). The asymmetric effect of tour guide service quality on tourist satisfaction. *Journal of Quality Assurance in Hospitality & Tourism, 19*(4), 521–542. doi:10.1080/1528008X.2018.1483283

Kusluvan, S. (2003). *Managing employee attitudes and behaviours in the tourism and hospitality*. Nova Science.

Kusluvan, S., & Kusluvan, Z. (2003). Perceptions and attitudes of undergraduate tourism students towards working in the tourism and hospitality industry in a developing economy. In S. Kusluvan (Ed.), *Managing employee attitudes and behaviours in the tourism and hospitality industry* (pp. 76–98). Nova Science.

Lackey, N. Q., & Pennisi, L. (2019). Ecotour guide training program methods and characteristics: a case study from the African bush. *Journal of Ecotourism*. Retrieved from https://www.tandfonline.com/doi/full/10.1080/14724049.2019.1686005

Lamsfus, C., & Alzua-Sorzabal, A. (2013). Theoritical framework for a tourism internet of things: Smart destinations. *Journal of Tourism and Human Mobility, 2*, 15–21.

Lamsfus, C., Martín, D., Alzua-Sorzabal, A., & Torres, M. E. (2015). Smart tourism destinations: an extended conception of smart cities focusing on human mobility. In I. Tussyadiah & A. Inversini (Eds.), *Information and communication technologies in tourism 2015* (pp. 363–375). Springer. doi:10.1007/978-3-319-14343-9_27

Landry, C., Greene, L., Matarasso, F., & Biancini, F. (1996). *The Art of urban regeneration: Urban renewal through cultural activity*. Comedia.

Larrinaga, C. (2005). A century of tourism in northern Spain: the development of high-quality provision between 1815 and 1914. In J. K. Walton (Ed.), *Histories of tourism representation, identity and conflict* (pp. 88–103). Channel View Publications. doi:10.21832/9781845410339-007

Laura, N. (2008). *Food tourism in India.* Retrieved December 10, 2013 from the website http: // traveltips.usatoday.com/food-tourism-india-18335.html

Leary, R. (1995). *Social psychology series. Self-presentation: Impression management and interpersonal behaviour.* Brown & Benchmark Publishers.

Leighton, D. S. R. (1970). The Internationalization of American business: The third industrial revolution. *Journal of Marketing*, *34*(3), 3–6.

Leymann, H. (1996). The content and development of mobbing at work. *European Journal of Work and Organizational Psychology*, *5*(2), 165–184. doi:10.1080/13594329608414853

Licciardi, G., & Amirtahmasebi, R. (Eds.). (2012). *The economics of uniqueness: investing in historic city cores and cultural heritage assets for sustainable development.* The World Bank. doi:10.1596/978-0-8213-9650-6

Lim, C. C., & Bendle, L. J. (2012). Arts tourism in Seoul: Tourist orientated performing arts as a sustainable niche market. *Journal of Sustainable Tourism*, *20*(5), 667–682. doi:10.1080/09669582.2011.636817

Lin, Y. C., Pearson, T. E., & Cai, L. A. (2011). Food as a form of destination identity: A tourism destination brand perspective. *Tourism and Hospitality Research*, *11*(1), 30–48. doi:10.1057/thr.2010.22

Liu, P., & Liu, Y. (2016). Smart tourism via smart phone. *Proceedings of International Conference on Communications*, 129-132).

Lloyd, R. (2002). Neo-bohemia: Art and neighborhood redevelopment in Chicago. *Journal of Urban Affairs*, *24*(5), 517–532. doi:10.1111/1467-9906.00141

Locke, K. (2001). *Grounded Theory in Management Research.* Sage publications.

Locock, L., Dopson, S., Chambers, D., & Gabbay, J. (2001). Understanding the role of opinion leaders in improving clinical effectiveness. *Social Science & Medicine*, *53*(6), 745–757. doi:10.1016/S0277-9536(00)00387-7 PMID:11511050

Lomine, L. (2005). Tourism in Augustan society (44 BC–AD 69). In J. K. Walton (Ed.), *Histories of tourism: Representation, identity and conflict* (pp. 69–87). Channel View Publications. doi:10.21832/9781845410339-006

Long, L. M. (1998). Culinary tourism: A folkloristic perspective on eating and otherness. *Southern Folklore*, *55*(3), 181.

Long, P. E., & Morpeth, N. (Eds.). (2016). *Tourism and the creative ındustries: Theories, policies and practice.* Routledge. doi:10.4324/9781315735870

Lopez de Avila, A. (2015). Smart destinations: XXI century tourism. *Proceedings of ENTER2015 Conference on Information and Communication Technologies in Tourism.*

Lowrie, K., & Frisch, M. (2007). *The arts community, arts village development and promotion of arts in woodbridge township.* National Center For Neighborhood And Brownfields Redevelopment.

Lucas, R., & Johnson, K. (2003). Managing Students As A Flexible Labor Resource In Hospitality and Tourism In Centraland Eastern Europe and the U. K. In S. Kusluvan (Ed.), *Managing Employee Attitudes And Behaviors In The Tourism and Hospitality Industry* (pp. 153–170). Nova Publishers.

Lu, J., Mao, Z., Wang, M., & Hu, L. (2015). Goodbye maps, hello apps? exploring the influential determinants of travel app adoption. *Current Issues in Tourism, 18*(11), 1059–1079. doi:10.1080/13683500.2015.1043248

Lynch, P., Molz, J. G., Mcintosh, A., Lugosi, P., & Lashley, C. (2011). Theorizing hospitality. *Hospitality & Society, 1*(1), 3–24. doi:10.1386/hosp.1.1.3_2

MacCannell, D. (1973). 'Staged Authenticity: Arrangement of Social Space in Tourist Settings'. *American Journal of Sociology, 79*(3), 589–603. doi:10.1086/225585

Macdonald, S. (2006). Mediating heritage: Tour guides at the former Nazi Party Rally Grounds, Nuremberg. *Tourist Studies, 6*(2), 119–138. doi:10.1177/1468797606071473

Mackenzie, J. M. (2005). Empires of travel: British guide books and cultural imperialism in the 19th and 20th centuries. In J. K. Walton (Ed.), *Histories of tourism representation, identity and conflict* (pp. 19–38). Channel View Publications. doi:10.21832/9781845410339-003

Madsen, E. S., Bilberg, A., & Grube Hansen, D. (2016, September). Industry 4.0 and digitalization call for vocational skills, applied industrial engineering, and less for pure academics. *Proceedings of the 5th P&OM World Conference P&OM.*

Magdola, P. M. (2000). The Campus Tour: Ritual and Community in Higher Education. *Anthropology & Education Quarterly, 31*(1), 24–46. doi:10.1525/aeq.2000.31.1.24

Magnini, V. P. (2009). Understanding and reducing work-family conflict in the hospitality industry. *Journal of Human Resources in Hospitality & Tourism, 8*(2), 119–136. doi:10.1080/15332840802269643

Magolda, P. M. (2001). What our Rituals Tell us about Community on Campus: A Look at the Campus Tour. *About Campus: Enriching the Student Learning Experience, 5*(6), 2–8. doi:10.1177/108648220100500602

Mak, A. H. N., Lumbers, M., & Eves, A. (2012). Globalization and Food Consumption in Tourism. *Annals of Tourism Research, 39*(1), 171–196. doi:10.1016/j.annals.2011.05.010

Mak, A. H. N., Wong, K. K. F., & Chang, R. C. Y. (2011). Critical issues affecting the service quality and professionalism of the tour guides in Hong Kong and Macau. *Tourism Management, 32*(6), 1442–1452. doi:10.1016/j.tourman.2011.01.003

Malaka, R., & Zipf, A. (2000). Deep map: challenging IT research in the framework of a tourist information system. In *Proceedings of the 7th International Conference in Information and Communication Technologies in Tourism* (pp.15-27). 10.1007/978-3-7091-6291-0_2

Mampaey, J., Huisman, J., & Seeber, M. (2015). Branding of Flemish higher education institutions: A strategic balance perspective. *Higher Education Research & Development, 34*(6), 1178–1191. doi:10.1080/07294360.2015.1024634

Mancini, M. (2001). *Conducing tours* (3rd ed.). Delmar Thomson Learning.

Mancini, M. (2000). *Conducting tours: A practical guide* (3rd ed.). Nelson Education.

Mandala, L. (2015). Arts, tourism & cultural diplomacy. In C. Lord (Ed.), Arts and America: Arts, culture, and the future of America's communities (pp.129-140). ABD: Americans for the Arts.

Mansfeld, Y., & Pizam, A. (2006). *Tourism security and safety from theory to practice.* Elsiever Butterworth-Heineman.

Mansfeld, Y., Ron, A., & Gev, D. (2000). *Muslim Tourism to Israel—Characterization, Trends and Potential.* University of Haifa: Center for Tourism, Pilgrimage, and Recreation Research. (in Hebrew)

Markham, A. N. (1998). Life online: Researching real experiences in virtual space. In C. Ellis & A. P. Bochner (Eds.), *Ethnographic Alternatives Book Series* (pp. 244–264). Altamira Press.

Markusen, A. (2006). Urban development and the politics of a creative class: Evidence from a study of artists. *Environment & Planning A, 38*(10), 1921–1940. doi:10.1068/a38179

Markusen, A., Gilmore, S., Johnson, A., Levi, T., & Martinez, A. (2006). *Crossover, how artists build careers across commercial, nonprofit and community work*. The Arts Economy Initiative, Project on Regional and Industrial Economics, Humphrey Institute of Public Affairs, University of Minnesota.

Markusen, A., & Schrock, G. (2006). The Artistic Dividend: Urban Artistic Specialisation and Economic Development Implications. *Urban Studies (Edinburgh, Scotland), 43*(10), 1661–1686. doi:10.1080/00420980600888478

Martín, H. S., & Herrero, A. (2011). Influence of the user's psychological factors on the online purchase intention in rural tourism: Integrating innovativeness to the UTAUT framework. *Tourism Management, 33*(2), 341–350. doi:10.1016/j.tourman.2011.04.003

Martin, W. H., & Mason, S. (2004). Leisure in an Islamic context. *World Leisure Journal, 46*(1), 4–13. doi:10.1080/04419057.2004.9674343

Masseno, M. D., & Santos, C. (2018). Smart tourism destinations priivacy risks on data protection. *Revista Eletrônica Sapere Aude, 1*(1), 125–149.

Mathews, V. (2010). Aestheticizing space: Art, gentrification and the city. *Geography Compass, 4*(6), 660–675. doi:10.1111/j.1749-8198.2010.00331.x

McBride, J. (2018). *The Economics of Hosting the Olympic Games*. Retrieved July 25, 2019, from https://www.cfr.org/backgrounder/economics-hosting-olympic-games

McCance, M. (2019). *UVA Today Series Presents Stories Explored by Commission on Segregation*. Retrieved from: https://news.virginia.edu/content/uva-today-series-presents-stories-explored-commission-segregation

McCartney, G., Butler, R., & Bennett, M. (2008). A strategic use of the communication mix in the destination image-formation process. *Journal of Travel Research, 47*(2), 183–196. doi:10.1177/0047287508321201

McDonald, M. A., Milne, G. R., & Hong, J. B. (2002). Motivational factors for evaluating sport spectator and participant market. *Sport Marketing Quarterly, 11*(2), 100–113.

McMahon, L. (2000). Bullying and harassment in the workplace. *International Journal of Contemporary Hospitality Management, 12*(6), 384–387. doi:10.1108/09596110010343666

Melewar, T. C., & Alwi, S. F. S. (2017). *Islamic marketing and branding: Theory and practice*. Routledge. doi:10.4324/9781315590035

Merriam-Webster. (2019). *Sözlük*. https://www.merriam-webster.com/dictionary/trek

Mesaritou, E., Coleman, S., & Eade, J. (2016). Introduction: Guiding the pilgrim. *Tourist Studies, 16*(1), 3–22. doi:10.1177/1468797615588425

Meyer, M., Johnson, D., & Ethington, C. (1997). Contrasting attributes of preventive health innovations. *Journal of Communication, 47*(2), 112–131. doi:10.1111/j.1460-2466.1997.tb02709.x

Miles, M. (2013). Art and culture: The global turn. In E. Grierson & K. Sharp (Eds.), *Re-imagining the city: Art, Globalization and Urban Spaces* (pp. 19–38). University of Chicago Press.

Miles, M., & Huberman, M. (1994). *Qualitative Data Analysis: an Expanded Sourcebook*. SAGE.

Miller, R. K., & Washington, K. D. (2006). *The 2007 travel & tourism market research handbook* (10th ed.). Richard K. Miller & Associates.

Ming, L. (2007). *Research in Campus Tourism: Actuality, Problems and Advice - a Literature Review of 10 Years Research in Campus Tourism in China.* doi: CNKI:SUN:ZTKB.0.2007-06-030

Ministry of Tourism and Antiquities. Jordan Government. (1998). *Legislations.* Retrieved from http://www.mota.gov.jo/Contents/Law_Regulations.aspx

Ministry of Tourism -MoT. (2019). *Tour guide Training Programs.* Retrieved from MoT website on 18 November, 2019, from http://tourism.gov.in/search/node/guide

Mohanty, P. (2014). Culinary Tourism In India: An Emerging Concept. *Abhinav Journal, 3*(5), 44–53.

Mohsin A. (2005). Tourist attitudes and destination marketing— the case of Australia's Northern Territory and Malaysia. *Tourism Management, 26*(5), 723– 732.

Mohsin, A., Ramli, N., & Alkhulayfi, B. A. (2016). *Halal Tourism: Emerging Opportunities, Tourism Management Perspectives* (Vol. 19). Part B.

Mok, C., Sparks, B., & Kadampully, J. (2013). *Service quality management in hospitality, tourism, and leisure.* Routledge. doi:10.4324/9780203047965

Moscarado, G. (1998). Interpretation and sustainable tourism: Functions, examples and principles. *Journal of Tourism Studies, 9*(1), 2–13.

Mulvaney, R., O'Neill, J., Cleveland, J., & Crouter, A. (2006). A model of work-family dynamics of hotel manager. *Annals of Tourism Research, 34*(1), 66–87. doi:10.1016/j.annals.2006.07.002

Muñoz, O. (1995). Interpretation guidelines. In D. Hawkins, M. Epler Wood, & S. Bittman (Eds.), *The Ecolodge Sourcebook.* Ecotourism Society.

Murgado, E. M. (2013). Turning food in to a gastronomic experience: Olive oil tourism. In N. Arcas, F. N. Arroyo López, J. Caballero, R. D'Andria, M. Fernández, E. R. Fernandez, A. Garrido, J. López-Miranda, M. Msallem, M. Parras, L. Rallo, & R. Zanoli (Eds.), *Present and future of the Mediterranean olive sector* (pp. 97–109). CIHEAM.

Murugadoss, K., Nagarathinam, S. R., Vinothkanna, S., & Soundarraj, H. (2016). A Study on Tourism Potential in Munnar Region, Kerala. *IJIRSET, 5*(12), 262–271.

Mussina, K. P., & Oryngazhiyeva, M. R. (2018). Smart cities as a tourist destination for smart travelers. *Innovative Economy: Perspectives of Development and Improvement, 6*(32), 12–16.

Mylod, E. J. (2013). *Latin Christian Pilgrimage in the Holy Land, 1187-1291* (Unpublished Dissertation). University of Leeds.

Mzakki, A. (2008). Islam as a symbolic commodity: Transmitting and consuming Islam through public sermons in Indonesia. In P. Kitiarsa (Ed.), *Religious commodification in Asia: Marketing gods* (pp. 205–219). Routledge.

Nagoor, C., & Kumar, N. (2010). Assessing the impact of the ASEAN-India FTA on the Tea Industry. *Economic and Political Weekly, 30*(44), 112–116.

Neirotti, D. L., Bosetti, H. A., & Teed, K. C. (2001). Motivation to attend the 1996 Summer Olympic Games. *Journal of Travel Research, 39*(3), 327–331. doi:10.1177/004728750103900315

Neumeier, S. & Pollermann, K. (2014). Rural Tourism as Promoter of Rural Development – Prospects and Limitations: Case Study Findings from a Pilot Project promoting Village Tourism. *European Countryside, 6*(4).

New Straits Times. (2014, Mar. 19). Attractive tour packages with Tabung Haji Travel. *New Straits Times.*

Nezakati, H., Chin, M. C., & Akhoundi, M. (2013). Factors Influencing People to Travel Abroad for Sport Tourism (Malaysia Evidence*). Journal of Applied Sciences (Faisalabad), 13*(24), 5708–5715. doi:10.3923/jas.2013.5708.5715

NFC-Technology. (n.d.). Retrieved from https://www.smart-tec.com/en/auto-id-world/nfc-technology

Nguyen, H. L. & University of Western Sydney. (2015). *The Impacts of Tour Guide Performance on Foreign Tourist Satisfaction and Destination Loyalty in Vietnam.* Sydney, Australia: University of Western Sydney.

Nicholson, R., & Pearce, D. (2000). Who Goes to Events: A Comparative Analysis of the Profile Characteristics of Visitors to Four South Island Events in New Zealand. *Journal of Vacation Marketing, 6*(3), 236–253. doi:10.1177/135676670000600304

Nickson, D. (2013). *Human resource management for hospitality, tourism and events.* Routledge. doi:10.4324/9780080966496

Nillson, M., & Mekonnen, T. (2016). Performing the "post-secular" in Santiago de Compestela. *Annals of Tourism Research, 57*, 18–30. doi:10.1016/j.annals.2015.11.001

Nilüfer Belediyesi. (2019). *Mysia Yolları.* http://www.nilufer.bel.tr/dosya_yoneticisi/mysia.pdf

Noe, F., Uysal, M., & Magnini, V. (2010). *Tourist customer service satisfaction an encounter approach.* Routledge. doi:10.4324/9780203852361

Nyaupane, P., & Timothy, D. J. (2013). Assessing visitors' preferences of various roles of tour guides in the Himalaya. *Tourism Analysis, 18*(1), 45–49. doi:10.3727/108354213X13613720283647

O'Gorman, K. (2010). Historical giants forefathers of modern hospitality and tourism. In K. O'Gorman (Ed.), *Tourism Giants* (pp. 3–17). CABI. doi:10.1079/9781845936525.0003

O'Mara, M. P. (2012). Beyond town and gown: University economic engagement and the legacy of the urban crisis. *The Journal of Technology Transfer, 37*(2), 234–250. doi:10.100710961-010-9185-4

OIC (Organization of Islamic Corporation). (n.d.). Retrieved from https://www.oic-oci.org/

Olsen, D. H. (2018). Religion, pilgrimage and tourism in the Middle East. In D. Timothy (Eds.), Routledge Handbook on Tourism in the Middle East and North Africa (pp. 109-124). London: Routledge.

Olsen, D. H. (2013). A scalar comparison of motivations and expectations of experience within the religious tourism market. *International Journal of Religious Tourism and Pilgrimage, 1*(1), 1–5.

Onetikk. (2013, January 31). Archive.Today: Himachali Cuisine, A Pleasantly Surprising Fusion [Web log post]. Retrieved from https://archive.is/20130216144406/http:// www.onetikk.com/blog/1358/himachali-cuisine-a-plesantly-suprising-fusion/

Ontario Ministry of Tourism. (2009). *OCHTP: Ontario Cultural and Heritage Tourism Product Research Paper.* Retrieved from Ontario Ministry of Tourism website on 05 April, 2018, from http://www.mtc.gov.on.ca/en/publications/Ontario_ Cultural_and_Heritage_Tourism.pdf

Orams, M. B. (1999). *Marine tourism: Development, impacts and management.* Routledge.

Orbasli, A. (2008). *Architectural Conservation. Principles and Practice.* Blackwell.

Ormsby, A., & Mannle, K. (2006). Ecotourism benefits and the role of local guides at Masoala National Park, Madagascar. *Journal of Sustainable Tourism*, *14*(3), 271–287. doi:10.1080/09669580608669059

Öter, Z. (2019). Guided Tour. In Encyclopedia of Tourism–Turkey: The Basic Concepts of Tourism and Hospitality Industry. Detay Publishing. (in Turkish)

Outman, J. L., & Outman, E. M. (2003). *Industrial revolution.* Thomson Learning.

Overend, D. (2012). Performing Sites: Illusion and Authenticity in the Spatial Stories of the Guided Tour. *Scandinavian Journal of Hospitality and Tourism*, *12*(1), 44–54. doi:10.1080/15022250.2012.678070

Oxford City Council. (2017). *Listed buildings - overview.* Retrieved 7 24, 2017, from https://www.oxford.gov.uk/info/20192/listed_buildings/867/listed_buildings_-_overview

Özel, M. K. (2011). Kent strüktürü ile tapınma yapıları arasındaki ilişki bağlamında Ayvalık Hamidiye Camisi. *MSGSÜ Mimarlık Fakültesi Dergisi*, *7*(11-12), 15–39.

Palumbo, F., Dominici, G., & Basile, G. (2014). The culture on the palm of your hand: How to design a user oriented mobile app for museums. In L. Aiello (Ed.), Management of cultural products: e-relationship marketing and accessibility perspectives (pp. 224-243). IGI Global.

Parsons, H., Houge Mackenzie, S., & Filep, S. (2019). Facilitating self-development: How tour guides broker spiritual tourist experiences. *Tourism Recreation Research*, *44*(2), 141–152. Advance online publication. doi:10.1080/02508281.2019.1582159

Patton, M. (2002). *Qualitative Evaluation and Research Methods* (3rd ed.). SAGE.

Pearce, P. (1984). Tourist- guide interaction. *Annals of Tourism Research*, *11*(1), 129–146. doi:10.1016/0160-7383(84)90100-2

Pengiran-Bagul, A. H. B. (2013). *Art tourism: An exploratory study in Sabah, Malaysia.* Retrieved October, 24, 2020, from http://www.ijbts-journal.com/images/main_1366796758/0083-Awangku.pdf

Pereira Brando Albino, S. (2015). *Tourism in university cities: the role of universities in place branding* (Unpublished Doctoral Dissertation). Instituto de Geografia e Ordenamento do Território. Lisbon: Universidade De Lisboa.

Perna, F., Custódio, M. J., & Oliveira, V. (2019). Local Communities and Sport Activities Expenditures and Image: Residents' Role in Sustainable Tourism and Recreation. *European Journal of Tourism. Hospitality and Recreation.*, *9*(1), 49–59. doi:10.2478/ejthr-2019-0006

Pevsner, N., & Williamson, E. (1985). *The Buildings of England. County Durham* (2nd ed.). Yale University Press.

Pew Research Centre. (2015). *The Future of World Religions: Population Growth Projections, Year 2010-2050.* https://www.pewforum.org/2015/04/02/religious-projections-2010-2050/

Pink, J. (Ed.). (2009). Muslim Societies in the Age of Mass Consumption: Politics, Culture and Identity between the Local and the Global. Tyne Newcastle: Cambridge Scholars.

Pizam, A. (1982). Tourism manpower: The state of the art. *Journal of Travel Research*, *21*(2), 5–9. doi:10.1177/004728758202100204

Pizam, A., & Reichel, A. (1996). The effect of nationality on tourist behavior: Israeli tour-guides' perceptions. *Journal of Hospitality & Leisure Marketing*, *4*(1), 23–49. doi:10.1300/J150v04n01_03

Plog, S. (2001). Why destination areas rise and fall in popularity, An update of a Cornell Quarterly Classic. *The Cornell Hotel and Restaurant Administration Quarterly*, *42*(3), 13–24. doi:10.1016/S0010-8804(01)81020-X

Plummer, R., Telfer, D., Hashimoto, A., & Summers, R. (2005). Beer tourism in Canada along the Waterloo-Wellington Ale Trail. *Tourism Management*, *26*(3), 447–458. doi:10.1016/j.tourman.2003.12.002

Pocock, D. (2013). *The Story of Durham*. The History Press.

Pond, K. (1993). *The professional guide*. VNR.

Popescu, A. I. (2012). Branding Cities as Educational Centres: The Role of Higher Education Institutions. *Management and Marketing. Challenges for the Knowledge Society*, *7*(3), 493–510.

Portalés, C., Lerma, J. L., & Pérez, C. (2009). Photogrammetry and augmented reality for cultural heritage applications. *The Photogrammetric Record*, *24*(128), 316–331. doi:10.1111/j.1477-9730.2009.00549.x

Poslad, S., Laamanen, H., Malaka, R., Nick, A., Buckle, P., & Zipf, A. (2001). Crumpet: creation of user-friendly mobile services personalised for tourism. In *Proceedings of the 3G Mobile Communication Technologies* (pp. 28- 32). 10.1049/cp:20010006

Poudel, S., & Nyaupane, G. P. (2013). The role of interpretive tour guiding in sustainable destination management: A comparison between guided and non-guided tourists. *Journal of Travel Research*, *52*(5), 659–672. doi:10.1177/0047287513478496

Prakash, M., & Chowdhury, N. (2010). Tour guides: Roles, challenges and desired competences. A review of literature. *International Journal of Hospitality and Tourism Systems*, *3*(1).

Pratt, A. C. (2008). Creative cities: The cultural industries and the creative class. *Geografiska Annaler. Series B, Human Geography*, *90*(2), 107–117. doi:10.1111/j.1468-0467.2008.00281.x

Puvaneswary, S. (2017). *Brunei takes serious look at Islamic Tourism*. https://www.ttgasia.com/2017/11/16/brunei-takes-serious-look-at-islamic-tourism/

Rabotic, B. (2010). Tourist guides in contemporary tourism. *International Conference on Tourism and Environment*, 1-10.

Rahman, N. A. A. (2019a). Prominent and Innovative Tourism Events in Peninsular Malaysia in Routledge Advances in Event Research. In *Tourism Events in Asia, Marketing and Development*. Routledge.

Rahman, N. A. A. (2019b). Halal Business and Halal Technology in Korea: History, Concepts and Research Opportunities. *Test Engineering and Management*, *81*(Nov- Dec), 5782–5790.

Rahman, N. A. A., Mohamad, M. F., Muda, J., Majid, Z. A., & Noh, H. M. (2018a). Linking halal requirement and branding: An examination of halal flight kitchen provider in Malaysia. *International Journal of Supply Chain Management*, *7*(3), 208–215.

Rahman, N.A., Abdul, M.F., Rahim, S.A., & Noh, H.M. (2018b). Implementing Air Cargo Halal Warehouse: Insight from Malaysia. *Journal of Islamic Marketing*, *9*(3).

Raj, R., & Morpeth, N. D. (Eds.). (2015). *Religious Tourism and Pilgrimage Management: An International Perspective*. CABI Publishing. doi:10.1079/9781780645230.0000

Rakic, T., & Lester, J. A. (2013). *Travel, tourism and art*. Ashgate.

Randall, C., & Rollins, R. B. (2009). Visitor perceptions of the role of tour guides in natural areas. *Journal of Sustainable Tourism*, *17*(3), 357–374. doi:10.1080/09669580802159727

Rand, G. D., Heath, E., & Alberts, N. (2003). The role of local and regional food in destination marketing: A South African situation analysis. *Journal of Travel & Tourism Marketing*, *14*(3-4), 97–112. doi:10.1300/J073v14n03_06

Raut, A., & Bhakay, J. (2012). Wine tourism in Maharashtra: Problems and solutions. *Asian Journal of Food and Agro-Industry*, *5*(2), 141–155.

Raviv, Y. (2015). *Falafel Nation: Cuisine and the making of national identity in Israel* (H. Hartman, Ed.). Nebraska Press. doi:10.2307/j.ctt1d989k8

Rawashdeh, A. A. (2012). Eco-tourism in Ajloun Forest Reserve: An Investigative Study. *Jordanian Journal for Social Science*, *5*, 151–170.

Rayner, C., & Hoel, H. (1997). A Summary Review of Literature Relating to Workplace Bullying. *Journal of Community & Applied Social Psychology*, *7*(3), 181–191. doi:10.1002/(SICI)1099-1298(199706)7:3<181::AID-CASP416>3.0.CO;2-Y

Razalli, M. R., Yusoff, R. Z., & Roslan, M. W. M. (2013). A framework of halal certification practices for hotel industry. *Asian Social Science*, *9*(11), 316.

Razzaq, S., Hall, C. M., & Prayaq, G. (2016). *The Capacity of New Zealand to accommodate the Halal Tourism market-or not*. Academic Press.

Reader, I. (2013). *Pilgrimage in the marketplace*. Routledge. doi:10.4324/9781315885704

Reisinger, Y., & Steiner, C. (2006). Reconceptualising interpretation: The role of tour guides in authentic tourism. *Current Issues in Tourism*, *9*(6), 481–498. doi:10.2167/cit280.0

Reknes, I., Pallesen, S., Magerøy, N., Moen, B. E., Bjorvatn, B., & Einarsen, S. (2014). Exposure to bullying behaviors as a predictor of mental health problems among Norwegian nurses: Results from the prospective SUSSH-survey. *International Journal of Nursing Studies*, *51*(3), 479–487. doi:10.1016/j.ijnurstu.2013.06.017 PubMed

Rey-Lopez, M., Barragans-Martinez, A., Peleteiro, A., Mikic-Fonte, F., & Burguillo, J. (2011). More tourism: mobile recommendations for tourism. In *Proceedings of the IEEE International Conference on Consumer Electronics* (pp. 347-348). Berlin, Germany: IEEE.

Reynolds, P. (1993). Food and Tourism: Towards an understanding of sustainable culture. *Journal of Sustainable Tourism*, *1*(1), 48–54. doi:10.1080/09669589309450700

Richards, G. (2007). Global Trends in cultural tourism. In G. Richards (Ed.), *Cultural tourism: global and local perspectives* (pp. 1–24). Haworth Hospitality Press.

Richardson, S., & Butler, G. (2012). Attitudes of Malaysian tourism and hospitality students' towards a career in the industry. *Asia Pacific Journal of Tourism Research*, *17*(3), 262–276. doi:10.1080/10941665.2011.625430

Rifkin, J. (2011). *The third industrial revolution: how lateral power is transforming energy, the economy, and the world.* Palgrave Macmillan.

Riley, M., Ladkin, A., & Szivas, E. (2002). *Tourism employment analysis and planning: Aspects of tourism* (Vol. 6). Channel View Publications. doi:10.21832/9781873150382

Rimmington, M., & Kozak, M. (1997). Developments in information technology: Imlications for the tourism industry and tourism marketing. *Anatolia: An International Journal of Tourism Hospitality Research*, *8*(3), 59–80. doi:10.1080/13032917.1997.9687121

Rinaldi, G. (2011). *Sport tourism: An exploration of the motivations and tourism experiences of Australian Football League interstate sport tourists travelling to Western Australia* (Thesis). Edith Covan University. Retrieved September 30, 2019, from https://ro.ecu.edu.au/theses_hons/34

Ringer, G. (2013). *Destinations: Cultural landscapes of tourism.* Routledge. doi:10.4324/9780203441381

Rinschede, G. (1992). Forms of religious tourism. *Annals of Tourism Research, 19*(1), 51–67. doi:10.1016/0160-7383(92)90106-Y

Ristova, C., Popeska, B., & Dimitrov, N. (2019). Possibilities for development of sport-recreational tourism of Kozuf and Plackovica Mountains. *The 11th International Scientific Conference "Modern Trends of Physical Education and Sport".*

Ritchie, B., & Adair, D. (2002). The Growing Recognition of Sport Tourism. *Current Issues in Tourism, 5*(1), 1–6. doi:10.1080/13683500208667903

Robert, H., and Eryurt, H. (2013). Culture Routes in Turkey. *e-dialogos, 3*, 6-7.

Roberts, B. R. (2015). *The third industrial revolution: iimplications for planning cities and regions.* Urban Frontiers Working Paper 1.

Roberts, M. (2003). *Durham. 1000 years of history.* Tempus Publishing Ltd.

Robinson, S. L., & Bennett, R. J. (1995). A Typology of Deviant Workplace Behaviors: A Multidimensional Scaling Study. *Academy of Management Journal, 38*(2), 555–572. doi:10.5465/256693

Rodrigo, P., & Turnbull, S. (2019). Halal Holidays: How is value perceived by Muslim tourist? *International Journal of Tourism Research, 21*(5), 675–692. doi:10.1002/jtr.2290

Rodrigues, J. M., Ramos, C. M., Cardoso, P. J., & Henriques, C. (2017). *Handbook of research on technological developments for cultural heritage and etourism applications.* IGI Global.

Rogers, E. M. (1962). *Diffusion of innovations.* The Free Press.

Rogers, E. M. (1995). *Diffusion of Innovations* (3rd ed.). Free Press.

Rojko, A. (2017). Industry 4.0 concept: Background and overview. *International Journal of Interactive Mobile Technologies, 11*(5), 77–90. doi:10.3991/ijim.v11i5.7072

Ross, G. F. (1993). Tourism and hospitality work interest and motivation among potential employees. *Hospitality Research Journal., 16*(2), 17–27. doi:10.1177/109634809301600204

Ross, G. F. (1997). Career stress responses among hospitality employees. *Annals of Tourism Research, 24*(1), 41–51. doi:10.1016/S0160-7383(96)00032-1

Roussou, M., Trahanias, P., Giannoulis, G., Kamarinos, G., Argyros, A., Tsakiris, D., ... Schulz, D. (2001, November). Experiences from the use of a robotic avatar in a museum setting. *Proceedings of the 2001 conference on Virtual reality, archeology, and cultural heritage*, 153-160. 10.1145/584993.585017

Rudnyckyj, D. (2009a). Market Islam in Indonesia. *Journal of the Royal Anthropological Institute, 15*(1), 182–200.

Rudnyckyj, D. (2009b). 'Spiritual Economies: Islam and Neoliberalism in Contemporary Indonesia'. *Cultural Anthropology, 24*(1), 104–141. doi:10.1111/j.1548-1360.2009.00028.x

Rudnyckyj, D. (2010). *Spiritual economies: Globalization and the afterlife of development.* Cornell University Press.

Ryan, C. (2016). *Halal Tourism, Tourism Management Perspective* (Vol. 19). Part B.

Sahin, S. (2018). Competencies of Tourist Guides. In S. Eser, S. Sahin, & C. Cakici (Eds.), *Tourism Guidance* (pp. 47–74). Detay Publishing. (in Turkish)

Sahoo, D. & Kurukshetra University. (2018). *Religious Gastronomy: A case study of Jagannath Dhaam in Puri.* Kurukshetra: Kurukshetra University.

Sahoo, D. (2014). Social Architecture of Ethnic food tourism in India. *Indian Journal of Tourism & Hospitality Management*, *5*(1), 53–66.

Sahoo, D. (2019). Religious Gastronomy as a Tourism Attraction: A case study of Hindu Temple Food Mahaprashada. *Research Review: International Journal of Multidisciplinary*, *4*(5), 919–927.

Sahu, P. (2018, July 23). Medium: Fabulous Trip to Mini-Israel(Kasol) [Web log post]. Retrieved from https://medium.com/@piyurisahu/fabulous-trip-to-mini-israel-kasol-ce17b530c399

Salazar, N. (2005). Tourism and glocalization "local" tour guiding. *Annals of Tourism Research*, *32*(3), 628–646. doi:10.1016/j.annals.2004.10.012

Salazar, N. B. (2006). Touristifying Tanzania: Local Guides, Global Discourse. *Annals of Tourism Research*, *33*(3), 833–852. doi:10.1016/j.annals.2006.03.017

Sandıkcı, Ö., & Rice, G. (Eds.). (2011). *Handbook of Islamic Marketing*. Edward Elgar Publishing. doi:10.4337/9780857936028

Sanjiv, K., & Suvamay, B. (2016). Tea tourism in Darjeeling. *International Journal of Advance Research and Innovative Ideas in Education*, *2*(4).

Sapra, A. (2019). *In Delhi Food walk's Chandigarh to Kasauli Food Tour* [Video file]. Retrieved on 12 December, 2019, from https://www.youtube.com/watch?v=9J8 V0mH1Dic&t=41s

Sapra, A. (2019). *In Delhi Food walk's Shimla Food Walk - Part 1/3 I Amazing Mandyali Dham (Himachali Thali)* [Video file]. Retrieved on 12 December, 2019, from https://www.youtube.com/watch?v=kwPE_SW3CR8

Sardar, Z. (2014). *Mecca: The sacred city*. Bloomsbury Publishing USA.

Sarı, İ. (2011). Frig Yürüyüş Yolu, 23-24 Aralık 2011. Kültür Turizmi Zirvesi Sunumu, Kültür Turizm Bakanlığı, Ankara.

Sarı, Y., & Kozak, M. (2005). Turizm pazarlamasina internetin etkisi: Destinasyon web siteleri için bir model önerisi. *Akdeniz İİBF Dergisi*, *9*, 248–271.

Sarma S. (2014). Tea Tourism in North East India. *Employment News, 39*(4).

Sato, G., Hirakawa, G., & Shibata, Y. (2017). Push typed tourist information system based on beacon and augumented reality technologies. In *Proceedings of the IEEE 31st International Conference on Advanced Information Networking and Applications* (pp. 298-303). 10.1109/AINA.2017.167

Saunders, C. E. (2012). *Art tourism, Guggenheim effect*. Retrieved October, 24, 2020, from https://confluence.cornell.edu/display/tam2011/Day+16+-+Charles+Saunders

Sawyer. (2019). *High performance collaborative robot*. Retrieved from https://www.rethinkrobotics. com/sawyer

Sayegh, B., Yagi, S., & Connell, K. (2011). *Christian Antiquities in Jordan*. Jordan Book Center.

Schat, A. C. H., & Kelloway, E. K. (2000). Effects of perceived control on the outcomes of workplace aggression and violence. *Journal of Occupational Health Psychology*, *5*(3), 386–402. doi:10.1037/1076-8998.5.3.386 PubMed

Scherle, N., & Nonnenmann, A. (2008). Swimming in cultural flows: Conceptualising tour guides as intercultural mediators and cosmopolitans. *Journal of Tourism and Cultural Change*, *6*(2), 120–137. doi:10.1080/14766820802364147

Schilderman, H. (2011a). Religious capital and public accountability: challenges to empirical theology. In L. Francis, & H. G. Ziebertz (Eds.), The Public Significance of Religion (pp. 41-63). Reiden: Brill. doi:10.1163/ej.9789004207066.i-495.24

Schilderman, H. (2011b). Varieties of Religious Solidarity. In L. Francis, & H. G. Ziebertz (Eds.), The Public Significance of Religion (pp. 169-190). Reiden: Brill. doi:10.1163/ej.9789004207066.i-495.66

Schlund, S., Hammerle, M., & Strolin, T. (2014). *How automation and digitization will change production*. Fraunhofer Institute for Industrial Engineering.

Schneider, B., & Bowen, D. E. (1985). Employee and Customer Perceptions of Service in Banks. Replication and Extension. *The Journal of Applied Psychology*, *70*(3), 423–433; Advance online publication. doi:10.1037/0021-9010.70.3.423

Schofield, P., & Fallon, P. (2012). Assessing the Viability of University Alumni as a Repeat Visitor Market. *Tourism Management*, *33*(6), 1373–1384. doi:10.1016/j.tourman.2011.12.021

Schon, D. A. (1963). *Reflection of concepts*. Humanities Press.

Schreiber, R. (1976). Sports interest. *A travel definition. The Travel Research Association 7th Annual Conference Proceedings*, 85–97.

Schuster, B. (1996). Rejection, Exclusion, and Harassment at Work and in Schools. *European Psychologist*, *1*(4), 293–317. doi:10.1027/1016-9040.1.4.293

Segittur. (2019). Retrieved from www.segittur.es

Sehgal, M. (2017, July 5). Why Israeli tourists in Kasol, called mini-Israel. *India Today*. Retrieved on 11 October, 2019 from https://www.indiatoday.in/india/story/india-pm-modi-israel-benjamin-netanyahu-kasol-mini-israel-himachal-kullu-israeli-tourists-1022439-2017-07-05

Semacraft. (2020). *Diffusion of Innovation Theory*. Retrieved March 12, 2020, from: http://semacraft.com/blog/wp-content/uploads/2010/11/blogDiffusionImage.jpg

Seo, B. K., Kim, K., & Park, J. I. (2010, November). Augmented reality-based on-site tour guide: a study in Gyeongbokgung. *Asian Conference on Computer Vision*, 276-285.

Serenari, C., Leung, Y. F., Attarian, A., & Franck, C. (2012). Understanding environmentally significant behavior among whitewater rafting and trekking guides in the Garhwal Himalaya, India. *Journal of Sustainable Tourism*, *20*(5), 757–772. doi:10.1080/09669582.2011.638383

Shabani, N., & Hassan, A. (2017). Augmented reality for tourism service promotion in Iran as an emerging market. In V. Nadda, S. S. Dadwal, & R. Rahimi (Eds.), *Promotional strategies and new service opportunities in emerging economies* (pp. 116–129). IGI Global. doi:10.4018/978-1-5225-2206-5.ch005

Shackley, M. (2001). *Managing sacred sites: Service provision and visitor experience*. Cengage Learning EMEA.

Shafaei, F., & Mohamed, B. (2015). Involvement and brand equity: A conceptual model for Muslim tourists. *International Journal of Culture, Tourism and Hospitality Research*, *9*(1), 54–67. doi:10.1108/IJCTHR-06-2014-0050

Shah, S. K. (2013). Prospects of Indian Tea Industry. *International Journal of Agricultural Sciences and Veterinary Medicine*, *1*(1), 18–25.

Shah, S. K., & Patel, V. A. (2016). Tea production in India: Challenges and opportunities. *Journal of Tea Science Research, 6*(5), 1–6.

Sharlene, M. (2012). *The Motivational Factors of the Role of Food in Tourism.* Paper presented at Conference: The many faces of the global tourism experience in the 21st century', 2012, TSVC. Retrieved from www.travelconference.co.uk / commentries.php?paper=182#.W3KzZc4zbX5

Sharma, P., Sharma, P., & Kaith, M. (2014). Traditional food, fairs and festivals of Himachal Pradesh. *International Journal of Asian Academic Research Associates, 1*(21), 384–390.

Sharma, S. K., & Bhowmick, S. (2016). Tea tourism in Darjeeling. *International Journal of Advance Research and Innovative Ideas in Education, 2*(4), 1120–1122.

Shaw, K., & Fincher, R. (2010). University students and the creative city. *Journal of Policy Research in Tourism, Leisure & Events, 2*(3), 199–220. doi:10.1080/19407963.2010.512203

Sheehan, R. (2014). Artists' positioning in and creating authentic identities through place and tourism in New Orleans's Jackson Square. *Journal of Tourism and Cultural Change, 12*(1), 50–67. doi:10.1080/14766825.2013.869228

Shen, J., Mei, T., & Gao, X. (2016). Landmark reranking for smart travel guide systems by combining and analyzing diverse media. *IEEE Transactions on Systems, Man, and Cybernetics. Systems, 46*(11), 1492–1504. doi:10.1109/TSMC.2016.2523948

Sheoran, S. K. (2017). Big data: A big boon for tourism sector. *International Journal of Research in Advanced Engineering and Technology, 3*, 10–13.

Shephard, K. L., & Royston-Airey, P. C. M. (2000). Exploring the role of part-time ecotourism guides in Central Southern England. *Journal of Sustainable Tourism, 8*(4), 324–332. doi:10.1080/09669580008667367

Shinde, K. A. (2010). Entrepreneurship and indigenous enterpreneurs in religious tourism in India. *International Journal of Tourism Research, 12*(5), 523–535. doi:10.1002/jtr.771

Shoup, J. (2007). *Culture and Customs of Jordan.* Greenwood Publishing.

Shryock, A. (2004) The new Jordanian hospitality: house, host, and guest in the culture of public display *Comparative Studies in Society and History, 46*, 35-62.

Shunnaq, M., & Shunnaq Ramadan, S. (2012). Camels in Asia and North Africa: Interdisciplinary Perspectives on their Past and Present Significance. In Camel Trekking in Jordan: Between Oriental Perception and Authentic Tourist Attraction. Verlag der Österreichischen Akademie der Wissenschaften (Austrian Academy of Sciences Press).

Sidali, K. L., Kastenholz, E., & Bianchi, R. (2015). Food tourism, niche markets and products in rural tourism: Combining the intimacy model and the experience economy as a rural development strategy. *Journal of Sustainable Tourism, 23*(8-9), 1179–1197. Advance online publication. doi:10.1080/09669582.2013.836210

Sinha, V. (2011). *Religion and commodification: 'Merchandizing' diasporic Hinduism.* Routledge. doi:10.4324/9780203842799

Smerecnik, K., & Andersen, P. (2011). The Diffusion of Environmental Sustainability Innovations in North American Hotels and Ski Resorts. *Journal of Sustainable Tourism, 19*(2), 171–196. doi:10.1080/09669582.2010.517316

Smith, R. (2015). *Smart tourism tools: linking technology with the touristic resources of city destinations* (Bachelor Thesis). NHTV University of Applied Sciences, International Tourism Management and Consultancy.

Smith, L. (2006). *Uses of Heritage.* Routledge. doi:10.4324/9780203602263

Smith, N. (1996). *The new urban frontier: Gentrification and revanchist city*. Routledge.

Smithsonian Museum. (n.d.). *Pepper the Robot*. Retrieved October 10, 2019, from https://www.si.edu/visit/pepper

Smith, T. (2012). *Thinking contemporary curating*. Independent Curators International.

Snelgrove, R., & Wood, L. (2010). Attracting and leveraging visitors at a charity cycling event. *Journal of Sport & Tourism, 15*(4), 269–285. doi:10.1080/14775085.2010.533918

Sobh, R., Belk, R. W., & Wilson, J. A. J. (2013). Islamic Arab hospitality and multiculturalism. *Marketing Theory, 13*(4), 443–463. doi:10.1177/1470593113499695

Son, A., & Xu, H. (2013). Religious food as a tourism attraction: The roles of Buddhist temple food in Western tourist experience. *Journal of Heritage Tourism, 8*(2), 248–258. doi:10.1080/1743873X.2013.767815

Sotiriadis, M. (2016). Events sponsorship as a business partnership: Suggesting the critical success factors, Special issue on Special Interest Tourism. *TOURISMOS: An International Multidisciplinary Journal of Tourism., 11*(5), 1–25.

Spencer, A. J., Buhalis, D., & Moital, M. (2012). A hierarchical model of technology adoption for small owner-managed travel firms: An organizational decision-making and leadership perspective. *Tourism Management, 33*(5), 1195–1208. doi:10.1016/j.tourman.2011.11.011

Sport Management Degrees. (2020). *Five Challenges facing the sports industry*. Retrieved January 24, 2019, from https://www.sports-management-degrees.com/lists/five-challenges-facing-the-sports-industry/

Sportyjob. (2015). *The contribution of sport to economic growth and employment in Europe*. Retrieved November 10, 2019, from https://www.sportyjob.com/blog/the-contribution-of-sport-to-economic-growth-and-employment-in-europe/

Stamboulis, Y., & Skayannis, P. (2003). Innovation strategies and technology for experience based tourism. *Tourism Management, 24*(1), 35–43. doi:10.1016/S0261-5177(02)00047-X

Standeven, J., & De Knop, P. (1999). *Sport Tourism*. Human Kinetics.

Stark, R., & Finke, R. (2000). *Acts of faith: Explaining the human side of religion*. University of California Press.

Statista. (2018). *Number of smartphone users worldwide from 2014 to 2020*. Retrieved from https://www.statista.com/statistics/

Stephenson, A. L., & Yerger, D. B. (2014). Does brand identification transform alumni into university advocates? *International Review on Public and Nonprofit Marketing, 11*(3), 243–262. doi:10.100712208-014-0119-y

Strauss, A., & Corbin, J. (1990). *Basics of qualitative research grounded theory procedures and techniques* (2nd ed.). Sage.

Strauss, A., & Corbin, J. (1998). *Basics of qualitative research: Techniques and procedures for developing grounded theory* (2nd ed.). Sage.

Stricker, R., Müller, S., Einhorn, E., Schröter, C., Volkhardt, M., Debes, K., & Gross, H. M. (2012). Konrad and Suse, two robots guiding visitors in a university *building*. In *Autonomous Mobile Systems 2012* (pp. 49–58). Springer. doi:10.1007/978-3-642-32217-4_6

Sugio, K. (2005). A consideration on the definition of the setting and management/protection measures for cultural routes. *15th ICOMOS General Assembly and International Symposium: 'Monuments and sites in their setting - conserving cultural heritage in changing townscapes and landscapes'*, 17 – 21.

Su, K., Li, J., & Fu, H. (2011). Smart city and the applications. *Proceedings of the International Conference on Electronics, Communications and Control*, 1028-1031.

Sun-Ah Ponting, S., Wearing, S., & Black, R. (2010) Exploring Practitioner Conceptualisations of Professionalism among Australian Ecotour Guides. In *CAUTHE, Proceedings of the 20th Annual CAUTHE Conference*. University of Tasmania.

Sun, Y., Song, H., Jara, A. J., & Bie, R. (2016). Internet of things and big data analytics for smart and connected communities. *IEEE Access: Practical Innovations, Open Solutions*, *4*, 766–773. doi:10.1109/ACCESS.2016.2529723

Suradin, M. (2018). Halal Tourism Promotion in Indonesia: An Analysis on Official Destination Websites. *Journal of Indonesian Tourism and Development Studies*, *6*(3), 143–158. doi:10.21776/ub.jitode.2018.006.03.01

Sweet, T. (2014). *Roslyn: The tour guide robot* (PhD Thesis). Computer Science & Engineering, University of Nevada, Reno, NV.

Szivas, E., & Riley, M. (1999). Tourism employment during economic transition. *Annals of Tourism Research*, *26*(4), 747–771. doi:10.1016/S0160-7383(99)00035-3

Tabung Haji Travel and Services. (n.d.). Retrieved from https://www.thts.com.my/

Talukdar, U., & Hazarika, C. (2017). Production and export of value added tea in India and its global competitiveness. *Economic Affairs*, *62*(4), 705–710. doi:10.5958/0976-4666.2017.00086.9

Tanrisever, C. (2016). The Necessity of cultural routes education for tourist guides in Turkey. *Bartın Üniversitesi İktisadi ve İdari Bilimler Fakültesi Dergisi*. http://iibfdergi.bartin.edu.tr/

Tanrısever, C. (2019). *Turist Rehberliğinde Uzmanlaşma Özel İlgi Turundan Özel Konulara (edt: Güzel, Ö, Köroğlu, Ö.)*. Nobel Yayınevi.

Tanwar, M., Tanwar, B., Tanwar, R. S., Kumar, V., & Goyal, A. (2018). Himachali Dham: Food, culture, and heritage. *Journal of Ethnic Foods*, *5*(2), 99–104. doi:10.1016/j.jef.2017.10.006

Tatar, C. F., Herman, G. V., & Gozner, M. (2018). Tourist Guides'. *Contribution to Sustainability in Romania, Geo Journal of Tourism and GeoSites*, *21*(1), 282–287.

Tavakol, M., & Dennick, R. (2011). Making Sense of Cronbach's Alpha. *International Journal of Medical Education*, *2*(1), 53–55. doi:10.5116/ijme.4dfb.8dfd PMID:28029643

Taylor, J. P. (2001). Authenticity and sincerity in tourism. *Annals of Tourism Research*, *28*(1), 7–26. doi:10.1016/S0160-7383(00)00004-9

Tea Board. (2015). Retrieved from http://www.teaboard.gov.in/pdf/stat/Monthly _Export_ of_ Tea.pdf

Tekin, Ö., Bideci, M., & Avcıkurt, C. (2017). Turist rehberliğinde yeni teknolojilerin kullanımı. In F. Ö. Güzel, V. Altıntaş, & İ. Şahin (Eds.), *Turist rehberliği araştırmaları öngörüler ve uygulamalar* (pp. 291–303). Detay Yayıncılık.

Temporal, P. (2011). *Islamic branding and marketing: Creating a global Islamic business*. John Wiley & Sons.

Tetik, N. (2006). Analysis of Expectations from Professional Tourist Guide and customers in Turkey: The Case of Kuşadası. In F. Irigüler & M. E. Güler (Eds.), *Tourist Guiding: "Cinderella" of the Tourism* (pp. 203-220). Retrieved December 16, 2019 from www.researchgate.net/publication/310957726_Tourist_Guiding_Cinderella_of_the_Tourism

The Straits Times. (2017). *Singapore is the 6th most popular travel destination for Muslim Millenials*. https://www.straitstimes.com/singapore/singapore-is-the-6th-most-popular-travel-destination-for-muslim-millennials-study

Thomas, J. (2019, February). *Typical Food of India*. Retrieved from https://traveltips.usatoday.com/main-food-india-11361.html

Thomas, K. W. (2002). Intrinsic Motitvation at Work: Building Energy & Commetmints. San Francisco: Berett-Koehler.

Tien, J. M. (2012). The next industrial revolution: Integrated services and goods. *Journal of Systems Science and Systems Engineering*, *21*(3), 257–296. doi:10.100711518-012-5194-1

Timeout. (2019). *Contribution of Sport to Economic Growth*. https://www.timeout.com/istanbul/tr/spor-ve-fitness/2016-yaz-olimpiyat-oyunlari

Timothy, D. J., & Iverson, T. (2006). Tourism and Islam: considerations of culture and duty. In *Tourism, religion and spiritual journeys* (pp. 202–221). Routledge. doi:10.4324/9780203001073

Timothy, D. J., & Ron, A. S. (2013). Understanding heritage cuisines and tourism: Identity, image, authenticity, and change. *Journal of Heritage Tourism*, *8*(2-3), 99–104. doi:10.1080/1743873X.2013.767818

Tolga, Ö., Korkmaz, H., & Atay, L. (2015). A research on professional attitudes of undergraduate tourist guidance students. *Journal of Travel and Hotel Management*, *12*(2), 26–41.

Topsakal, Y., Yüzbaşıoğlu, N. & Çuhadar, M. (2018a). Industrial revolutions and tourism: Turkey tourism 4.0 swot analysis and proposal for adaptation process. *Süleyman Demirel University the Journal of Faculty of Economics and Administrative Sciences, 23*(Special Issue), 1623-1638.

Topsakal, Y., Yüzbaşıoğlu, N., & Akıncı, Z. (2018c). Smart tourism: cannot tourism enterprises unable to generate smart solutions be able to compete. In *Proceedings of the 9th International Conference of Strategic Research on Scientific Studies and Education* (pp. 259-272). Antalya, Turkey: Academic Press.

Topsakal, Y., Yüzbaşıoğlu, N., & Bahar, M. (2018b). Proposal for the hotel sector on the wat to be 'smart' at the age of Industry 4.0. In *Proceedings of the Second International Congress on Future of Tourism: Innovatıon, Entrepreneurship and Sustainability* (pp. 252-256), Mersin, Turkey: Academic Press.

Topsakal, Y., Yüzbaşıoğlu, N., & Çelik, P. (2018d). New generation tourist cards: tourist card proposal for the Antalya destination. In *Proceedings of the International Antalya Congress* (pp. 1321-1333). Antalya, Turkey: Academic Press.

Topsakal, Y. (2018). Disabled-friendly mobile services in the context of smart tourism: Recommendations for Turkey tourism. *Journal of Tourism Intelligence and Smartness*, *1*(1), 1–13.

Topsakal, Y. (2019). Big data and tourism. In O. İçöz & M. Uysal (Eds.), *Tourism Encyclopedia - Turkey: Tourism and Hospitality Industry Basic Terms* (p. 66). Detay Publicaiton.

Topsakal, Y. (2019a). Mobile applciation in tourism. In O. İçöz & M. Uysal (Eds.), *Tourism Encyclopedia - Turkey: Tourism and Hospitality Industry Basic Terms* (p. 614). Detay Publicaiton.

Topsakal, Y. (2019b). Location based applications. In O. İçöz & M. Uysal (Eds.), *Tourism Encyclopedia - Turkey: Tourism and Hospitality Industry Basic Terms* (p. 291). Detay Publicaiton.

Topsakal, Y., & Çelik, P. (2017). A new strategy for tourism: smart destinations. In G. Haşit, Ç. Hakkı, & M. E. Merter (Eds.), *Socio-economic strategies II: management* (pp. 96–106). IJOPEC.

Tourism Policy, H. P. (2019). *The Himachal Pradesh Tourism Policy-2019*. Retrieved from Himachal tourism website on 14 November, 2019, from https://himachal tourism.gov.in/wp-content/uploads/2019/09/Himachal-Pradesh-Tourism-Policy-2019.pdf

Tourism Statistics, H. P. (2019). *Himachal Tourism Statistics-2019*. Retrieved from Himachal tourism website on 16 November, 2019, from https://himachaltourism. gov.in/wp-content/uploads/2019/03/edited_Tourist-statastics.pdf

Trahanias, P., Argyros, A., Tsakiris, D., Cremers, A., Schulz, D., Burgard, W., ... Kamarinos, G. (2000, May). Tourbot-interactive museum tele-presence through robotic avatars. *Proc. of the 9th International World Wide Web Conference*.

Trail, G. T., Robinson, M. J., Dick, R. J., & Gillentine, A. J. (2003). Motives and points of attachment: Fans versus spectators in intercollegiate athletics. *Sport Marketing Quarterly*, *12*(4), 217–227.

Trail, G., & James, J. (2001). The motivation scale for sport consumption: Assessment of the scale's psychometric properties. *Journal of Sport Behavior*, *24*(1), 108–127.

TravelM. Q. (2020). Retrieved from https://mqtravel.co.id/

Trépanier, S.-G., Fernet, C., & Austin, S. (2015). A longitudinal investigation of workplace bullying, basic need satisfaction, and employee functioning. *Journal of Occupational Health Psychology*, *20*(1), 105–116. doi:10.1037/a0037726 PubMed

Tribe, J. (2001). Research Paradigms and the Tourism Curriculum. *Journal of Travel Research*, *392*(4), 442–448. doi:10.1177/004728750103900411

Tribe, J., & Lewis, A. (2003). Attitudes of the young to careers in hospitality and tourism: Review and recommendations. In S. Kusluvan (Ed.), *Managing employee attitudes and behaviours in the tourism and hospitality industry* (pp. 67–76). Nova Science.

Trujillo Flores, M. M., Rivas Tovar, L. A., & Vilchis, F. L. (2014). Mobbing: A theoretical model quantifying factors affecting the role of women executives in the institutions of public education in Mexico. *Contaduría y Administración*, *59*(1), 195–228. doi:10.1016/S0186-1042(14)71249-5

Tsai, C. T. L. (2013). Culinary tourism and night markets in Taiwan. *International Journal of Business and Information*, *8*(2).

Tsaur, S., Dai, Y., & Liu, J. (2018). SOCO's impact on service outcomes of tour guides: The moderating effect of customers' shopping orientation. *Current Issues in Tourism*, *21*(8), 917–933. doi:10.1080/13683500.2015.1118444

Tucker, H. (2016). Empathy and tourism: Limits and Possibilities. *Annals of Tourism Research*, *57*, 31–43. doi:10.1016/j.annals.2015.12.001

Tucker, H., & Carnegie, E. (2014). World heritage and the contradictions of 'universal value'. *Annals of Tourism Research*, *47*(July), 63–76. doi:10.1016/j.annals.2014.04.003

Tung, V. W. S., & Law, R. (2017). The potential for tourism and hospitality experience research in human-robot interactions. *International Journal of Contemporary Hospitality Management*, *29*(10), 2498–2513. doi:10.1108/IJCHM-09-2016-0520

TUREB. (2019, Nov 11). *Association of Tourist Guides in Turkey*. Retrieved from http://www.tureb.org.tr/en

Turner, B. S. (2008). Introduction: The price of piety. *Contemporary Islam*, *2*(1), 1–6. doi:10.100711562-007-0034-7

Tuzunkan, D. (2018). Mobbing in Tourism Enterprises. *The Case of Turkish Riviera.*, *13*(4), 1813–1819.

Uçar, H. (2014). Ayvalık'ta Hagia Triada Kilisesi mimari analizi. *Trakya University Journal of Engineering Sciences*, *15*(1), 7–18.

United Nations World Tourism Organisation. (2011). *Religious Tourism in Asia & the Pacific*. Retrieved from UNWTO website on 27 May, 2018, from https://cf.cdn.unwto.org/sites/all/files/pdf/110325_religious_tourism_excerpt.pdf

United Nations World Tourism Organisation. (2012). *Global Report on Food Tourism. AM Report: Volume Four*. Retrieved from UNWTO website on 02 January, 2018, from http://cf.cdn.unwto.org/sites/all/files/docpdf/ amreports4-foodtourism.pdf

United Nations World Tourism Organisation. (2017). *Second Global Report on Gastronomy Tourism. Affiliate members Report: Volume Sixteen*. Retrieved from UNWTO website on 21 March, 2018, from http://cf.cdn.unwto.org/sites/all/files/pdf/ gastronomy_report_web.pdf

UNWTO Would Tourism Organization. (2019). Weidenfeld A, Ron A. 2008. Religious needs in the tourism industry. Anatolia. *International Journal of Tourism and Hospitality Research, 19*(2), 18–22.

Upadhyay, Y., & Sharma, D. (2014). Culinary preferences of foreign tourists in India. *Journal of Vacation Marketing, 20*(1), 29–39. doi:10.1177/1356766713486143

Urry, J. (1990). The consumption of tourism. *Sociology, 24*(1), 23–35. doi:10.1177/0038038590024001004

Valente, T. W. (1996). Social Network Thresholds in the Diffusion of Innovations. *Social Networks, 18*(1), 69–89. doi:10.1016/0378-8733(95)00256-1

van der Wal, S. (2008). *Sustainability issues in the tea sector a comparative analysis of six leading producing countries* (J. Turner, Ed.).

Van Dijik, P. A., Smith, L. D., & Cooper, B. K. (2007). Are you for real? An evaluation of the relationship between emotional labour and visitor outcomes. *Tourism Management, 32*(1), 39–45. doi:10.1016/j.tourman.2009.11.001

Van Nieuwkerk, K. (2008). Creating an Islamic cultural sphere: Contested notions of art, leisure and entertainment. An Introduction. *Contemporary Islam, 2*(3), 169–176. doi:10.100711562-008-0059-6

Van Nieuwkerk, K. (2013). *Performing piety: singers and actors in Egypt's Islamic revival.* University of Texas Press.

Van Rheenen, D., Cernaianu, S., & Sobry, C. (2017). Defining sport tourism: A content analysis of an evolving epistemology. *Journal of Sport & Tourism, 21*(2), 75–93. doi:10.1080/14775085.2016.1229212

Vargo, S. L., & Lusch, R. F. (2008). From goods to service (s): Divergences and convergences of logics. *Industrial Marketing Management, 37*(3), 254–259. doi:10.1016/j.indmarman.2007.07.004

Varsha, K. (2018). *Visit Panchgani And Be A Part Of The Strawberry Festival.* Retrieved on 5 November, 2019, from https://www.yatrablog.com/strawberry-festival-in-panchgani

Vasudeva, S. (2014, June 19). *Indian Food Tours: The Growing Trend of 2014.* Retrieved from http://cooks.ndtv.com/article/show/indian-food-tours-the-growing-trend-of-2014-539675

Verter, B. (2003). Spiritual capital: Theorizing religion with Bourdieu against Bourdieu. *Sociological Theory, 21*(2), 150–174. doi:10.1111/1467-9558.00182

Visit Scotland. (2016). *Scotland of Food & Drink Survey report: Whisky Tourism – Facts and Insights, March 2015.* Retrieved from Visit Scotland website on 15 January, 2018, from http://www.visitscotland.org/pdf/Whisky%20 Tourism%20%20Facts %20and%20Insights2.pdf

Vukonic, B. (1996). *Tourism and Religion.* Pergamon, Elsevier Science Ltd, Tourism Social Science Services.

Walker, K., & Weiler, B. (2017). A new model for guide training and transformative outcomes: A case study in sustainable marine-wildlife ecotourism. *Journal of Ecotourism, 16*(3), 269–290. doi:10.1080/14724049.2016.1245736

Walton, J. (Ed.). (2005). *Histories of tourism: Representation, identity and conflict.* Channel View Publications. doi:10.21832/9781845410339

Wang, D., Park, S., & Fesenmaier, D. (2012). The role of smartphones in mediating the tourism experience. *Journal of Travel Research, 51*(4), 371–387. doi:10.1177/0047287511426341

Wang, D., Xiang, Z., & Fesenmaier, D. (2014). Smartphone use in everyday life and travel. *Journal of Travel Research, 55*(1), 1–12.

Wang, F., Yuan, Y., Wang, X., & Qin, R. (2018). Societies 5.0: A new paradigm for computational social systems research. *IEEE Transactions on Computational Social Systems*, *5*(1), 2–8.

Wang, X., Li, X. R., Zhen, F., & Zhang, J. H. (2016). How smart is your tourist attraction? Measuring tourist preferences of smart tourism attractions via a FCEM-AHP and IPA approach. *Tourism Management*, *54*, 309–320. doi:10.1016/j.tourman.2015.12.003

Wang, Y., Yu, Q., & Fesenmaier, R. D. (2002). Defining the virtual tourist community: Implications for tourism marketing. *Tourism Management*, *23*(4), 407–417. doi:10.1016/S0261-5177(01)00093-0

Wan, Y. K. P., Wong, I. A., & Kong, W. H. (2014). Student career prospect and industry commitment: The roles of industry attitude, perceived social status, and salary expectations. *Tourism Management*, *40*, 1–14. doi:10.1016/j.tourman.2013.05.004

Ward, J. (2016). Down by the sea: Visual arts, artists and coastal regeneration. *International Journal of Cultural Policy*, *24*(4), 1–18.

Washburn, D., Sindhu, U., Balaouras, S., Dines, R. A., Hayes, N., & Nelson, L. E. (2010). *Helping CIOs understand smart city initiatives.* Retrieved from http://www.uwforum.org/upload/board/forrester_help_cios_smart_city.pdf

Watson, C. (2013, May). *How the arts and culture tourism spur economic development.* Retrieved October, 24, 2020, from https://www.westerncity.com/article/how-arts-and-cultural-tourism-spur-economic-development

Watson, T. (1995). *Sociology, work, and industry.* Routledge.

Webb, D. (2002). *Medieval European Pilgrimage, c.700–c.1500.* Palgrave. doi:10.1007/978-1-4039-1380-7

Weed, M. (2014). After 20 years, what are the big questions for sports tourism research? [Editorial]. *Journal of Sport & Tourism*, *19*(1), 1–4. doi:10.1080/14775085.2015.1032505

Weiler, B., & Black, R. (2015). *Tour guiding research: Insights, issues and implications* (Vol. 62). Channel View Publications.

Weiler, B., & Black, R. (2015). *Tour guiding research: Insights, issues, and implications.* Channel View Publications.

Weiler, B., & Davis, D. (1993). An exploratory investigation into the roles of the nature-based tour leader. *Tourism Management*, *14*(2), 91–98. doi:10.1016/0261-5177(93)90041-I

Weiler, B., & Ham, S. (2002). Tour guide training: A model for sustainable capacity building in developing countries. *Journal of Sustainable Tourism*, *10*(1), 52–69. doi:10.1080/09669580208667152

West, M. A., & Farr, J. L. (1990). Innovation at work. In M. A. West & J. L. Farr (Eds.), *Innovation and creativity at work: Psychological and organizational strategies* (pp. 309–333). Wiley.

WFTGA. (2003). What is a Tourist Guide? Retrieved January 5, 2020, from World Federation of Tourist Guide Associations website: http://www.wftga.org/tourist-guiding/what-tourist-guide

WFTGA. (2018). *What is a Tourist Guide?* Retrieved December 05, 2019, from www.wftga.org/tourist-guiding/what-touris-guide

WFTGA. (2019, Nov 11). *World Federation of Tourist Guide Associations.* Retrieved from http://www.wftga.org/tourist-guiding/what-tourist-guide

Whitehead, R. (2019). *Sun, Sea and Saudi: How countries around the world are courting Muslim tourist.* https://cnaluxury.channelnewsasia.com/experiences/how-countries-are-courting-muslim-tourists-11571148

Winter, T. (2013). Clarifying the critical in critical heritage studies. *International Journal of Heritage Studies, 19*(6), 532–545. doi:10.1080/13527258.2012.720997

Winterton, J., & Winterton, R. (2002). *Developing managerial competence*. Routledge. doi:10.4324/9780203020395

Woodward, S. (2013). Campus tourism, universities and destination development. In M. Smith & G. Richards (Eds.), *Routledge Handbook of Cultural Tourism* (pp. 265–272). Routledge.

World Federation of Tourist Guide Associations. (n.d.). *What is a Tourist Guide?* Retrieved October 15, 2019, from http://www.wftga.org/tourist-guiding/what-tourist-guide

World Federation of Tourist Guides Association. (2019). *World-wide survey of tourist guides*. Retrieved from: http://wftga.org/membership/surveys-tourist-guiding-worldwide

World Tourism Organization. (2019). International Tourism Highlights, 2019 Edition. doi:10.18111/9789284421152

World Travel and Tourism Council. (2018). *Travel & Tourism Economic Impact 2018*. World Travel and Tourism Council.

World Travel and Tourism Council. (2019). *Travel & Tourism Economic Impact 2019 World*. London UK: WTTC. Retrieved from: https://www.wttc.org/-/media/files/reports/economic-impact-research/regions-2019/world2019.pdf

WST. (2020). *Sport Tourism*. Retrieved January 24, 2019, from https://www.wst-show.com/en/sport-tourism

WTTC. (2019). *Economic Impact Research*. Retrieved November 10, 2019, from https://www.wttc.org/-/media/files/reports/economic-impact-research/regions-2019/world2019.pdf

WWF. (2019). *Dünya Doğayı Koruma Vakfı*. https://www.worldwildlife.org/

Xinhua, L., Juntao, C., Jinmei, Z., & Lin, L. (2016). A Study on Sports Tourism Competitiveness Based on Factor Analysis Method. *12th International Conference on Computational Intelligence and Security*, 16-19.

Yao, G. (2012). A study on the construction framework of intelligent tourism. [Social Sciences]. *Journal of Nanjing University of Posts and Telecommunications, 14*(2), 13–16.

Yasuda, S. (2018). Entrepreneurship for Religious Tourism in Mumbai, India. In S. Yasuda, R. Raj & K. Griffin (Eds.), Religious tourism in Asia: Tradition and change through case studies and narratives (pp. 21-29). Wallington: CABI.

Yasuda, S. (2013). Commitment for Strategy: Religious Entrepreneur Networks in Syrian Shi'ite Religious Tourism. *The Journal of Sophia Asian Studies, 31*, 35–49.

Yetgina, D., & Benligiray, S. (2019). The effect of economic anxiety and occupational burnout levels of tour guides on their occupational commitment. *Asia Pacific Journal of Tourism Research, 24*(4), 333–347. doi:10.1080/10941665.2018.1564681

Yıldırım, A., & Şimşek, H. (2018). *Qualitative research methods in the social sciences*. Seçkin Publishing.

Yıldırım, G., & Özbek, Ö. (2019). A Study on Sexual Harassment to Professional Tour Guides. *Turist Rehberliği Dergisi, 2*(2), 72–87.

Yildiz, Z., & Cekic, S. (2015). Sport Tourism and its History and Contribution of Olympic Games to Touristic Promotion. *International Journal of Science Culture and Sport,* (SI4), 326-337.

Yıldız, Ş. (2018). Profesyonel turist rehberlerinin teknolojik rehber algısı belirlemeye yönelik bir araştırma: İstanbul örneği. *Journal of Social And Humanities Sciences Research, 5*(31), 4686–4697.

Yıldız, Ş. (2019). The rise of robot guides in tourist guidance profession. *Suleyman Demirel University Visionary Journal, 10*(23), 164–177.

Yin, R. K. (2003). *Case study research: Design and methods* (3rd ed.). Thousand Oaks: Sage Publications Inc.

Yoyotte, J. (1960). Les Pelerinages Dans I'Egypte Ancienne. In J. Yoyotte (Ed.), Les Pelerinages, 'Sources Orientales' iii (pp. 17–74). Academic Press.

Yurcu, G., & Akinci, Z. (2019). The mediating role of mobbing in the relationship between personal value and depression: A case study of hospitality employees. *Journal of Human Resources in Hospitality & Tourism, 18*(4), 471–503. doi:10.1080/15332845.2019.1626852

Yu, X., & Weiler, B. (2006). Service ethics for ecotourism guides. In B. Prideaux, G. Moscardo, & E. L. James (Eds.), *Managing tourism and hospitality services: Theory and international applications* (pp. 181–194). CAB International. doi:10.1079/9781845930127.0181

Yüzbaşıoğlu, N., Çelik, P., Topsakal, Y., & Bahar, M. (2018). Industry 4.0 and smart tourism: Developing a smart tourist guidance application for Antalya destinastion. *Proceedings of the Innovation and Global Issues in Social Sciences, III*, 707–718.

Zailani, S., Ali, S. M., Iranmanesh, M., Moghavvemi, S., & Musa, G. (2016). Predicting Muslim medical tourists' satisfaction with Malaysian Islamic friendly hospitals. *Tourism Management, 57*, 159–167. doi:10.1016/j.tourman.2016.05.009

Zapf, D. (1999). Organisational, work group related and personal causes of mobbing/bullying at work. *International Journal of Manpower, 20*(1/2), 70–85. doi:10.1108/01437729910268669

Zhang, L. J. (2004). (Master's). Zhejiang University. Retrieved from Wangfang Database (Y600349).

Zhang, H. Q., & Chow, I. (2004). Application of importance-performance model in tour guides performance: Evidence from mainland Chinese outbound visitors in Hong Kong. *Tourism Management, 25*(1), 81–91. doi:10.1016/S0261-5177(03)00064-5

Zhang, L., Li, N., & Liu, M. (2012). On the basic concept of smarter tourism and its theoretical system. *Luyou Xuekan, 27*(5), 66–73.

Ziyarah Tours. (2020). Retrieved from https://www.ziyarahtours.net/

Zuahar, J. (2004). Historical perspectives of sports tourism. *Journal of Sport & Tourism, 9*(1), 5–101. doi:10.1080/1477508042000179348

About the Contributors

Gulsun Yildirim has B.A. degree in Tourism Management from Adnan Menderes University and ELT from Pamukkale University. After graduating from B.A. she has M.A degree from Eskişehir Osmangazi University in Tourism Management and Ph.D. degree from Selcuk University/Gazi University in Tourism Management. She worked as a tour guide for three years. She also worked as a lecturer in Ardesen Vocational School of Recep Tayyip Erdogan University between 2012-2019. She has been working as Assistant Professor in the department of Gastronomy and Culinary Arts in the Tourism Faculty of Recep Tayyip Erdoğan University. Her Research Interests are Rural Tourism, Agri-tourism, Sustainable Tourism, Tourism Marketing, Qualitative Research Methods, Tour Guiding, Women Empowerment. She is a reviewer on IGI Global's International Journal of Tourism and Hospitality Management in the Digital Age (IJTHMDA). Her work appears in book chapters released by Routledge.

Ozlem Ozbek worked as a research assistant in Konya Necmettin Erbakan University Tourism Faculty, Tourism Guidance Department between 2012-2017. She completed her Ph.D in 2017 with her thesis on augmented reality marketing. She has been working as an assistant professor at Bandirma Onyedi Eylul University since 2017. Ozbek has experience in book reviewer at the Emerald publishing and her researches has been published in book projects such as Routhledge and IGI Global. By profession, she is a tour guide. Ozlem Ozbek's major research interests are digital marketing, art therapy and tourism, touristic product diversification and service quality of tour guides.

Ceyhun Caglar Kilinc started M.A. in Business in Eastern Mediterranean University (EMU) after graduating from the Department of Business Administration in Selcuk University. After completing the M.A. program, he worked as lecturer in the Beysehir Vocational School during 2004-2009 academic year. He started Ph.D program in Business in Selcuk University 2005-2009 academic year. He had been an Assistant Professor of Business in Tourism Faculty of Selcuk University between 2010-2014. He gained the title of an Associated Professor of Marketing in 2015. He is still working in Akdeniz University Manavgat Tourism Faculty. His expertise is on marketing and he also is studying on tourism marketing, service quality and tourism. Dr. Kilinc is also a member of some international academic journals worldwide and now is the co-editor of the International Journal of Economic and Management Perspectives.

* * *

Nor Aida Abdul Rahman is an Associate Professor at Universiti Kuala Lumpur, Malaysian Institute of Aviation Technology (UniKL MIAT) in Subang, Selangor, Malaysia. Currently, she served as a Head of Aviation Management at Universiti Kuala Lumpur. She has worked as internal and external trainer in management, supply chain, Halal logistics and postgraduate research. She has written three books on postgraduate research. Her research work has appeared in several reputable academic journals, book chapter and refereed conference proceedings. She earned PhD degree in Management (supply chain management) from Brunel University, London, UK. She is also serving as External Academic Advisor in college, a chartered member for Chartered Institute of Logistics and Transport Malaysia (CILTM), HRDF Certified Trainer, Chairman (Academic Committee) for Malaysian Association of Transportation, Logistics and Supply Chain Schools (MyATLAS), Vice President (Research Journal) for Institute for Research in Management and Engineering UK (INRME), JAKIM Halal certified Trainer, UniKL Halal Professional Board and a member of Academy of Marketing, UK.

Areej Shabib Aloudat is Associate Professor at Yarmouk University, Jordan and has worked there since 2010. Since 2015 she has been the Dean's Assistant for Quality Assurance and Accreditation and her responsibilities include curriculum and study plan development at both under and post-graduate levels. She leads the Faculty team for the Tourism Management Program that has been awarded TedQual certification. Areej holds a PhD in tourism from the University of Bedfordshire, UK. Her research interests include tour guiding, women's empowerment in tourism, tourism and peace, and grounded theory.

Sakib Bin Amin is an Associate Professor in the Department of Economics and the Director of the Accreditation Project Team (APT) at North South University, Bangladesh. He holds a Ph.D. in Economics from Durham University (UK). Dr. Amin was also the receiver of the Commonwealth Rutherford Fellowship (2017-2018) for conducting his postdoctoral research at Durham University (UK). He is a fellow of the Higher Education Academy (HEA). His research focuses on Energy and Tourism Policy in the Developing Countries.

Mehmet Bahar is Assistant Prof in the Department of Culinary Art at the Cappadocia University where he has been a faculty member since 2018. Bahar completed his Master Degree at Akdeniz University and his bachelor degree at Aydın Adnan Menderes University. He got Ph.D. degree at Akdeniz University in 2017. His research interests lie in the area of Industry 4.0, smart tourism, culinary art.

Foued Benghadbane received the Ph.D. in urban planning at the University of Constantine (Algeria) in 2009. He is currently teaching in the Department of Urban Techniques Management of the University "Larbi Ben M'hidi" Oum El Bouaghi- Algeria. His research interests in the Geography of Tourism, Geography of Services, GIS and Remote Sensing, which is evident through the publications on these aspects, He directed several national projects of the CNEPRU type which dealt with many topics, including transport, urban heritage and sustainable development.

Rosemary Black holds a PhD from Monash University and is Associate Professor in Interpretation and Ecotourism at Charles Sturt University. She teaches and undertakes research in sustainable tourism, tour guiding, environmental communication and social dimensions of natural resource management. Prior to becoming an academic she worked in protected area management, adventure travel and community conservation. She has published six books and more than 60 refereed publications. Rosemary undertakes applied research with industry partners including protected area management, tourism agencies and community-based organizations. Rosemary is on the editorial board of three international tourism journals.

Birsen Bulut-Solak graduated from Ankara University with a BSc in Dairy Technology. She obtained a MSc and a PhD in Food Engineering from Selcuk University (TR). She was awarded an International Post-Doctoral Research Fellowship by the Scientific and Technical Research Council of Turkey (TUBITAK). She fulfilled her post-doctoral research in Food and Nutrition Sciences at University College Cork (IRL) under the supervision of Dr. James Anthony O'Mahony between 2013 and 2015. She was a university lecturer in the Programme of Dairy Technology at Selcuk University between 2002 and 2013. Dr. Bulut-Solak is currently an Assistant Professor in the Department of Gastronomy and Culinary Arts at Selcuk University, Konya (TR). She has also been an ambassador in Global Harmonization Initiative for Turkey since 2016.She is currently involved in three significant researches that are concerned with food safety, consumer acceptance of traditional foods and the determinant of food choice. She also works closely with others internationally (e.g. University College Cork and University of Illinois).

Elizabeth Carnegie is a former museum professional who now leads the MSc Creative and Cultural Industries programme at Sheffield University Management School, where she lectures on heritage and cultural property management.

Sabriye Çelik Uğuz who was born in 1973, is a graduate in tourism management and completed her master's and doctorate degrees in economics. She is also a graduate of photography department in the field of fine arts. She has been working as Ph.D. Associate Professor and head of department of tourism and hotel management at Balıkesir University, School of Burhaniye Applied Sciences. Since 2003, she has worked in the projects of the Ministry of Culture and Tourism, Istanbul Archeology Museum, UNESCO, European Union, History Foundation and similar public and private institutions. In 2012, she was a visiting lecturer at Primorska University/Portoroz-Slovenia and worked on "Festival Tourism". She participated in Erasmus Training Mobility Manchester/England/2013, in Ersamus Teaching Mobility Krakow/Poland/2015 and Kaunas/Lithuania/2018. Many articles, chapters, and papers were published in national and international journal, book and congresses. Her main fields of study are tourism, economic and art.

Jean-Michel Chapuis is an Associate Professor at University of Paris Panthéon-Sorbonne. Jean Michel Chapuis is a Doctor in Management Science, from the University of Burgundy and graduated in economics. His research that explores the dynamics of prices in tourism, hotels and restaurants is published in international journals and conferences. He is a reviewer for International Journal of Hospitality Management and Tourism Management. Currently, four Ph.D. students work under his supervision on medical tourism, destination management and hotel management. Dr. Chapuis is since 2013 a member of the board of directors of the oldest French Institute of Tourism. He has led the national board of Higher Education for Tourism as the chairman, nominated by the French Minister of Higher Education & Research Innovation 2016-2019. Pr Chapuis has launched several tourism trainings tailored to the needs of local and international companies such as bachelor in resorts management, and a master of science in hospitality management. Different workshops dealing with forecasting or length of stay controls are very appreciated by hotel managers.

Eda Rukiye Donbak received her bachelor degree from Anadolu University, School of Tourism and Hotel Management in 2000, In 2017, she received her philosophy of doctorate from Atatürk University, Tourism Management and Hotel Department. Her academic career started in 2011 as lecturer at Adıyaman University, School of Tourism and Hotel Management, Tourism Guidance and Travel Management Department. Since 2007 she is licensed tourist guide and 2018, she is founder and head of Tourism Guidance Department of Adıyaman University Tourism Faculty. Her research interests are tourist guiding, ecotourism, strategic management, strategic alliances, organizational ecology, organizational behaviour, organizational competencies and competitiveness.

Ramazan Eren is an assistant professor at the Department of Tourism Management at Akdeniz University. His research interests include gastronomy tourism, tourism and gastronomy education, food image and rural tourism.

Sally Everett joined King's Business School, King's College London as their inaugural Professor of Business Education in January 2019 and has since became the College lead for inclusive education and was most recently appointed the Vice Dean (Education) in September. Sally was previously the Deputy Dean for the Faculty of Business and Law at Anglia Ruskin University (2013 – 2018) and institutional lead for their work on inclusive learning. She is a National Teaching Fellow and a Principal Fellow of the Higher Education Academy. Before joining ARU, Sally was the Head of Department for tourism, marketing and hospitality (and previously a Senior Lecturer in Tourism) at the University of Bedfordshire (2009-2013). Sally is also a widely published tourism researcher and over the last ten years has produced a significant body of work on food and drink tourism, including her textbook: 'Food and Drink Tourism: Principles and Practices' (Sage Publications, 2016). Her other research interests include social and cultural tourism geographies, heritage tourism and inclusive curriculum design. Recent publications include work on community resistance, slow towns, inclusive teaching, food festivals, and destination marketing.

Onur Icoz is Assistant Professor of Tourism Management at the Department of Travel Management, Faculty of Tourism, Adnan Menderes University, Kusadasi, Turkey. He obtained his PhD in Tourism Management from Dokuz Eylul University in 2013. His research interests include Marketing in Tourism, Social Media Marketing, Sport Tourism and Destination Branding.

Orhan Icoz is Professor of Tourism Management at the Department of Tourism Guidance, School of Applied Sciences, Yasar University, Izmir, Turkey. He obtained his PhD in Tourism Management from Dokuz Eylul University in 1987. His research interests include Economics of Travel and Tourism, Tourism and Hospitality Marketing, Tourism Planning and Policy, Tourism and Transportation.

Abdullah Karaman is the head of Department of Travel and Tourism Management Guidance at Selcuk University, Faculty of Tourism. He has been involved in many national and international projects as executive and coordinator. He has many researches on various topics related to tourism.

Sawsan Haider Khreis is an assistant Professor in tourism management. she received her PhD in tourism management (Highest honours) from the University Paris 1, Panthéon. La Sorbonne, Paris/ France in 2012. Her actual position is an assistant professor in the department of tourism and travel department at Yarmouk University. She is also the General Secretary of Arab Tourism Experts Union. she wrote many articles in well-known scientific journals. she participates all the time in many summits and conferences all around the world as a keynote speaker.

Sawsan Khreis, an Assistant Professor at University of Yarmouk, is a doctor in tourism management, from the University Paris 1, Panthéon, La Sorbonne. Currently, two Ph.D. students work under her supervision. Dr. Khreis is since 2016 the General secretary of Arab tourism experts. She published many articles in several journals.

Oğuz Nebioğlu is an Assistant Professor at Alanya Alaaddin Keykubat University, Tourism Faculty, Department of Tourism Guidance.

Debasis Sahoo received his Bachelor's degree (B. Sc. in Hospitality & Hotel Administration) from the National Council of Hotel Management & Catering Technology, India in 2004. He was hired as a Chef by the Oberoi International Hotels Group in 2004 and later worked with Taj & ITC group of hotels in the Pan Asian restaurants. Subsequently he completed his 'Masters & Ph. D' in Tourism Management and is currently working as Assistant Professor in department of tourism and travel at the Central University of Himachal Pradesh, Dharamshala. He writes and presents widely on issues and prospects related to gastronomy tourism, religious gastronomy as well as sustainable issues in the field of tourism environment.

Kürşad Sayin was born in Ankara in 1972. He graduated from 9 September University, Tourism faculty in 1995. From 1990 to 2005 he worked as a staff and manager at many hotels in the tourism sector. He has been working at Selçuk University since 2005. He graduated from the Ph.D. in Tourism administration in 2019. He serves as the head of the Hotel, Restaurant and Catering Services Department. He is married and has two children.

Mohammed Shunnaq is a socio-cultural Anthropologist. Dedicated to discipline and like to do research in different areas like Tourism and local community, city and life and in Ethno-archaeology. He is the dean of faculty of Tourism and hotel management at Yarmouk University, Irbid, Jordan.

Halil Sunar works as a lecturer at Giresun University, Tirebolu Mehmet Bayrak Vocational High School. He conducts research in the fields of Media Psychology, Educational Psychology, Behavioral Science Quantitative Social Research, Qualitative Social Research, Communication and Media, Vocational Education, Educational Assessment, Educational Management.

Canan Tanrisever is an Associate Professor and Head of Tourism and Hotel Management Department at Kastamonu University in Turkey since 2010. She also worked for University of South Carolina in the USA between 2012-2014 as an adjunct professor. She graduated her Ph. D and Master Degree in Landscape Architecture from Ankara University in 2013. She has bachelor degree from Tourism Management and Guidance in Baskent University. From 1996 to 2007 she worked in many seven and five-star hotels like Ciragan Palace, Rixos, Dedeman, Holiday Inn, travel agencies and Turkish Airlines. Since 2007 she has been working in academia. She has published over 30 articles and 21 book chapters and books. She has attended 36 international conferences and has been in 50 countries in 6 continents. She has given lectures in 4 countries like USA, Hong Kong, Japan and Switzerland. She is also a Professional International Tour Guide. Her research interests are tour guidance, cultural studies and heritage sites.

Yunus Topsakal is a Doctor in the Department of Tourism Management at the Adana Alparslan Türkeş Science and Technology where he has been a faculty member since 2013. Topsakal completed his Master Degree at Akdeniz University and his bachelor degree at I.D. Bilkent University. He got Ph.D. degree at Akdeniz University in 2017. His research interests lie in the area of Industry 4.0, smart tourism, digital tourism, innovation, destination competitiveness. He has collaborated actively with researchers in several other disciplines of statistic, economics, and agriculture.

Simon Woodward is an experienced heritage professional who, for the last 11 years, has lectured on heritage and cultural tourism management on the MSc Responsible Tourism Management programme at Leeds Beckett University. He also works as a consultant to a broad range of organisations including the World Bank, UNWTO and other public and private sector bodies.

Shin Yasuda is Associate Professor at Takasaki City University of Eonocmis in Takasaki City, Gunma Prefecture in Japan. Yasuda received his Bachelor of Arts degree in history from Sophia University in Tokyo, Japan in 2007. He obtained his master's and Ph.D. degrees in area study from Graduate School of Asian and African Area Studies in Kyoto University, Japan in 2010 and 2012. He is presently working at Department of Tourism Policy, Faculty of Regional Policy in Takasaki City University of Economics in Takasaki City, Gunma Prefecture in Japan. His research interests include Islamic tourism, religious tourism in Islamic countries, history of tourism in the Middle Eastern countries, and theoretical studies in religious tourism, anthropology of tourism and sociology of tourism. He edited the book Religious Tourism in Asia: Tradition and Change through Case Studies and Narratives (CABI publishing, 2018) with Razaq Raj and Kevin Griffin.

Nedim Yüzbaşioğlu is a Professor in the Department of Tourism Management at the University of Akdeniz where he has been a faculty member since 2009. Yüzbaşıoğlu completed his Ph.D. at Cardiff University. His research interests lie in the area of strategic management, SMEs in tourism industry, tourism policy and planning, destination competitiveness. In recent years, he has focused on social innovation, medical tourism and intellectual capital. He has collaborated actively with researchers in several other disciplines of statistic, economics, and agriculture. Yüzbaşıoğlu is the instructor for strategic management and tourism policy on undergraduate topics.

Index

www.ingramcontent.com/pod-product-compliance
Lightning Source LLC
LaVergne TN
LVHW062135300926
849161LV00010B/2533